Modern Optics and Photonics of Nano- and Microsystems

Modern Optics and Photonics of Nano- and Microsystems

Yu. N. Kulchin

CISP

CRC Press
Taylor & Francis Group
Boca Raton   London   New York

CRC Press is an imprint of the
Taylor & Francis Group, an **informa** business

CRC Press
Taylor & Francis Group
6000 Broken Sound Parkway NW, Suite 300
Boca Raton, FL 33487-2742

First issued in paperback 2020

© 2018 by CISP
CRC Press is an imprint of Taylor & Francis Group, an Informa business

No claim to original U.S. Government works

ISBN-13: 978-0-367-57160-3 (pbk)
ISBN-13: 978-0-8153-6976-9 (hbk)

Visit the Taylor & Francis Web site at
http://www.taylorandfrancis.com

and the CRC Press Web site at
http://www.crcpress.com

Introduction

Three decades, covering the end of the XX century and the beginning of the XXI century, are characterized by rapid progress in the development and implementation of technologies based on the achievements of photonics. Photonics is an interdisciplinary scientific field, which is based on fundamental and applied research related to the study of the properties of optical radiation, processes and technologies of generation and control of light characteristics, the phenomena of interaction of radiation with the matter, the processes of transmission and processing of optical signals, fibre optic communication and development of complex information systems.

Optical sciences, which are one of the earliest and most important areas of research in the field of physical sciences, are now actively penetrating into diverse fields of research, including chemistry, biology, medicine and engineering. In this regard, the world trends in the development of optics, optical instrumentation and optical material science have undergone significant changes in recent years. The changes touched upon the development of a new generation of optical materials, the discovery of new optical phenomena and effects, which formed the basis for the creation of fundamentally new optical elements, devices and systems.

Today, photonics is seen as a priority direction for the development of science and technology in many leading countries of the world, as it serves as a key element for solving many social problems: from energy generation and its effective use, to ensuring public health, security, adequate response to climate change, etc.

Considering the foregoing, modern ideas about the development of optics and photonics, as fundamental foundations for the development of scientific and industrial instrument making in the ranges of X-ray, ultraviolet, visible, infrared and terahertz electromagnetic radiation, do not allow a contrast boundary between them.

The experimental and theoretical results achieved in optics and photonics allowed us to develop new technologies that significantly transformed our life. First of all, it concerns the rapid development of communication systems and information technologies, as well as the implementation of laser technologies in industry and medicine. The number of new scientific results obtained in the field of optics and photonics is rapidly growing. Only since the 80-ies of the last century seven Nobel Prizes in Physics were received for the research performed in these areas. In this connection, it is extremely difficult to state all the results achieved in one monograph. This monograph is devoted to several key achievements of modern photonics and optics, which are presented in the following 10 chapters:

1. Fundamentals of nonlinear optics;
2. Phenomena of filamentation and generation of a supercontinuum in the propagation of laser pulses in a nonlinear medium;
3. Photonic crystals;
4. Nonlinear optics of optical fibres;
5. Fibre lasers;
6. Photonics of self-organizing nanostructured biomineral objects and their biomimetic analogs;
7. Dynamic holography and optical Novelty filters;
8. Adaptive optoelectronic systems SMART-GRID monitoring of physical fields and objects;
9. Laser cooling, trapping and controlling atoms;
10. Photonics of nanostructures.

The material of these chapters is devoted to the consideration of the processes of linear and nonlinear propagation of laser pulses in volume and light-guiding media, new sources of coherent radiation-fibre lasers, photonic crystal objects created by living Nature, and new photonic information systems using dynamic spatial filters and distributed fibre-optic measurement systems, as well as achievements in the field of ultra-low laser cooling of atoms and studies of the optical and nonlinear optical properties of heterogeneous nanostructured systems .

The content and construction of the monograph are very close to the corresponding course "Modern optics", which the author has read over the past two years to graduate and postgraduate students of the National Research Nuclear University "MEPhI" and the Institute of Automation and Control Processes, Far Eastern Branch of the Russian Academy of Sciences.

Since a significant part of the material presented in the monograph has not yet been reflected in the curriculum literature and remains to be described only in original articles and monographs, the author hopes that this monograph will be useful both to specialists and students and graduate students making their first steps in such fascinating areas of research which are modern optics and photonics.

Fundamentals of Nonlinear Optics

Introduction

When in ordinary life we talk about the propagation of light in matter, it is initially assumed that the characteristics of the medium do not change under its influence. Indeed, the electric fields in atoms and molecules, which are the main elements determining the structure and optical properties of matter, are very large. In this connection, the light sources that existed in the pre-laser era could provide electric fields in the light electromagnetic wave many orders of magnitude smaller than the intra-atomic ones. Therefore, the influence of the electromagnetic field of the lightwave on the properties of the medium turned out to be negligibly small and did not manifest itself under real conditions. Since in this case the response of the medium to the external optical action is proportional to the electric field strength in the wave, the theoretical description of the phenomena arising in the interaction of light with matter is usually called linear optics.

The discovery of optical quantum generators (lasers) made it possible to generate optical fields with a strength comparable to that of an intra-atomic field. The effect of such optical radiation on the medium leads to a change in its optical properties. This means that the wave in the medium experiences both self-action and can influence the propagation of other waves in the medium. Naturally, this leads to the emergence of new phenomena that have not been observed before and requires the development of new theoretical approaches. As a result, a new field of science arose, which was called the nonlinear optics.

The term 'nonlinear optics' was first introduced by the Russian scientist S.I. Vavilov in 1925, who managed to observe a decrease in the absorption of light by uranium glasses when optical radiation passes through them with high intensity. In the early 60s of the same century, after the creation of lasers, nonlinear optical phenomena became not only observable, but also turned into a serious tool for studying matter, and became the basis for the creation of completely new laser devices. Thus, the subject of nonlinear optics are the processes of interaction of light with matter, the nature of which depends on its intensity. Such processes include phenomena of resonant medium clarification, two-photon or multiphoton light absorption, optical breakdown of the medium, generation of optical harmonics, 'straightening' of light, stimulated light scattering, self-focusing of light beams, self-modulation of pulses, and a number of other effects manifested in fields of laser radiation. The classification of all these nonlinear optical effects is presented in detail in [1–5].

Incoherent nonlinear optical effects. Such optical characteristics of the medium, as the refractive index and the absorption coefficient, do not depend on the intensity of light if the reaction of the medium to the electric field of the lightwave is described by linear material equations. In a constant or slowly varying external electric field with a sufficiently low value of the field strength, the polarization vector of the medium **P**, as a rule, depends linearly on the field strength vector **E** of the lightwave and is described by the following material equation [6]:

in the CGS system

$$P = \chi E, \tag{1.1}$$

in the International System of Units system

$$P = \varepsilon_0 \div E, \tag{1.2}$$

where ε_0 is the dielectric constant. (*Further, the formulas in this chapter are given only in the CGS system in which $\varepsilon_0 = 1$*).

The proportionality coefficient χ is called the dielectric susceptibility or the polarizability of the medium and depends on the properties of the medium: chemical composition, concentration, structure, aggregate state of the medium, and also on temperature, mechanical stress, etc.

However, this material equation is approximate since it is valid only for the electric field of the lightwave E small compared with the intensity of intra-atomic electric fields E_a. To estimate the value of E_a we can assume that the order of magnitude is $E_a \sim e/a^2$, where a is the size of the atom, e is the electron charge. As $a \sim 10^{-8}$ cm, then the intra-atomic electric field will be of the order of $E_a \sim 10^8$ V/cm.

In light beams from non-laser sources, the field strengths do not exceed $0.1-10$ V/cm and the linear material equations (1.1), (1.2) are satisfied with great accuracy. However, in intense laser beams, the field strength E reaches values of 10^6-10^7 V/cm. In such fields, the model of a harmonic oscillator for describing the behaviour of an optical electron in an atom is no longer applicable and the relationship between the polarizability of the medium and the intensity of the light-wave field becomes nonlinear. As a result, the optical characteristics of the medium depend on the radiation intensity, which does not lead to any small corrections, but to fundamentally new effects that are not observed in linear optics.

Nonlinear optics significantly expands our understanding of the interaction of light with matter. For example, in the earliest experiments on nonlinear optics carried out by S.I. Vavilov and V.L. Levshin, a slight decrease (up to 1.5%) of the uranium glass absorption coefficient with increasing light intensity was observed. The appearance of this nonlinear saturation effect is due to the alignment of populations of two energy levels between which quantum transitions occur with the absorption and emission of light.

Subsequently, the decrease in the fraction of the absorbed power with increasing light intensity, that is, the clarification??? of the nonlinear absorbing medium during passage of strong light beams, has found application in laser technology where it is used to modulate the quality of optical resonators for the purpose of generating ultrashort high-power pulses (see, for example, Chapter 5). In this case, a cell with a nonlinear absorbing medium is an automatic shutter opening under the influence of a powerful light beam. It is very important that such a shutter has a low inertia, since after passing a powerful light pulse, the medium quickly becomes again opaque for the low-intensity light. In an active medium with population inversion, the saturation effect leads to a decrease in the gain with increasing light intensity and, thus, to the establishment of a steady-state lasing regime in lasers. If the medium is opaque for the low-intensity light due to the saturation effect becomes transparent in strong light fields, then the opposite situation is observed in the optical transparency region of the medium. Here, as a result of multiphoton absorption of radiation, intense light can be absorbed much more strongly than weak light. At high radiation densities, a system with energy levels E_1 and E_2 can absorb in an elementary act two photons with frequencies ω_1 and ω_2, such that $\hbar\omega_1 + \hbar\omega_2 = E_2 - E_1$. In particular, in the first case, when $\omega_1 = \omega_2 = (E_2 - E_1)/2\hbar$, the probability of absorption of two photons is proportional to the product of the intensities of the beams with frequencies ω_1 and ω_2 (or the square of their intensity for $\omega_1 = \omega_2$).

Nonlinear incoherent effects include the multiphoton process of radiation absorption. For example, in the act of interaction of light with the medium, it is possible to simultaneously absorb three photons or more. Multiphoton absorption is used in nonlinear laser spectroscopy and allows obtaining information on the energy levels of quantum systems.

In experiments with focused laser beams, such high light-field intensities are attained when the processes in which the atom absorbs eight or more photons become available to observation. As a result, photoionization of the atom with he light of low frequency can occur, that is, in the intense light fields, the red boundary of the photoelectric effect disappears.

One of the most striking examples of nonlinear incoherent effects is the self-focusing of light. The effect is that the medium acquires focusing (lens) properties in the field of a powerful laser beam. As a result, the light beam 'collapses', turning into a thin luminous filament, or breaks up into several such filaments (the phenomenon is considered in Chapter 2). The mechanism of self-focusing is associated with a change in the refractive index of the medium under the action of a powerful lightwave. The reasons for this may be different. For example, the electrostriction in a light field leads to the appearance of a pressure that changes the density of the medium in the region occupied by the light beam, and, consequently, also changes the refractive index of the medium. In a liquid, a strong light field leads to the orientation of anisotropically polarizable molecules due to the interaction of light with the induced dipole moment, while the medium becomes anisotropic, and the average refractive index for the orienting field increases. This effect is commonly called the high-frequency Kerr effect. The change in the refractive index here, as in the well-known static Kerr effect, is due to the 'alignment' of molecules over the field.

Finally, the change in the density, and, consequently, in the refractive index of the medium can be associated with heating caused by the dissipation of the energy of a powerful lightwave.

The self-focusing of light was theoretically predicted by G.A. Askar'yan in 1962, and experimentally first observed by N.P. Pilipetskii and A.R. Rustamov in 1965. In their experiments, narrow luminous filaments were photographed in organic liquids irradiated by a focused beam of a ruby laser.

The self-defocusing effect is inverse to self-focusing. Under the action of high-power laser radiation, a nonlinear addition to the absorption index appears in the medium. In the event that this additive is positive, self-focusing takes place (a nonlinear positive quasilens is formed). With a negative addition, the refractive index on the beam axis decreases and a nonlinear negative quasilens is formed, which leads to self-defocusing of the radiation. Thus, when laser radiation passes through a liquid, even its weak absorption leads to thermal expansion of the liquid, a decrease in its density, and, as a consequence, its refractive index. Ultimately, this leads to the formation of a thermal defocusing lens, which increases the angular divergence of the laser beam.

The same mechanisms cause another effect of nonlinear self-action of light – self-modulation of the light pulse. In self-modulation of a pulse, which, for example, can occur in an optical fibre, the frequency spectrum of the pulse drastically broadens, which makes it possible, by subsequent compression, to obtain extremely short light pulses. This effect is used in the systems of generation of femtosecond laser pulses.

We note that the main feature of the incoherent nonlinear optical effects considered above is their independence from the phase of laser radiation.

Nonlinear optical effects due to high coherence of radiation. It is known that in addition to high intensity, laser radiation has a high degree of coherence. The latter is manifested in nonlinear optical effects, in which the phase relations play a decisive role. Such effects include: harmonic generation, frequency mixing, stimulated Raman scattering of light, and violation of the superposition principle for strong lightwaves in a medium.

Generation of harmonics. Frequency mixing. This phenomenon, for example, consists in doubling the frequency of light in the propagation of a powerful laser beam in a crystal. The mechanism of the process is related to the nonlinearity of an elementary atomic oscillator. The doubling of the frequency of light in the crystal was the first coherent nonlinear optical effect discovered shortly after the creation of the laser in 1961. In the experiments of the American scientist P. Franken, the radiation of a ruby laser having a wavelength of λ = 694.3 nm was focused into a quartz crystal. The radiation emerging from the crystal was unfolded into the spectrum with the aid of a dispersion prism and was focused on a photographic plate. Experience has shown that, in addition to the light at the frequency of the laser (ω), the light comes out of the crystal at a doubled frequency (the second harmonic) having a wavelength of $\lambda_2 = \lambda/2$ ($\omega_2 = 2\omega$) = 347.15 nm. Despite the fact that the radiation second harmonic in P. Franken's experiment was extremely weak, this experience played a fundamental role, initiating the development of nonlinear optics, and subsequent experiments showed that using other crystals, the efficiency of second harmonic generation can be dramatically increased. Methods for transforming a significant fraction of the laser radiation to the second harmonic have been developed; in some cases it is possible to obtain the lasing efficiency close to 100%.

In addition to generating the second harmonic, it is possible to generate total and difference radiation frequencies of two or more lasers: $\omega_s = \omega_1 \pm \omega_2$, as well as higher harmonics and total and difference frequencies in higher-order processes $\omega_s = n\omega_1 \pm m\omega_2$ (where n and m are integers). Currently, this effect is widely used to convert the frequency of laser radiation and the development of spectral ranges in which the creation of lasers is difficult.

Stimulated Raman scattering of light. This effect (abbreviated SRS) consists in the fact that in the field of a powerful laser beam the medium generates intense radiation shifted in frequency relative to the laser by an amount equal to the frequency of molecular vibrations. The mechanism of the process is the same as in the case of spontaneous scattering – modulation of light by molecular vibrations. However, unlike spontaneous scattering, which is very weak and directed in all directions, stimulated scattering resembles laser generation. The power and directivity of the stimulated scattering are commensurable with the analogous parameters of the laser beam. The reason for this is that stimulated scattering does not occur on chaotic thermal molecular vibrations, but on vibrations excited and phased by light in a large volume of the medium. The transition of spontaneous scattering to stimulated scattering occurs when the intensity of the exciting light exceeds a certain definite quantity called the SRS threshold.

For the first time stimulated Raman scattering was observed in 1962 by American scientists E. Woodbury and V. Nga in the study of the Q-switching regime of a ruby laser with the aid of a Kerr cell with nitrobenzene. They found the appearance of an infrared component in the laser radiation, the frequency of which was 1345 cm^{-1} lower than the frequency of the main laser radiation. Since the frequency shift coincided with one of the natural oscillation frequencies of the nitrobenzene molecule, it was suggested that the appearance of the infrared component is associated with the Raman scattering of light in nitrobenzene, and the high intensity of radiation is due to the forced character of the process, in which molecular vibrations are strongly swinging with light. This assumption was confirmed in subsequent experiments with various liquids, as well as with gases and solids. Experience shows that the stimulated Raman scattering is observed only at a sufficiently high light intensity achieved by focusing the beam into a cuvette. Since stimulated scattering has a high intensity, commensurate with the intensity of the laser beam, this circumstance makes it possible to create effective laser transducers radiation based on the SRS process. Currently, such converters are used to convert the radiation frequency, as well as to compress (shorten the duration and increase the power) of laser pulses. In addition, based on the SRS process, it is possible to carry out a coherent summation of the emissions of several laser modules. The excitation of coherent molecular vibrations with the aid of a pair of lightwaves (the method of biharmonic pumping) is also widely used in the spectroscopy of coherent anti-Stokes scattering of light (CARS).

Violation of the superposition principle for strong lightwaves in a medium. The principle of superposition is that different lightwaves, differing in frequency, direction of propagation or polarization, propagate and interact with the environment independently of each other. In nonlinear optics this is not so. As we have seen, new spectral components of the field appear in nonlinear optical processes, different lightwaves interact strongly with each other, energy is exchanged between them until the complete transformation of one wave into another. A typical example of this kind is the generation of a second optical harmonic. Thus, for nonlinear-optical processes of interaction of radiation with matter, a violation of the superposition principle is characteristic.

1.1. Polarization of dielectrics in a constant electric field

Any electromagnetic process in a medium is described by the Maxwell equations [6]

$$
\begin{cases}
\operatorname{rot} \mathbf{E} = -\dfrac{\partial \mathbf{B}}{\partial t}, \\[2mm]
\operatorname{div} \mathbf{B} = 0, \\[2mm]
\operatorname{rot} \mathbf{H} = \mathbf{j} + \dfrac{\partial \mathbf{D}}{\partial t}, \\[2mm]
\operatorname{div} \mathbf{D} = \rho,
\end{cases}
\tag{1.3}
$$

which in the operator form have the form

$$
\begin{cases}
\nabla \times \mathbf{E} = -\dfrac{\partial \mathbf{B}}{\partial t}, \\[2mm]
\nabla \cdot \mathbf{B} = 0, \\[2mm]
\nabla \times \mathbf{H} = \mathbf{j} + \dfrac{\partial \mathbf{D}}{\partial t}, \\[2mm]
\nabla \cdot \mathbf{D} = \rho,
\end{cases}
\tag{1.4}
$$

where $\mathbf{E}(\mathbf{r},t)$, $\mathbf{H}(\mathbf{r},t)$ are the vectors of electric and magnetic field strengths at a point with a radius vector $\mathbf{r}$ at the time t; ρ is the density of electrical charges; $\mathbf{j}$ is the vector of current density; $\mathbf{D}$ and $\mathbf{B}$ are the vectors of electrical and magnetic inductions.

However, these equations are not sufficient to solve the problem; we need material equations that establish additional connections between these vectors [2–6]:

$$
\begin{cases}
\mathbf{D} = \mathbf{E} + \mathbf{P} = \varepsilon \mathbf{E}, \\
\mathbf{B} = \mathbf{H} + \mathbf{M} = \mu \mathbf{H}, \\
\mathbf{j} = \sigma \mathbf{E}.
\end{cases}
\tag{1.5}
$$

We recall that in the International System of Units the equations (1.5) will have the form

$$
\begin{cases}
\mathbf{D} = \varepsilon_0 \mathbf{E} + \mathbf{P} = \varepsilon_0 \varepsilon \mathbf{E}, \\
\mathbf{B} = \mu_0 \mathbf{H} + \mathbf{M} = \mu_0 \mu \mathbf{H}, \\
\mathbf{j} = \sigma \mathbf{E},
\end{cases}
\tag{1.6}
$$

where ε_0 and μ_0 are the dielectric and magnetic constants, ε and μ are the dielectric and magnetic permeabilities of the medium, and σ is the electric conductivity of the medium.

Equations (1.5) establish a connection between the vector of macroscopic polarization of the medium **P**, the vector of the macroscopic magnetization of the medium **M** and vectors **D** and **B**, as well as between the current density **j** and the electric field strength **E**. Further, we will not take into account the magnetic properties of the medium.

In the isotropic case, the macroscopic polarization of the medium depends on the electric field strength. The coefficient of proportionality in such a dependence is the dielectric susceptibility of the medium $\chi(\mathbf{E})$, which in general also depends on **E**. If we take this dependence into account, then, using (1.1) and (1.5), we obtain:

$$\begin{cases} \mathbf{P} = \chi(\mathbf{E})\mathbf{E}, \\ \mathbf{D} = (1 + \chi(\mathbf{E}))\mathbf{E} = \varepsilon(\mathbf{E})\mathbf{E}. \end{cases} \tag{1.7}$$

The value $\varepsilon(\mathbf{E}) = 1 + \chi(\mathbf{E})$ is called the dielectric permittivity of the medium.

In weak fields, the dielectric susceptibility of the medium and the permittivity are constants that do not depend on the electric field strength and, consequently, the reaction of the medium to the external field is linear:

$$\begin{cases} \mathbf{P} = \chi\mathbf{E}, \\ \mathbf{D} = (1 + \chi)\mathbf{E} = \varepsilon\mathbf{E}. \end{cases} \tag{1.8}$$

Nonlinear effects are manifested only when the fields are strong enough and the quantities χ and ε can no longer be regarded as independent of the field strength. To illustrate the appearance of the nonlinear dependence of the quantities χ and ε, we calculate them in the framework of a simple classical problem. To do this, let us consider a gas consisting of the simplest atoms (two point charges of the quantity e: a nucleus and an electron) without a constant electric dipole moment. In the absence of an external field, the position of the mass centred for point charges in the atom is the same. If the atomic gas is placed in a constant electric field, then the charges in each atom will move a certain distance r. For simplicity, we assume that the electron displacement coincides with the direction of the external electric field. Then we can ignore the vector nature of the quantities involved in the problem, and operate with scalar quantities. Thus, atoms will acquire a dipole moment: $d = e \cdot r$.

If the environment consists of N atoms, then its macroscopic polarization will be

$$P = N \cdot d = N \cdot e \cdot r. \tag{1.9}$$

Two forces act on an electron in an atom: one acts on the side of the electric field with force $F_E = eE$, and the second – elastic, due to the potential field in which the electron moves, returning the electron to its former position, with force: $\mathbf{F}_{el} = -kr - qr^3$ (this force in the general case depends nonlinearly on the electron displacement) [4], where k and q are the linear and nonlinear elastic deformation constants. Equating the forces F_E and F_y, we obtain an equation for determining the amount of displacement of an electron in an external field:

$$eE = kr + qr^3. \tag{1.10}$$

Determining the value r from (1.9) and substituting its value in (1.10), we obtain a nonlinear equation for the polarization of the medium:

$$\frac{e^2 N}{k} E = P + \frac{q}{ke^2 N^2} P^3. \tag{1.11}$$

We represent the polarization of the medium in the form $P = P_L + \delta P$, where P_L is a linear component of the polarization of the medium, independent of the magnitude of the electric field, and δP is a small nonlinear addition to polarization. We solve equation (1.11) with respect to P, assuming that the power term of P^3 is small. In this connection, equation (1.11) can be represented as a system of two equations: one for the terms of the zeroth and the other for the terms of the first order of smallness:

$$\begin{cases} P_L = \dfrac{e^2 N}{k} E, \\[2mm] \delta P = -\dfrac{q}{ke^2 N^2} P_L^3. \end{cases} \tag{1.12}$$

Using the system (1.12), we obtain an expression for the polarization of the medium:

$$P = P_L + \delta P = \left(\frac{e^2 N}{k} - \frac{qe^4 N}{k^4} E^2 \right) E. \tag{1.13}$$

Comparing the solution obtained with (1.7), we obtain

$$\chi(E) = \frac{e^2 N}{k} - \frac{qe^4 N}{k^4} E. \tag{1.14}$$

As can be seen, the dielectric susceptibility of the medium is a function of the electric field strength. If the magnitude of the field is sufficiently weak (much less than the intra-atomic field), then the second term in (1.14) can be neglected (this means that the displacement is small and in the expression for the term it can be neglected) and the dielectric susceptibility of the medium becomes a constant.

Before that, we considered the case of an isotropic medium. When the medium is anisotropic, the dielectric susceptibility and permeability of the medium, instead of scalars, become tensors of the second rank, and the connection between the vectors **P, D, E** has the form:

$$\begin{cases} P_i = \sum_{j=1}^{3} \chi_{ij}(E) E_j, \\ D_i = \sum_{j=1}^{3} (d_{ij} + \chi_{ij}(E)) E_j = \sum_{j=1}^{3} \varepsilon_{ij}(E) E_j, \end{cases} \tag{1.15}$$

where d_{ij} is the unit tensor, $\varepsilon_{ij}(E)$ — in the general case, the tensor of the permittivity of the medium, which depends on the magnitude of the electric field strength.

In the Cartesian coordinate system, the components of the vector **D** are represented as:

$$\begin{cases} D_x = \varepsilon_{xx} E_x + \varepsilon_{xy} E_y + \varepsilon_{xz} E_z \\ D_y = \varepsilon_{yx} E_x + \varepsilon_{yy} E_y + \varepsilon_{yz} E_z \\ D_z = \varepsilon_{zx} E_x + \varepsilon_{zy} E_y + \varepsilon_{zz} E_z \end{cases} \tag{1.16}$$

or

$$\mathbf{D} = \begin{pmatrix} \varepsilon_{xx} & \varepsilon_{xy} & \varepsilon_{xz} \\ \varepsilon_{yx} & \varepsilon_{yy} & \varepsilon_{yz} \\ \varepsilon_{zx} & \varepsilon_{zy} & \varepsilon_{zz} \end{pmatrix} \mathbf{E}. \tag{1.17}$$

Using these relations, knowing the parameters of the external electric field and the dielectric susceptibility tensor of a particular anisotropic medium (usually determined by experimental methods), one can always calculate its polarization.

1.2. Polarization of an isotropic dielectric in a light field

Let us consider the polarization of a dielectric in a high-frequency light field using the same simple gas model. Since the intensity of the electric field now depends on time, it is necessary to solve the dynamic, and not static, problem for the motion of an electron. The equation of motion of the electron is written in the form

$$\begin{cases} m\dfrac{d^2r}{dt^2} = F_E + F_{\text{ef}} + F_{\text{f}}, \\[2mm] F_{\text{f}} = -m\gamma_{\text{f}}\dfrac{dr}{dt}, \end{cases} \tag{1.18}$$

where F_{f} is the friction force proportional to the velocity of motion of the electron, γ_{f} is the coefficient of friction; F_E is a force acting on the side of the external electric field; F_{ef} is the elastic force. For the case of a weak electric field, we can use the first approximation for the elastic force: $F_{\text{ef}} = -kr$. Substituting this expression in (1.18), we obtain

$$\begin{cases} m\dfrac{d^2r}{dt^2} = eE(t) - kr - m\gamma_{\text{f}}\dfrac{dr}{dt}, \\ \text{or} \\ \dfrac{d^2r}{dt^2} + \gamma_{\text{f}}\dfrac{dr}{dt} + \dfrac{k}{m}r = eE(t). \end{cases} \tag{1.19}$$

We multiply the last equation in (1.19) by eN, taking into account the expression for the polarization of the medium (1.9) and the expression for the natural frequency of the electron oscillations $\omega_0^2 = k/m$ we obtain an equation for the polarization of the medium:

$$\frac{d^2P}{dt^2} + \gamma_f\frac{dP}{dt} + \omega_0^2 P = \frac{e^2N}{m}E(t). \tag{1.20}$$

If the lightwave changes in accordance with the harmonic law: $E(t) = E_0\cos(\omega t)$, the solution for polarization will be found in the form $P(t) = P_0\cos(\omega t + \varphi)$. Substituting this expression into (1.20) we obtain

$$-\omega^2 P_0\cos(\omega t + \varphi) - \gamma_T\omega P_0\sin(\omega t + \varphi) + \omega_0^2 P_0\cos(\omega t + \varphi) = \frac{e^2 N}{m}E_0\cos(\omega t)$$

or

$$\left(\omega_0^2 - \omega^2\right)P_0(\cos\omega t\cos\varphi - \sin\omega t\sin\varphi) -$$
$$-\gamma_f\omega P_0(\cos\omega t\sin\varphi + \sin\omega t\cos\varphi) = \frac{e^2}{m}E_0\cos\omega t. \tag{1.21}$$

Equating separately the terms at $\cos\omega t$ and $\sin\omega t$, we obtain a system of equations

$$\begin{cases} -(\omega_0^2 - \omega^2)\sin\varphi - \gamma_T\omega\cos\varphi = 0, \\ (\omega_0^2 - \omega^2)P_0\cos\varphi - \gamma_f\omega P_0\sin\varphi = \frac{e^2 N}{m}E_0. \end{cases} \tag{1.22}$$

From the first equality, we determine the phase of the oscillations of polarization of the medium:

$$\text{tg}\,\varphi = -\frac{\gamma_f\omega}{\omega_0^2 - \omega^2}. \tag{1.23}$$

Using the second equality in (1.22) and the expression (1.23), we obtain the expression for the polarization amplitude of the medium:

$$P_0 = \frac{\dfrac{e^2 N}{m}E_0\sqrt{1 + \text{tg}\,\varphi}}{(\omega_0^2 - \omega^2) - \gamma_f\omega\,\text{tg}\,\varphi} = \frac{e^2 N}{m}\frac{E_0}{\sqrt{(\omega_0^2 - \omega^2)^2 - (\gamma_f\omega)^2}}. \tag{1.24}$$

As a result, the expression for the polarization of the medium has the following form

$$P = \frac{e^2 N}{m}\frac{E_0\cos(\omega t + \varphi)}{\sqrt{(\omega_0^2 - \omega^2)^2 - (\gamma_f\omega)^2}}. \tag{1.25}$$

As indicated by the expression (1.25), the polarization of the dielectric medium changes with the frequency of the light field wave ω and the polarization amplitude of the medium depends on the relationship between the intrinsic frequency of the oscillations of the electrons in the atom ω_0 and the frequency of the incident light radiation ω.

In most cases, some limiting cases are of special interest. In particular, when the frequency of the lightwave coincides with the intrinsic frequency of the oscillations of the electrons of the medium (resonance case): $\omega = \omega_0$, the polarization amplitude of the medium is maximum and equal to:

$$P_0 = \frac{e^2 N}{m \gamma_f \omega_0} E_0. \tag{1.26}$$

Another limiting case may be the large difference of the frequencies of the lightwave and the resonance frequency of the oscillations of the electrons of the medium: $\omega \gg \omega_0$ or $|\omega - \omega_0| \gg \gamma_T$. Consequently, from (1.23) it follows that

$$|\operatorname{tg}\varphi| = \frac{\gamma_f \omega}{\omega_0^2 - \omega^2} = \frac{\gamma_f \omega}{(\omega_0 - \omega)(\omega_0 + \omega)} \approx$$
$$\frac{\gamma_f \omega}{2\omega(\omega_0 - \omega)} = \frac{\gamma_f}{2(\omega_0 - \omega)} \ll 1. \tag{1.27}$$

This means that in this case the phase for the polarization of the medium is close to 0 and the polarization oscillations take place almost completely in the phase with the oscillations of the lightwave. Using equation (1.25), we derive the expression for the ionisation of the dielectric medium

$$P(\omega,t) = \frac{e^2 N}{m} \frac{E_0 \cos(\omega t)}{|\omega_0^2 - \omega^2|} = \frac{e^2 N}{m} \frac{E(t)}{|\omega_0^2 - \omega^2|} = \chi(\omega)E(t). \tag{1.28}$$

As indicated by (1.28), the dielectric susceptibility of the medium depends on the frequency of light radiation.

Another limiting cases may be $\omega = 0$ which indicates the effect of a direct electric field on the medium, consequently

$$\chi(0) = \frac{e^2 N}{m\omega_0^2} = \frac{e^2}{k} N, \tag{1.29}$$

which coincides with the previously derived equation (1.14) at $E = 0$.

Until now, it was assumed that the electron is subjected to the effect of a field with a low strength. Taking this into account, we used the linear approximation for small displacements of the electron in the light field wave. However, if it is assumed that the strength of the light field and the displacement of the electrode can be sufficiently large, then we should use another approximation for the elastic medium: $F_y = -kr - qr^3$, where q is the nonlinearity coefficient. As a result, the equation of motion of the electron in the light field wave has the form different from (1.19):

$$m\frac{d^2 r}{dt^2} = eE(t) - kr - qr^3 - m\gamma_f \frac{dr}{dt}. \tag{1.30}$$

The equation, describing the variation of polarization of the medium, also changes

$$\frac{d^2 P}{dt^2} + \gamma_f \frac{dP}{dt} + \frac{q}{me^2 N^2} P^3 + \omega_0^2 P = \frac{e^2 N}{m} E(t). \tag{1.31}$$

As in the previous case, it will be assumed that the field $E(t)$ changes in accordance with the harmonic law. The solution of the equation (1.31) as previously will be obtained in the form $P = P_L + \delta P$. After substituting this expression into (1.31) one obtains

$$\frac{d^2 P_L}{dt^2} + \frac{d^2 \delta P}{dt^2} + \gamma_f \frac{dP_L}{dt} + \gamma_f \frac{d\delta P}{dt} + \frac{q}{me^2 N^2}$$
$$(P_L^3 + \delta P^3 + 3P_L^2 \delta P + 3P_L \delta P^2) + \omega_0^2 P_L + +\omega_0^2 \delta P = \frac{e^2 N}{m} E(t). \tag{1.32}$$

Grouping in (1.32) the terms of the zeroth and first order of smallness, and ignoring the terms with the multipliers γ_T, δP^2, δP^3 and other terms of the second and third order of smallness, we obtain the following system of the equations

$$
\begin{cases}
\dfrac{d^2 P_L}{dt^2} + \omega_0^2 P_L = \dfrac{e^2 N}{m} E(t), \\[2ex]
\dfrac{d^2 \delta p}{dt^2} + \omega_0^2 \delta P + \dfrac{q}{me^2 N^2} P_L^3 = 0.
\end{cases}
\tag{1.33}
$$

Away from the resonance, the solution of the first equation in (1.33) has the form (1.28). Substituting this equation into the second equation of the system (1.33) gives

$$
\frac{d^2 \delta P}{dt^2} + \omega_0^2 \delta P = -\frac{q}{me^2 N^2} P_L^3 = -\frac{q\chi(\omega)}{me^2 N^2} E^3(t).
\tag{1.34}
$$

Since the strength of the field of the lightwave changes according to the harmonic law, then

$$
E^3(t) = \frac{1}{4} E_0^3 (3\cos\omega t + \cos 3\omega t).
\tag{1.35}
$$

The equation (1.34) is the equation of the harmonic oscillators subjected to the effect of the external force (the right hand part of the equation) consisting of two components, one of which changes with the frequency ω and the other with frequency 3ω. Therefore, the solution of (1.34) will be obtained in the form

$$
\delta P = \delta P_\omega \cos\omega t + \delta P_{3\omega} \cos 3\omega t.
\tag{1.36}
$$

Substituting (1.36) into (1.34) the obtain the following values

$$
\delta P_\omega = -\frac{3}{4} \frac{q\chi^3(\omega)}{me^2 N^2 (\omega_0^2 - \omega^2)} E_0^3
\tag{1.37}
$$

and

$$
\delta P_{3\omega} = -\frac{1}{4} \frac{q\chi^3(\omega)}{me^2 N^2 (\omega_0^2 - 9\omega^2)} E_0^3.
\tag{1.38}
$$

According to (1.28)

$$\chi(\omega) = \frac{e^e N}{m} \frac{1}{\left|\omega_0^2 - \omega^2\right|},$$ (1.39)

and therefore, substituting (1.36), (1.37) and (1.39) into the expression $P = P_L + \delta P$, we obtain the general solution:

$$P(\omega,t) = P_L(\omega,t) + \delta P(\omega,t) = \chi(\omega) E_0 \cos \omega t -$$

$$-\frac{3}{4} \frac{q\chi^3(\omega)}{me^2 N^2 (\omega_0^2 - \omega^2)} E_0^3 \cos \omega t$$

$$-\frac{1}{4} \frac{q\chi^3(\omega)}{me^2 N^2 (\omega_0^2 - 9\omega^2)} E_0^3 \cos 3\omega$$

$$= \left[\chi(\omega) - \frac{3}{4} \frac{q\chi^3(\omega)}{me^2 N^2 (\omega_0^2 - \omega^2)} E_0^2 \right]$$

$$E_0 \cos \omega t + \left[-\frac{1}{4} \frac{q\chi^2(\omega)}{me^2 N^2 (\omega_0^2 - 9\omega^2)} E_0^2 \right] \times$$

$$\times E_0 \cos 3\omega t = \chi(\omega, E_0) E_0 \cos \omega t + \chi(3\omega, E_0) E_0 \cos 3\omega t,$$

(1.40)

where

$$\chi(\omega, E_0) = \chi(\omega) - \frac{3}{4} \frac{q\chi^3(\omega)}{me^2 N^2 (\omega_0^2 - \omega^2)} E_0^2,$$ (1.41)

and

$$\chi(3\omega, E_0) = -\frac{1}{4} \frac{q\chi^3(\omega)}{me^2 N^2 (\omega_0^2 - 9\omega^2)} E_0^2.$$ (1.42)

Thus, the polarization of the dielectric in the strong light field depends not only on the strength and frequency of the incident radiation but also on its third harmonics. Since the charge carrying out oscillations with some frequency emits an electromagnetic wave with the same frequency, then in the effect of the nonlinear influence on the medium of the strong light field the radiation spectrum should contain electromagnetic waves with the frequencies ω and 3ω, and in a general case also higher harmonics.

1.3. Interaction of high intensity electromagnetic fields with the nonlinear medium

As shown previously, the formation of the anharmonism of the oscillations of the electrons under the effect of the strong light field leads to the formation of high frequency orders in the polarization spectra of the medium and of the electromagnetic radiation determined by this. The polarization of any medium at time t can be described as an exponential series of the value of the strength of the electric field of the lightwave [2–4];

$$\mathbf{P}(t) \propto \chi^{(1)}\mathbf{E}(t) + \chi^{(2)}\mathbf{E}^2(t) + \chi^{(3)}\mathbf{E}^3(t) + \dots . \qquad (1.43)$$

Here the coefficients of the series $\chi^{(n)}$ are the n-th order of the dielectric susceptibility of the medium and are usually interpreted as the n-th order of the nonlinearity, corresponding to this medium. Usually, $\chi^{(n)}$ is regarded as the tensor of the $n + 1$ order which depends both on the nature of interaction of radiation with the medium and on the symmetry of the material of the nonlinear medium.

Sometimes, the dependence of the polarization vector of the medium $\mathbf{P}$ on the strength of the electric field $\mathbf{E}$ is denoted as follows

$$\mathbf{P}(t) = \chi^{(1)} : \mathbf{E}(t) + \chi^{(2)} : \mathbf{E}^2(t) + \chi^{(3)} : \mathbf{E}^3(t) + \dots , \qquad (1.44)$$

or in the International System of Units

$$\mathbf{P}(t) = \varepsilon_0 \chi^{(1)} : \mathbf{E}(t) + \varepsilon_0 \chi^{(2)} : \mathbf{E}^2(t) + \varepsilon_0 \chi^{(3)} : \mathbf{E}^3(t) + \dots , \qquad (1.45)$$

where ε_0 is the electrical constant, and $\chi(n)$ is as previously the component of the n-th order of the dielectric susceptibility of the matter. The symbol ':' indicates the scalar product of the matrices. Taking the above considerations into account we can write (1.44) in the detailed form for every i-th component of the polarization vector $\mathbf{P}$ in the following form

$$P_i = \sum_{j=1}^{3} x_{ij}^{(1)} E_j + \sum_{j=1}^{3}\sum_{k=1}^{3} x_{ijk}^{(2)} E_j E_k + \sum_{J=1}^{3}\sum_{k=1}^{3}\sum_{l=1}^{3} \chi_{ijkl}^{(3)} E_j E_k E_l + \dots , \qquad (1.46)$$

where i, j, k and $l = 1$, 2, 3. In most cases, it is assumed that $P_1 = P_x$, i.e., the component of field polarization is parallel to the x axis, $E_2 = E_y$, and so on.

For the isotropic medium with no free charges, the Maxwell equations maybe written in the form [2–5]

$$\text{rot rot } \mathbf{E} + \frac{n^2}{c^2}\frac{\partial^2}{\partial t^2}\mathbf{E} = -\frac{1}{c^2}\frac{\partial^2}{\partial t^2}\delta\mathbf{P}, \tag{1.47}$$

or in the operator form

$$\nabla\times\nabla\times\mathbf{E} + \frac{n^2}{c^2}\frac{\partial^2}{\partial t^2}\mathbf{E} = -\frac{1}{c^2}\frac{\partial^2}{\partial t^2}\delta\mathbf{P}, \tag{1.48}$$

where $\delta\mathbf{P}$ is the nonlinear part of polarization of the medium in (1.43) and $n = \sqrt{\varepsilon}$ is the refractive index of the medium which is determined by the linear part in the expression (1.43) for $\mathbf{P}$, c is the speed of light in vacuum.

For the linear medium, taking into account the rules of action with the operators [6]

$$\text{rot rot}(...) = \text{div}(\text{div}(...)) - \text{rot}^2(...) \tag{1.49}$$

or

$$\nabla\times\nabla\times\mathbf{E} = \nabla\cdot(\nabla\cdot\mathbf{E}) - \nabla^2\mathbf{E}, \tag{1.50}$$

the Gauss law in the absence of the free charges in the medium [2]

$$\text{div } \mathbf{D} = 0 \tag{3.6}$$

or

$$\nabla\cdot\mathbf{D} = 0. \tag{1.51}$$

Taking into account the relationships (1.3) for the medium without nonlinearity we have the following form of the wave equation

$$\text{rot}^2\mathbf{E} - \frac{n^2}{c^2}\frac{\partial^2}{\partial t^2}\mathbf{E} = 0 \tag{1.52}$$

or

$$\nabla^2\mathbf{E} - \frac{n^2}{c^2}\frac{\partial^2}{\partial t^2}\mathbf{E} = 0. \tag{1.53}$$

Taking into account the relationship between $\mathbf{E}$ and $\mathbf{D}$ in the form (1.8), the Gauss law cannot be used directly in the form (1.51) for the nonlinear medium, because the expression

$$\text{div } \mathbf{E} = 0 \qquad (1.54)$$

or

$$\nabla \cdot \mathbf{E} = 0 \qquad (1.55)$$

can hold only for isotropic media.

Nevertheless, even when this relationship is not equal accurately to 0, it is quite often almost negligible and, consequently, may be disregarded. Consequently, the wave equation for the nonlinear medium can be reduced to the form:

$$\text{rot}^2\mathbf{E} - \frac{n^2}{c^2}\frac{\partial^2}{\partial t^2}\mathbf{E} = \frac{1}{c^2}\frac{\partial^2}{\partial t^2}\delta\mathbf{P} \qquad (1.56)$$

or

$$\nabla^2\mathbf{E} - \frac{n^2}{c^2}\frac{\partial^2}{\partial t^2}\mathbf{E} = \frac{1}{c^2}\frac{\partial^2}{\partial t^2}\delta\mathbf{P}. \qquad (1.57)$$

The nonlinear wave equation (1.56) is a nonuniform differential equation. The solution of this equation is obtained using the Green functions. Physically, the equation contains the solution of the homogeneous equation (1.52) in the inhomogeneous medium

$$\frac{1}{c^2}\frac{\partial^2}{\partial t^2}\delta\mathbf{P},$$

which has the meaning of the source of electromagnetic forces. As a result of this, the nonlinear interaction is the result of pumping energy to the medium as a result of the process of bonding of the different place.

Usually, the *n*-th order is obtained by mixing *n* + 1 waves. For example, if we observe the second-order of nonlinearity (the case of the three-wave mixing), the nonlinear component of polarization *P*, according to (1.44), has the form

$$\delta P = \chi^{(2)}E^2(t). \qquad (1.58)$$

If it is assumed that $E(t)$ is the result of composition of two waves with different frequencies ω_1 and ω_2, then $E(t)$ can be written in the following form

$$E(t) = E_1 e^{-i\omega_1 t} + E_2 e^{-i\omega_2 t} + \text{c.c.,} \tag{1.59}$$

where the c.c. denotes the complexly conjugate term. Substitution of (1.59) into (1.58) for the nonlinear component of polarization of the medium leads to the expression

$$\delta P = \chi^{(2)} E^2(t) = \chi^{(2)} [|E_1|^2 e^{-i2\omega_1 t}$$
$$+ |E_2|^2 e^{-i2\omega_2 t} + 2E_1 E_2 e^{-i(\omega_1+\omega_2)t} + \tag{1.60}$$
$$2E_1 E_2^* e^{-i(\omega_1-\omega_2)t} + 2(|E_1|^2 + |E_2|^2 + \text{c.c.}].$$

It may be seen that the equation (1.60) contains the following frequency components: $2\omega_1$, $2\omega_2$, $\omega_1 + \omega_2$, $\omega_1 - \omega_2$ and 0. Thus, the three-wave mixing process corresponds to the following nonlinear optical effects: second harmonic generation, generation of waves at the total frequency, generation of waves at the difference frequency and optical rectification, respectively.

1.3.1. Second harmonic generation

In this section we examine as an example the process of second harmonic generation (SHG) of electromagnetic radiation which is very important for applications in practice. As indicated by (1.60), the effect of generation of total frequencies in the nonlinear medium may be described as when radiations at frequency ω_1 and ω_2 arrive in the medium, the electromagnetic wave with the frequency ω_3 forms at the output:

$$\omega_1 + \omega_2 = \omega_3. \tag{1.61}$$

If $\omega_1 = \omega_2 = \omega$, then the waves of a certain frequency travel into the medium, and the radiation at the total frequency is their second harmonics

$$\omega_3 = \omega + \omega = 2\omega. \tag{1.62}$$

The vector $\chi^{(2)}$ (the tensor of the third rank) is referred to as the quadratic nonlinear polarizability of matter. The essential condition for the second harmonic generation is that $\chi^{(2)}$ should differ from zero. This occurs in anisotropic media with no centre of symmetry. Actually, if the matter is isotropic or has a centre of symmetry, then the variation of the direction of the applied electrical field **E** should cause the polarization **P** to change its sign. To satisfy this requirement, the terms containing the odd powers in the decomposition (1.44) should not be present, i.e., the value of $\chi^{(2)}$ should be equal to 0. In addition to this, the medium should not show any absorption for all the interacting waves.

1.3.2. The phase synchronism condition

The generation of radiation at the total (or difference) frequency is most efficient if the wave with the frequency ω_3, arriving at the given element of the volume of the medium from the preceding elements, is located in the required phase with radiation at the same frequency which is generated in this element of the volume. The intensity of generation in this case increases by several orders of magnitude because it builds up along the entire length of the non-linear medium [2–5]. This favourable relationship of the phases is realised if the following equality is fulfilled for the wave vectors of the interacting waves $\mathbf{k}_i$:

$$\mathbf{k}_1 + \mathbf{k}_2 = \mathbf{k}_3. \tag{1.63}$$

The equation (1.63) is the *condition of phase (wave or spatial) synchronism.*

It may easily be noted that the equations (1.61) and (1.63) for the interacting light quanta denote the fulfilment of the laws of conservation of energy $E = \hbar\omega$ and the pulse $\mathbf{p} = \hbar\mathbf{k}$.

The phase synchronism condition may be fulfilled for the waves with different polarizations at specific directions of propagation of these waves in the anisotropic crystals. The formation of synchronism can be illustrated on the example of the surfaces of wave vectors for the most important practical case of a negative uniaxial crystal for which the refractory index of the extraordinary wave is smaller than the refractory index of the ordinary wave $(n_e^\omega < n_0^\omega)$ [7 – 8].

Figure 1 shows the cross-sections of such surfaces by the plane *XZ*, when the *Z* axis is parallel to the optical axis *C*.

Suppose that in the process of interaction in a crystal: $(\omega_1 + \omega_2 = \omega_3)$ waves with frequencies ω_1 and ω_2 have a linear

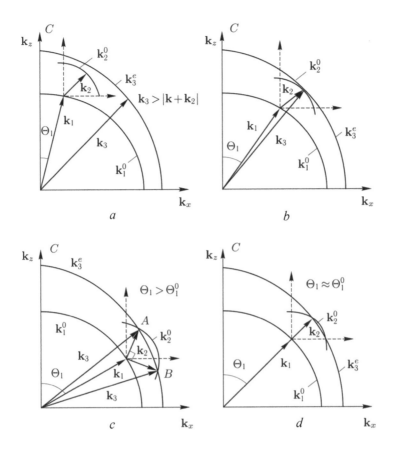

Fig. 1. Mutual position of the surfaces of the wave vectors at the frequencies $\omega_{1,2,3}$ (the type of interaction – 00 → e): a – there is no synchronism at any $\mathbf{k}_2$; b – the case of tangential synchronism; c – the case of critical vector synchronism; d – the case of the one-dimensional critical synchronism. C is the optical axis of the crystal.

polarization and propagate in the crystal as ordinary waves, i.e. vectors of the electric field strength of the waves $\mathbf{E}_1$ and $\mathbf{E}_2$ are orthogonal to the C-axis. To satisfy the synchronism condition in a negative crystal, the wave ω_3 must necessarily be extraordinary, i.e., linearly polarized, with a vector $\mathbf{E}_3$ lying in the plane drawn through the C-axis and to the orthogonal vectors $\mathbf{E}_1$ and $\mathbf{E}_2$. This type of interaction is denoted by *ooe*.

First of all, we construct the surface of the wave vector $\mathbf{k}_1$ (the sphere for $n_0^{\omega_1}$) and fix some direction of it. Taking the end of this vector for the beginning of a new coordinate system with the axes parallel to the original, we construct the surface of vector $\mathbf{k}_2$ (also a sphere for $n_0^{\omega_2}$)). Finally, we construct in the initial coordinate system a surface of the vector $\mathbf{k}_3$ corresponding to an extraordinary wave (an ellipsoid of rotation for $n_0^{\omega_3}$).

Depending on the angle θ_1, different situations are possible. While it is sufficiently small (the case shown in Fig. 1 *a*), $k_3 > |\mathbf{k}_1 + \mathbf{k}_2|$ for any direction $\mathbf{k}_2$, in view of the normal dispersion in the transparency region ($n_0^{\omega_3}$, $n_e^{\omega_3} > n_0^{\omega_1}$, $n_0^{\omega_2}$). However, due to the curvature of the surface $\mathbf{k}_3$, for sufficiently large θ_1, the surfaces $\mathbf{k}_2$ and $\mathbf{k}_3$ touch (the case shown in Fig. 1 *b*), if, of course, the degree of anisotropy is sufficiently large. For a point of tangency, the triangle of vectors $\mathbf{k}_{1,2,3}$ is closed, and condition (1.63) is satisfied. With further increase of θ_1, the tangency is replaced by the intersection at two points *A* and *B* (Fig. 1 *c*). In this case we speak of a critical vector synchronism, and in the case of Fig. 1 *b* – non-critical (tangential). Near it there is a collinear (one-dimensional) critical synchronism, shown in Fig. 1 *d*. In the case of harmonic generation, this kind of synchronism is usually used, and in some crystals it is possible to realize synchronism in the direction $\theta_1 = 90°$ under certain conditions, which in this geometry is already tangential and therefore less dependent on the crystal axis angular detuning, the divergence of the laser beams, and so forth.

1.3.3. Generation of the second optical harmonic (SHG)

As already mentioned above, SHG is a special case for interaction of the form (1.61) when $\omega_1 = \omega_2 = \omega$. For this case the condition of critical collinear synchronism can be written as follows:

$$k_\omega + k_\omega = k_{2\omega}, \qquad (1.64)$$

where k_ω and $k_{2\omega}$ are the moduli of vectors $\mathbf{k}_\omega$ and $\mathbf{k}_{2\omega}$ respectively. The value k is called the wave number. If the wave propagates in a medium with a refractive index n, then $k = 2\pi n/\lambda$. Taking this into account, from (1.64) we get $n_\omega + n_\omega = 2n_{2\omega}$, that is, to perform synchronism with SHG it is necessary that the following condition be satisfied in a nonlinear medium:

$$n_\omega = n_{2\omega}. \qquad (1.65)$$

In some optically anisotropic crystals, one can choose a direction of propagation of the lightwave for which the refractive index,

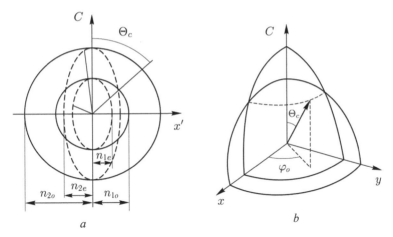

Fig. 2. Cross sections of refractive index surfaces for negative uniaxial crystal.

for example for an ordinary fundamental frequency beam, will be equal to the refractive index of the extraordinary ray of the second harmonics.

Figure 2 shows schematically the section of the surfaces of refractive indices for a negative uniaxial crystal (the optical axis is denoted by C). For this crystal at a given fixed frequency $n_e < n_0$. Current values of extraordinary refractive indices (for an arbitrary angle θ between the wave vector and the optical axis of the crystal) will be denoted by the subscript e, i.e. n^e.

As can be seen from Fig. 2, $n_{ie} = n^e_i\big|_{\theta=90°}$ and $n_{io} = n^e_i\big|_{\theta=0°}$, where $i = 1, 2$.

There are a number of crystals that do not have a centre of symmetry, for which equality $n^\omega_{1o} = n_e(\theta)^{2\omega}$ is fulfilled when waves propagate with frequencies respectively ω and 2ω at a certain angle θ_c to the optical axis of the crystal, as shown in Fig. 2. Consequently, the synchronism condition (1.65) is satisfied in this direction.

Knowing the main values of the refractive indices n_{1o}, n_{2o}, n_{1e}, we can calculate the synchronism angle θ_c. Since the cross-section of the surface of the refractive indices of a uniaxial crystal by the plane $x'C$ (see Fig. 2 a) is a circle and an ellipse, then the equality

$$\begin{cases} n^0_1(\theta) = n_{10}, \\ n^e_2(\theta) = \dfrac{n_{2e}}{\sqrt{1-\varepsilon^2_2\cos^2(\theta)}}, \end{cases} \tag{1.66}$$

holds, where the eccentricity of the ellipse of the refractive indices of the crystal is

$$\varepsilon_2 = \sqrt{1 - (n_{2e}/n_{20})^2}. \tag{1.67}$$

Substituting (1.67) into the synchronism condition (1.65), we obtain for the wave interaction action according to the scheme *ooe*:

$$\cos^2\left(\theta_c^{(ooe)}\right) = \frac{1}{\varepsilon_2^2}\left[1 - \left(\frac{n_{2e}}{n_{20}}\right)^2\right]. \tag{1.68}$$

As can be seen, if two ordinary linearly polarized main radiation waves with coincident planes of polarization interact with an orthogonally linearly polarized extraordinary second-harmonic wave, there is a definite angle at which this interaction will be most effective.

For many nonlinear optical crystals, synchronous SHG is also possible in the interaction of ordinary and extraordinary waves of the fundamental radiation with an extraordinary second-harmonic wave, i.e., the interaction *ooe*. The condition for collinear phase matching in this case is

$$k_{1o} + k_1^e = k_2^e. \tag{1.69}$$

According to this relation, for the refractive indices of interacting waves we obtain the expression

$$\frac{(n_{1o} + n_1^e)}{2} = n_2^e. \tag{1.70}$$

Using (1.70), we can find the value of the synchronism angle:

$$\cos^2\theta_c^{(oee)} \approx 2\frac{\dfrac{n_{1o} + n_{1e}}{2} - n_{2e}}{n_{2e}\varepsilon_2^2 - \dfrac{n_{1e}\varepsilon_1^2}{2}}, \tag{1.71}$$

where

$$\varepsilon_1 = \sqrt{1 - (n_{1e}/n_{1o})^2}. \tag{1.72}$$

From the practical point of view, it should be noted that in the SHG process the synchronism direction in the crystal is characterized by a certain, rather small, angular width, which can be represented as θ_c $\pm \bar{o}$, where $\bar{o}$ is the angle of detuning. This means that in directions with angles exceeding $\theta_c + \bar{o}$ and $\theta_c - \bar{o}$ there is no synchronous interaction of the waves. Usually the angular synchronism width $2\bar{o}$ is a few minutes. In addition, the phase synchronism condition also determines a certain maximum frequency interval ($\Delta\omega$) for interacting waves. The characteristic width of the band of interacting waves, or the required monochromaticity of the initial waves, usually does not exceed several angströms. This implies the requirements for laser radiation at the fundamental frequency:

- the divergence must be less than the angular width of the synchronism $2\bar{o}$);
- the monochromaticity of the radiation should not be greater than its frequency width $\Delta\omega$.

Otherwise, not all the power of the laser beam will participate in the process of adding the wave frequencies.

Considering the intersection surfaces of the indicatrices of the refractive indices for an ordinary beam of the fundamental frequency and an extraordinary second-harmonic wave (see Fig. 2 *b*), one can see the intersection of a sphere with an ellipsoid of revolution. In the angle θ_c the intersection of these surfaces represents a point, and in the angle φ_o is a circle. Accordingly, the efficiency of generation of the second harmonic as the crystal rotates through the angle θ depends strongly on this angle, but practically does not depend on the angle φ_o.

It follows from (1.66) that the prerequisite for effective SHG is fulfillment of the condition

$$\Delta k = k_{2\omega} - 2k_{\omega} = 0. \tag{1.73}$$

Otherwise, if $\Delta k \neq 0$, then a doubled frequency wave generated in some plane (z_{in}), reaching another plane (z_{out}), will not be in phase with a wave of doubled frequency generated in this plane. As a result of the interference of such waves, the two adjacent maxima of this interference process will be spaced apart by a distance, called the 'coherent length', equal to:

$$L_c = \frac{2\pi}{\Delta k} = \frac{2\pi}{k_{2\omega} - 2k_{\omega}} = \frac{\lambda}{2(n_{2\omega} - n_{\omega})}. \tag{1.74}$$

The 'coherent length', in essence, is the maximum length of the crystal, which can be used for SHG. In particular, if $\lambda = 1$ μm, and $n_{2\omega} - n_{\omega}$ 0.01, then $L_c = 100$ μm.

1.4. The Kerr effect in a nonlinear medium

As a rule, it is customary to distinguish between the static and dynamic Kerr effects.

1.4.1. Static Kerr effect

The electro-optical or static Kerr effect is a special case where a slowly varying electric field is applied to the electrodes between which the medium is placed. Under the influence of this electric field birefringence occurs in the material, which manifests itself in the difference in the refractive indices of the medium for light fields polarized parallel or perpendicular to the applied field. The difference in the values of the refractive indices Δn is described by the following equation [3,4]:

$$\Delta n = \lambda_0 K E^2, \tag{1.75}$$

where λ_0 is the wavelength of the light radiation in vacuum, K is the Kerr constant, and E is the magnitude of the applied electric field. The values of the Kerr constant usually lie in the range from 10^{-18} to 10^{-14} m/V^2 for crystalline materials and from 10^{-22} to 10^{-15} m/V^2 for liquids [9].

This difference in refractive index values is due to the fact that the material acts as a wave (phase) plate when light is introduced into it in a direction perpendicular to the applied electric field. Therefore, if the material is placed between two crossed polarizers, the radiation will not be passed through this system, provided that the external electric field is turned off. On the contrary, virtually all radiation will pass through the system at the optimum value of the applied electric field. The large values of the Kerr constant make it possible to observe the passage of light even at sufficiently low values of the applied electric field strength.

Some polar liquids, for example nitrotoluene ($C_7H_7NO_2$) or nitrobenzene ($C_6H_5NO_2$), exhibit high Kerr values. A glass cell filled with a similar liquid is called a Kerr cell. Such cells are often used as light modulators, since the Kerr effect reacts very quickly to changes in the electric field. As a result, light can be modulated by such devices with a frequency of up to 10 GHz. Since the Kerr effect is actually very weak, this requires the use of large control voltages up to 30 kV. This strongly distinguishes the Kerr effect from the Pockels effect, which operates with substantially lower voltages. Another feature of the Kerr cells is that the most accessible material – nitrobenzene – is poisonous. Some transparent crystals can also be used to create Kerr cells, but they have much smaller values of the Kerr constant.

For nonlinear materials, the electrical polarizability of the medium **P** depends on the magnitude of the electric field **E** and is described by the expressions (1.44) or (1.46). For a linear optical medium, only the first term in equation (1.44) is important and the polarization depends linearly on the electric field strength. For materials demonstrating the non-vanishing Kerr effect, in (1.44) the third term in the expansion containing $\chi^{(3)}$ is essential. The even member in (1.44) is usually discarded because of the inverse symmetry of Kerr media.

Suppose that the electric field in the medium **E** is created by a lightwave with frequency ω and amplitude $\mathbf{E}_0$ together with a constant electric field $\mathbf{E}_f$:

$$\mathbf{E} = \mathbf{E}_f + \mathbf{E}_0 \cos(\omega t). \qquad (1.76)$$

Substituting (1.76) into (1.44), taking into account the remarks made above, we can obtain an expression for the polarization vector of the medium **P**. For the static Kerr effect, when $|\mathbf{E}_f| \gg |\mathbf{E}_0|$, all members can be neglected, except for the linear part and the term with $\chi^{(3)}|\mathbf{E}_f|^2\mathbf{E}_0$. As a result, we obtain the expression

$$\mathbf{P} \approx \left(\chi^{(1)} + 3\chi^{(3)} |\mathbf{E}_f|^2 \right) \mathbf{E}_0 \cos(\omega t), \qquad (1.77)$$

which is analogous to the case of a linear connection between the polarization and the electric field of a lightwave and contains a nonlinear susceptibility term proportional to the square of the amplitude of the applied external constant field. Thus, for nonsymmetric media, the nonlinear addition to the refractive index will be proportional to the square of the amplitude of the applied external constant field $|\mathbf{E}_f|^2$ and, as in (1.75), will have the form:

$$\Delta n = \lambda_0 K \left| \mathbf{E}_f \right|^2. \qquad (1.78)$$

This means that a constant electric field induces birefringence in the medium. As a result, the Kerr cell, to which a constant transverse electric field is applied, can act as a phase plate rotating the plane of polarization of a lightwave propagating along it. Therefore, a cell, in combination with a polarizer, can be used as a gate or a light modulator.

The value of the Kerr constant depends on the medium used and, for example, for water is $9.4 \ 10^{-14} \, \text{m/V}^2$, and for nitrobenzene $4.4. \ 10^{-12} \, \text{m/V}^2$. For crystals, the susceptibility of the medium is in most cases a tensor, and the Kerr effect depends on the modification of this tensor.

1.4.2. Dynamic (optical) Kerr effect

The optical Kerr effect is manifested when an electric field is created by a light radiation propagating in the medium. In this case, changes of the refractive index are proportional to the local values of the light intensity. These changes in the refractive index lead to nonlinear optical effects: self-focusing, phase self-modulation, modulation instability, which are the basis for Kerr mode-locking. These effects are observed or are significant only at very high radiation intensities, that is, when using lasers. In the case of the optical or dynamic Kerr effect, intense light radiation in a medium is capable of self-modulating its refractive index without the application of an external electric field. In this case, the electric field in the medium has the form

$$\mathbf{E} = \mathbf{E}_0 \cos(\omega t), \tag{1.79}$$

where $\mathbf{E}_0$ is the vector of the lightwave amplitude.

Substituting (1.79) into (1.44) and leaving only the linear term and the third-order term $\chi^{(3)} |\mathbf{E}_0|^2$, we obtain the following expression for the polarization vector of the medium

$$\mathbf{P} \approx \left(\chi^{(1)} + \frac{3}{4} \chi^{(3)} |\mathbf{E}_0|^2 \right) \mathbf{E}_0 \cos(\omega t). \tag{1.80}$$

Taking into account, as in Section 1, that for nonlinear polarization of an optical medium

$$\mathbf{P} = \mathbf{P}_L + \delta\mathbf{P} = \chi\mathbf{E}_0\cos(\omega t) = (\chi_L + \chi_{NL})\mathbf{E}_0\cos(\omega t) =$$
$$= \left(\chi^{(1)} + \frac{3}{4}\chi^{(3)}|\mathbf{E}_0|^2\right)\mathbf{E}_0\cos(\omega t). \tag{1.81}$$

As can be seen, the dielectric susceptibility of the medium is a superposition of the linear term (χ_L) with a nonlinear addition (χ_{NL})

$$\chi = (\chi_L + \chi_{NL}) = \left(\chi^{(1)} + \frac{3}{4}\chi^{(3)}|\mathbf{E}_0|^2\right), \tag{1.82}$$

from which it follows that

$$n = (1 + \chi)^{1/2} = (1 + \chi_L + \chi_{NL})^{1/2} \approx n_0\left(1 + \frac{1}{2n_0^2}\chi_{NL}\right), \tag{1.83}$$

where $n_0 = (1+\chi_L)^{1/2}$ is the linear part of the refractive index environment. Using the representation in the form of a Taylor series and taking into account that $\chi_{NL} \ll n_0^2$ we obtain the following dependence of the refractive index of the medium on the intensity of the radiation introduced into it:

$$n = n_0 + \frac{3}{8n_0}\chi^{(3)}|\mathbf{E}_0|^2 = n_0 + n_2 I_0, \tag{1.84}$$

where n_2 is the nonlinear refractive index, and I_0 is the amplitude value of the intensity of the lightwave. As can be seen, the refractive index of the medium varies in proportion to the intensity of the light propagating in the medium.

Actually, the value of n_2 is very small and for most materials, including for ordinary glasses, is of the order of 10^{-20} m²/W. Nevertheless, the use of laser radiation with an intensity of more than 1 GW/cm² can lead to a significant change in the refractive index of the medium as a result of the dynamic Kerr effect.

The optical Kerr effect is manifested in a number of fundamentally important nonlinear optical effects. These include: temporary self-modulation of the radiation phase, self-induced phase and frequency shifts for pulses of laser radiation propagating in the medium. These processes, along with dispersion, can lead to the emergence of a soliton mode of radiation propagation [10].

1.5. Kerr self-focusing of light in a nonlinear medium

At the present time, it is fairly well known that, for powerful light emission, the spatial distribution of intensity in a light beam causes a spatial modulation of the refractive index of the medium in accordance with the spatial distribution of the beam intensity. As a result of self-focusing, the intensity of light on the beam axis increases significantly, which leads to the appearance of another nonlinear optical process, which is called the multiphoton ionization of the medium. This process begins to play an important role when the intensity of light in the beam becomes very large. In turn, the ionization of the medium, caused by the large value of the field strength of the lightwave, leads to a decrease in the refractive index of the medium, that is, to the formation of a negative lens.

As a result, the last light beam will be defocused. Thus, the propagation of high-power laser pulses in a medium with a Kerr nonlinearity of the type (1.84) is associated with alternation of focusing and defocusing of radiation.

Self-focusing very often accompanies the process of the propagation of femtosecond laser radiation in a variety of solid, liquid, and gaseous media. Depending on the properties of the medium and the intensity of the laser radiation, several mechanisms simultaneously accompany the process of changing the refractive index of the medium, the consequence of which is the self-focusing process, but the self-focusing caused by the Kerr nonlinearity of the medium and the self-defocusing caused by the plasma generation will always be determining processes.

The phenomenon of self-focusing was predicted by the Soviet physicist G.A. Askar'yan in 1961 and was first observed by N. F. Pilipetsky and A.R. Rustamov in 1965. The foundations of a mathematically rigorous description of the process were laid by V.I. Talanov.

Of particular interest is the consideration of spatially bounded beams [1–3]. In this case, there is a so-called nonlinear refraction: in a field of a bounded beam, a homogeneous nonlinear medium becomes inhomogeneous: a powerful beam of light emanating through a substance changes its properties, which, in turn, bends the path of the beam itself. Therefore, depending on whether the refractive index is increasing or decreasing in the beam field (that is, from the

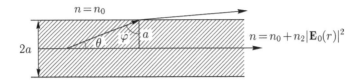

Fig. 3. Propagation of a light beam in a medium with Kerr nonlinearity.

sign of nonlinearity), the energy concentration near the beam axis or, conversely, its scattering is observed.

Consider a light beam of radius $r = a$ with an amplitude depending on the distance along the normal from the axis, which propagates in a medium with a cubic (Kerr) nonlinearity (Fig. 3). According to (1.82) and (1.83), for the refractive index and the dielectric constant of the medium, we can write the following expressions

$$\begin{cases} n = n_0 + n_2 \mid \mathbf{E}_0(r)\mid^2, \\ \varepsilon = \varepsilon_0 + \varepsilon_2 \mid \mathbf{E}_0(r)\mid^2, \end{cases} \tag{1.85}$$

which after a number of transformations

$$n = \varepsilon^{1/2} \approx \varepsilon_0^{1/2}\left(1 + \frac{1}{2}\frac{\varepsilon_2}{\varepsilon_0}\mid \mathbf{E}_0(r)\mid^2\right) = n_0 + n_2 \mid E_0(r)\mid^2 =$$

$$= n_0(1 + \frac{n_2}{n_0}\mid \mathbf{E}_0(R)\mid^2),$$

allow us to obtain the following expressions for the linear part of the dielectric constant of the medium and the coefficient of the Kerr nonlinearity of the dielectric constant of the medium:

$$\begin{cases} \varepsilon_0 = n_0^2, \\ \varepsilon_2 = 2\frac{n_2\varepsilon_0}{n_0}. \end{cases} \tag{1.86}$$

In the general case, when one can not neglect the nonlinear absorption in the medium, the coefficient of nonlinearity of the dielectric constant of the medium is complex:

$$\varepsilon_2 = \varepsilon_2' + i\varepsilon_2''. \tag{1.87}$$

In the absence of nonlinear absorption in the medium (when

$\varepsilon_2' = 0$) provided $\varepsilon_2' > 0$ and the Gaussian profile of the transverse distribution of the intensity of the light beam $I_0(r) = |E_0(r)|^2$ the medium will be self-focusing, since the refractive index of the medium on the beam axis, where the beam intensity is higher, increases. At the boundary of the light beam, due to the decrease in the magnitude of the refractive index, complete internal reflection of the rays is possible. Taking into account the essence of the circuit in Fig. 3 and expressions (1.85), we can calculate the value of the critical angle θ_0, at which the total internal reflection occurs:

$$\theta_0 = \arccos\left(\frac{n_0}{n_0 + n_2 \, | \, E_0(r)|^2}\right). \tag{1.88}$$

Thus, all rays propagating at an angle $\theta > \theta_0$ leave the beam, and those for which $\theta < \theta_0$ deviate to the beam axis. Divergence of rays in a spatially-bounded beam with a flat the phase front is determined by the diffraction angle:

$$\theta_d = \frac{0.61 \, \lambda_0}{n_0 a}, \tag{1.89}$$

where λ_0 is the wavelength of light in vacuum. Depending on the ratio of the angles θ_0 and θ_d the following three physical situations are possible:

1. If $\theta_0 < \theta_d$ the beam spreads out, but more slowly than in a linear medium;

2. If is $\theta_0 = \theta_d$ diffraction spreading is completely compensated by nonlinear refraction.

Since we are dealing with a nonlinear effect, the self-focusing of light radiation is observed only when a certain threshold value of its power is exceeded [11]:

$$P_c = \kappa \frac{\lambda_0^2}{4\pi n_0 n_2}, \tag{1.90}$$

where λ_0 is the wavelength of the radiation wave in a vacuum and κ is a constant, depending on the intensity distribution in the light ray. Despite the fact that there is no general analytic expression for κ, its value can be calculated for most optical ray profiles. The lower limit of the value is $\kappa \approx 1.86225$, it corresponds to Taun's rays, whereas for the Gaussian ray $\kappa \approx 1.8962$. For air $n_0 \approx 1$ and

$n_2 \approx 4 \cdot 10^{-23}$ m²/W for $\lambda_0 = 800$ nm and the critical power value for self-focusing is 2.4 GW, which corresponds to a pulse energy of 0.3 mJ for a pulse of 100 fs duration. For quartz $n_0 \approx 1.453$ W and $n_2 \approx 2.4 \cdot 10^{-20}$ m²/W and the critical power value is $P_c \approx 1.6$ MW.

The regime in which an original waveguide is formed in the medium, along which the light beam propagates without divergence, is called the self-channelization of the beam. The critical radiation power at which this effect occurs can be estimated as [9, 11]

$$P_c = \frac{(1.22)^2 \lambda_0^2 c}{256 n_2}. \tag{1.91}$$

3. If $\theta_0 > \theta_d$ the rays deviate to the beam axis and self-focusing of the light beam occurs. In this case, the nonlinear medium acts like a collecting lens, the focal length of which can be estimated by introducing the so-called diffraction length:

$$L_{df} = \frac{k_0 a^2}{2} \approx \frac{a}{\theta_d}, \tag{1.92}$$

Proceeding from this, the condition $\theta_0 = \theta_d$ is equivalent to

$$L_{df}^0 = \frac{a}{2} \left(\frac{n_0}{n_2 |E_0|^2} \right)^{1/2}, \tag{1.93}$$

where L_{df}^0 is called the self-focusing length.

In the case of high radiation power, when $P \gg P_c$, the behaviour of the light beam can be described in the approximation of geometrical optics, and the focal length is $z_f \approx L_{df}^0$.

We note that in the case of nonlinear absorption in the medium, $\varepsilon_2' < 0$ a decrease in the refractive index will occur in the medium, the result of which will be the self-defocusing of the propagating radiation.

The solution of the system of Maxwell's equations for a medium with a nonlinearity (1.3) in the approximation of a slow change in the amplitude of a wave (in comparison with the period of oscillations of a lightwave) in [9] also reveals the presence of foci for a propagating lightwave whose position is approximately described by expression

$$z_f \approx \xi \frac{k_0 a^2}{(P/P_c)^{1/2} - 1}, \tag{1.94}$$

where ξ is some constant.

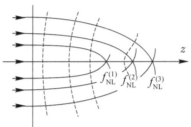

Fig. 4. Multifocus structure of a light beam in a nonlinear medium.

The process of increasing intensity in the focus region of a lightwave when it self-focusing is limited to nonlinear absorption effects associated with the complex part of the nonlinearity coefficient ε_2'', due to such nonlinear optical effects as multiphoton absorption, energy transfer to the components of stimulated scattering, optical breakdown, etc.

As a result, a number of foci are formed on the beam axis, corresponding to successive focusing of different annular zones of the beam (Fig. 4). At each focus, a power of the order of the critical flows in (and is partially absorbed). The total number of foci is limited by the initial beam power, and also by the absorption value. The presence of self-focusing can also lead to instability of the beam, that is, to an exponential increase in small spatial intensity fluctuations.

As a result, the light beam is broken into separate strands of power P_c and the size [9, 11]

$$a_{fil} = \frac{1}{k_0} \left(\frac{n_0}{n_2 \, |E_0(r)|^2} \right)^{1/2}. \tag{1.95}$$

1.6. Plasma self-focusing

Modern advances in laser technologies have made it possible to observe self-focusing during the interaction of intense laser radiation with a plasma [12, 13]. The possibility of plasma focusing of laser radiation was mentioned for the first time by the Soviet physicist G.A. Askar'yan [11]. Self-focusing of light radiation in a plasma can be observed as a result of the manifestation of the following effects: thermal expansion, relativistic and ponderomotive interactions of charges [14, 15].

Thermal self-focusing is due to the collision heating of charges in the plasma irradiated by electromagnetic radiation. An increase in the plasma temperature leads to a hydrodynamic expansion of its volume and, as a consequence, to an increase in the refractive index [16]. Relativistic self-focusing is caused by an increase in the mass of electrons propagating with a velocity close to the speed of light, which leads to the following change in the refractive index n_{rel} [17, 18]:

$$n_{rel} = (1 - \omega_p^2 / \omega^2)^{1/2}, \tag{1.96}$$

where ω is the circular frequency of the lightwave, and ω_p is the plasma frequency:

$$\omega_p = \left(\frac{ne^2}{K_i m_e \varepsilon_0} \right)^{1/2}, \tag{1.97}$$

where m_e and e is the mass and charge of an electron, K_i is the ionization parameter [19], n i concentration.

Ponderomotive self-focusing is due to the presence of a ponderomotive force that pushes electrons from the region where laser radiation is more intense, which leads to an increase in the refractive index and, accordingly, to the focusing of light [19].

The description of the contribution of all the above-mentioned processes to the phenomenon of self-focusing is a rather difficult problem, but there is a certain threshold value of the laser radiation power at which plasma self-focusing begins to manifest itself

$$P_c = \frac{mc^5 \omega^2}{e^2 \omega_p^2} \approx 17 \left(\frac{\omega}{\omega_p} \right)^2, \text{ GW} \tag{1.98}$$

where m is the electron mass, c is the speed of light in vacuum, ω is the circular frequency of the radiation, e is the charge of the electron, and ω_p is the plasma frequency. For an electron density of 10^{19} cm^{-3} and a radiation wavelength of 800 nm, the critical power for laser radiation at which plasma self-focusing will be observed will be about 3 TW. This value of power is easily achieved in modern lasers, the real power of which is already close to the petawatt value (PW). For example, with a pulse duration of 50 fs and an energy of 1 J, the peak pulse power is 20 TW.

Therefore, self-focusing in plasma can compensate for natural diffraction and lead to channeling of the laser beam. Such an effect can play a positive role in a number of cases, in particular, it can contribute to an increase in the interaction length of laser radiation with the medium. This is important, for example, for the process of laser particle acceleration, laser cutting and laser ablation, and also for the generation of high harmonics.

1.7. Phase self-modulation of light radiation

Phase self-modulation (PSM) of light radiation is a nonlinear optical effect, consisting in the dependence of the phase of the light pulse on its intensity due to the Kerr effect [1–5].

$$n(t) = n_0 + n_2 I(t),$$ (1.99)

where $I(t)$ is the time-dependent intensity of the laser radiation, which is capable of modulating the phase of the lightwave.

According to (1.99), after the laser pulse passing through the distance $z = L$ in the medium, the latter acquires a nonlinear phase shift

$$\Phi(t) = \omega_0 t - kL = \omega_0 t - \frac{2\pi}{\lambda_0} n(I)L,$$ (1.100)

where ω_0 and λ_0 are the central frequency and the corresponding wavelength of the light pulse in vacuum.

By analogy with expression (1.100), due to the dependence of the refractive index of the medium on the radiation intensity, the time dependence of the field intensity of the laser pulse leads to the appearance of a time-dependent change in the radiation frequency:

$$\omega(t) = \frac{\partial \Phi(t)}{\partial t} = \omega_0 - \frac{2\pi L}{\lambda_0} \frac{\partial n(I)}{\partial t} = \omega_0 - \frac{2\pi L}{\lambda_0} n_2 \frac{\partial I}{\partial t},$$ (1.101)

which, in turn, depends on the sign of n_2.

As can be seen, the leading edge of the pulse shifts to low frequencies, and the back edge to high frequencies. Near the maximum of the pulse intensity, the phase/frequency change is almost linear.

When an ultrashort monochromatic pulse of a Gaussian form with a central frequency ω_0 propagates in the Kerr medium, its intensity as a function of time can be represented in the form

$$I(t) = I_0 \exp\left(-\frac{t^2}{\tau^2}\right),\qquad (1.102)$$

where I_0 is the amplitude value of the intensity, and 2τ is the half-width of the pulse.
Then

$$n_2 \frac{\partial I(t)}{\partial t} = -n_2 I_0 \frac{2t}{\tau^2} \exp\left(-\frac{t^2}{\tau^2}\right).\qquad (1.103)$$

Taking into account (1.103) and assuming $n_2 > 0$, according to (1.101), for the instantaneous pulse frequency we obtain expression

$$\omega(t) = \omega_0 + \frac{4\pi L n_2 I_0}{\lambda_0 \tau^2} t \exp\left(-\frac{t^2}{\tau^2}\right).\qquad (1.104)$$

According to (1.104), near the intensity maximum ($t \cong 0$), the pulse frequency changes practically linearly, which can be represented in the form

$$\omega(t) = \omega_0 + \alpha_1 t,\qquad (1.105)$$

where

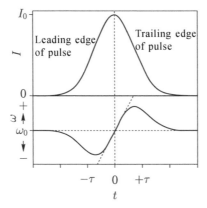

Fig. 5. Time form of a pulse propagating in a nonlinear medium (upper curve). Dependence of the change in the pulse frequency with PSM (lower curve).

$$\alpha_1 = \left(\frac{\partial \omega}{\partial t}\right)_{t=0} = \frac{4\pi L n_2 I_0}{\lambda_0 \tau^2}. \tag{1.106}$$

Figure 5 shows the plot of the intensity of the light pulse as a function of time and the corresponding graph of the dependence of the instantaneous pulse frequency on time. The last graph illustrates the increase in the radiation frequency (blue shift) for the trailing edge of the pulse ($t > 0$) and the decrease in the emission frequency (red shift) for the leading edge of the pulse ($t < 0$).

The effect of accelerating the trailing edge and slowing down the front caused by the phenomenon of PSM results in the squeezing of the light pulses. In pulses of sufficient power, a balance can be observed between the nonlinearity compressing the pulse and the dispersion, leading to pulse broadening. The signal thus obtained is an optical soliton.

Phase self-modulation can play a positive and negative role in the transmission of information over fibre-optic communication lines. Its negative aspects include the possibility of pulse broadening and the impact on its stability. On the other hand, the change in the pulse spectrum can be used for optical switching and obtaining shorter signals. With this help, it is also possible to improve the amplification of radio frequencies in microwave optical communication lines.

References

1. Slabko V.V., Zakarlyuka A.V., Lyamkina N.E., Nonlinear optics: abstracts lectures, Krasnoyarsk: IPK SFU, 2008.
2. Matveev A.N., Optics, Moscow, Vysshaya shkola, 1985.
3. Akhmanov S.A., Nikitin S.Yu., Physical optics, Moscow, Izd-vo MGU, 1998.
4. Evtikhiev N.N., et al., Information optics, Moscow Energy Institute, Moscow, 2000.
5. Dmitriev V.G., Tarasov L.V., Applied nonlinear optics, Moscow, Fizmatlit, 2004. 512 p.
6. Born M., Wolf E., Fundamentals of optics, Moscow, Nauka, 1973.
7. Sivukhin D.V., General course of physics. Optics. V. 4, Moscow, Nauka, Nauka, 1980.
8. Landsberg G.S., Optics, Moscow, Fizmatlit, 2003.
9. Akhmanov S.A., et al., Usp. Fiz. Nauk, 1967, V. 93, No. 9, pp. 19–70.
10. Agrawal G., Non-linear fiber optics, Moscow, Mir, 1996.
11. Askar'an G.A., Usp. Fiz. Nauk, 1973. V. 111, No. 10, pp. 249–260.
12. Cerullo G., Silverstri S., Rev. Sci. Instrum., 2003, No. 74. pp. 1–23.
13. Borisov A.B., et al., Phys. Rev. Lett., 1992, V. 68, pp. 2309–2312.
14. Mori W.B., Phys. Rev. Lett., 1988, V. 60, pp. 1298–1301.
15. Perkins F.W., Valeo E.J., Phys. Rev. Lett., 1974, V. 32, pp. 1234–1237.
16. Max C.E., et al., Phys. Rev. Lett., 1974, V. 33, pp. 209–212.

2

Filamentation Phenomena and Generation of Supercontinuum in Propagation of Laser Pulses in a Nonlinear Medium

Introduction

The phenomenon of filamentation of high-power pulsed laser radiation consists in localizing the energy of the light field in a thin extended filament under the action of self-focusing in a medium and nonlinearity in a self-induced laser plasma that limits the collapse of the beam [1]. As a result of self-focusing, the intensity of the laser pulse increases and its width decreases, but 'collapse' does not occur due to the defocusing effect of the electron plasma created by multiphoton ionization of the molecules of the medium. The formation of plasma is due to the fact that the photons of laser radiation knock out electrons from the molecules of the medium. Since the plasma has a smaller refractive index than the medium, it behaves formally as a scattering lens and begins to defocus the beam, reducing its intensity in the centre. After passing through the plasma region, the laser beam continues its movement and the situation repeats. As a result, balancing the processes of self-focusing and defocusing, the laser beam is able to overcome considerable distances without diffraction divergence. In the zone of maximum intensity, a focus, moving along the axis of propagation of the pulse, is recorded, the trace of which is called a *filament*, and the process of formation of such structures is called *filamentation*.

It should also be noted that, in addition to high intensity, the laser pulse should also have a short duration (femtosecond). Otherwise, instead of multiphoton ionization of the medium, a multistage ionization process may begin in which the electron concentration becomes so large that they begin to ionize the molecules even far from the region occupied by the propagating laser beam. This leads to an imbalance between self-focusing and defocusing, and the laser beam ceases to be focused and quickly diverges. Filamentation is observed in gaseous, liquid and solid transparent dielectrics and is accompanied by the formation of plasma channels, the super-broadening of the frequency and angular momentum spectra, and nonlinear optical effects. Self-focusing (see section 1.3) of radiation is the main physical cause of the formation of extended light filaments. The phenomenon of self-focusing of electromagnetic waves in general form was predicted in 1962 by G.A. Askar'yan.

Unique experiments on the non-stationary self-focusing of pulsed radiation were performed by A.M. Prokhorov and V.V. Korobkin, W. Lin and S.L. Chin [2]. Self-focusing of a powerful picosecond laser pulse in air at a distance of 25 m from the exit from the laser system was recorded by N.G. Basov, P.G. Kryukov, Yu.V. Senatsky, and S.V. Chekalin [3].

Self-focusing of pulsed laser radiation is a threshold effect, for its observation it is necessary to exceed the radiation power over the critical power of self-focusing, which varies from several megawatts in a condensed medium to a gigawatt level in gases of atmospheric density.

With the development of high-power femtosecond laser systems, it became possible to observe self-focusing and filamentation of collimated radiation in atmospheric air. The first experiments on filamentation in the air of radiation from a titanium–sapphire laser system were performed in the laboratory of Prof. J. Moore at the University of Michigan (USA), in the laboratory of Prof. A. Mysyrovich at the Polytechnic School Palezo (France) and at the Centre for Optics, Photonics and Lasers of the Laval University (Quebec, Canada), by a joint Canadian–Russian group of the International Training and Education Laser Centre of the M.V. Lomonosov Moscow State University and the Center for Optics, Photonics and Lasers at the Laval University under the guidance of professors V.P. Kandidov and S.L. Chin. In these experiments with pulses of 100–250 fs duration and a peak power of about 10 GW at a wavelength of 775–800 nm, filament formation with a length of

10–50 m and the formation of coloured conical emission rings in the visible spectral range were recorded [4].

The filament of a laser pulse is a self-forming guiding system in which an increase in the intensity of the light field occurs, together with filtration and emission at the fundamental mode axis, just as in optical fibres [4]. In a long filament with a high intensity of the light field, the intensity of nonlinear optical interaction with the medium increases. One of the manifestations of this interaction is the **superbroadening of the pulse spectrum, or the generation of a supercontinuum** whose spectral band, for example, for femtosecond pulses at a wavelength of 0.8 μm, can extend from the visible to the near infrared range [4].

$$\frac{U_i}{\hbar\omega_0} > 2, \tag{2.1}$$

where U_i is the width of the band gap of the material, and ω_0 is the radiation frequency.

The width of the anti-Stokes band of the supercontinuum increases with the width of the forbidden band exceeding the threshold value. It has also been experimentally established that the broadening of the supercontinuum spectrum into the blue region does not depend on the parameters of the medium and radiation, but is determined only by the relation (2.1).

A key physical factor for the generation of a supercontinuum by ais the presence of an additive to the refractive index, depending on the intensity of the laser radiation, in a medium with Kerr nonlinearity (1.84).

The appearance of an intensity-dependent additive to the refractive index in the case of short laser pulses leads to a physically significant phase modulation of the laser field – phase self-modulation (PSM) (see section 7, chapter 1).

Taking into account the expressions (1.99), (1.100), we represent the nonlinear phase shift of a pulse passing a distance L in a medium with a Kerr nonlinearity, in the form

$$\Phi_{nl}(t) = \frac{\omega}{c} n_2 L I(t). \tag{2.2}$$

According to equations (1.99), (1.100) the time dependence of the intensity of the field in the light pulse leads to the formation of the time-dependence deviation of the laser pulse frequency:

$$\Delta\omega = \frac{\partial\Phi_{nl}(t)}{\partial t} = \frac{\omega}{c}n_2 L\frac{\partial I}{\partial t}. \tag{2.3}$$

Using (1.102) and (2.3), one can estimate the maximum spectral broadening of the pulse induced by the phase self-modulation effect as:

$$\Delta\omega = \frac{\omega}{c}n_2 L\frac{I_0}{\tau}, \tag{2.4}$$

where I_0 is the peak intensity of the light pulse, τ is the pulse time.

Thus, the effect of phase self-modulation of a radiation pulse in a nonlinear medium leads to a broadening of its spectrum in proportion to the distance travelled.

To obtain more accurate expressions for the time envelope of the pulse $E(r, z, t)$ and the nonlinear phase shift induced by the PSM phenomenon, we usually use the elementary PSM theory, based on the approximation of slowly varying amplitudes and taking into account the dispersion in the first order of the power-law expansion of the propagation constant $\beta(\omega)$ for light pulses propagating in the volume of a nonlinear material [6]:

$$\beta(\omega) \approx \beta(\omega_0) + u^{-1}(\omega - \omega_0), \tag{2.5}$$

where ω_0 is the central frequency of the light pulse, and $u = (\partial\beta/\partial\omega|_{\omega=\omega_0})^{-1}$ is the group velocity.

In the accompanying coordinate system $(r, z, t = t'-z/u)$ [6], where t' is the actual time, the evolution of the nonlinear phase shift for the light pulse can be presented in the form

$$\delta\varphi_{nl}(t) = \frac{\omega}{c}n_2 z I_0(r,t), \tag{2.6}$$

where $I_0(r, t)$ is the envelope of the intensity of the light pulse.

Using (2.6), we obtain an expression for the deviation of the instantaneous field frequency from the central frequency ω_0:

$$\delta\omega_{nl} = \frac{\partial(\delta\varphi_{nl})}{\partial t} = \frac{\omega}{c}n_2 z\frac{\partial I_0(r,t)}{\partial t}. \tag{2.7}$$

For example, for a pulse with a quadratic dependence of the envelope on the running time

$$I_0(t') \approx I_0(0)\left(1 - \frac{t^2}{\tau_0^2}\right),$$ (2.8)

the frequency shift becomes

$$\delta\omega_{nl} = 2\frac{\omega}{c}n_2\frac{I_0}{\tau_0^2}tz.$$ (2.9)

The power density spectrum of the pulse with phase self-modulation of the form (2.8) is calculated by the formula

$$S(\omega) = \left|\int_0^\infty I(t)\exp[i\omega t + \delta\varphi_{nl}]dt\right|^2.$$ (2.10)

Since we have only considered the first order in material dispersion, the time-dependent self-action of the light pulse should lead to a symmetric broadening of its spectrum. However, as follows from the results of experimental studies, the asymmetry of the spectral broadening arises already at moderate intensities of laser radiation, which is associated with the inclusion of a number of physical mechanisms for the interaction of radiation with matter. The three most important mechanisms are related to spatial self-action, the formation of the shock front of the pulse envelope, and the finite time of the nonlinear-optical response of the medium. Below we give only a brief description of each of these factors.

The spatial self-action is related to the Kerr nonlinearity of the medium. In a manner similar to the time profile of the envelope intensity of the laser field $I_0(t)$ leading to the modulation of the pulse phase, the inhomogeneity of the intensity of the laser field in the beam $I_0(r)$ on the transverse coordinate r forms a nonlinear lens $n(r) = n_0 + n_2I_0(r)$, which, in turn, leads to self-focusing or self-defocusing of the beam as a function of the sign of n_2. The increase in the intensity of laser radiation in a self-focusing beam leads to an increase in the efficiency of nonlinear optical interactions. As was shown above, self-focusing is accompanied by an uncontrolled change in the intensity and phase of the laser pulse and leads to a complex spatial dynamics of the laser beam, one of the interesting manifestations of which is the decay of the laser beam into thin filaments. Due to the high intensity of laser radiation, filaments are sources of broadband optical radiation. However, the control of the properties of such radiation is an intractable task.

The formation of the shock front of the envelope of the light pulse is due to the dependence of the group velocity of the pulse on the intensity. In an environment with $n_2 > 0$, the nonlinearity leads to a positive additive to the group velocity. The pulse maximum under these conditions propagates more slowly than its fronts. The rear edge of the pulse becomes more steep, and the leading edge is more gently sloping. In the frequency representation, such a transformation of the light pulse leads to an asymmetry of its intensity spectrum. For $n_2 > 0$, the maximum of the pulse spectrum shifts to the low-frequency region, and the short-wave wing of the spectrum turns out to be much longer than the long-wavelength one.

The effects associated with the finite time of the nonlinear response of the medium become especially noticeable for pulses of short duration, at which the nonlinearity of the medium can no longer be considered instantaneous. The delay of the nonlinear response is equivalent to the dispersion of the nonlinearity of the medium in the frequency representation. A short pulse propagating in a medium with a delayed nonlinearity undergoes a low-frequency shift. The spectral broadening induced by the retarded nonlinearity is thus equivalent to the Raman scattering. In view of the extreme complexity of the problem, all the above-mentioned effects can approximately be taken into account only in numerical calculations.

Numerical studies of the space-time and spectral transformations of femtosecond radiation in condensed media use the approximation of a slowly varying amplitude [7]. According to this method, the equation for the complex amplitude $E(r, z, t)$ in the travelling coordinate system on the carrier frequency ω_0 has the form [6]

$$2ik_0 \frac{\partial E(r,z,t)}{\partial z} = \hat{T}\Delta_\perp E(r,z,t) +$$

$$+ \int_{-\infty}^{\infty} \frac{1}{1+\Omega/\omega_0} \cdot [k^2(\omega_0 + \Omega) - (k_0 + k'_\omega \Omega)^2]\tilde{E}(r,z,\Omega)\exp(i\Omega t)d\Omega +$$

$$+ \frac{2k_0^2}{n_0}\hat{T}\left[\Delta n_K(r,z,t)E(r,z,t)\right] + \frac{2k_0^2}{n_0}\hat{T}^{-1}\left[\Delta n_p(r,z,t)E(r,z,t)\right] -$$

$$- ik_0\hat{T}\left[\sigma N_e(r,z,t)E(r,z,t)\right] - ik_0\alpha(r,z,t)E(r,z,t),$$

(2.11)

where $\tilde{E}(r,z,t)$ – the time-dependent Fourier-image of the envelope; $\Omega = (\omega - \omega_0)$ is the frequency shift in the spectrum of the harmonic pulse at a frequency ω from the central frequency ω_0; $k_0 = \omega_0 n(\omega_0)$ /c_0 is the wave number; c_0 is the speed of light in a vacuum; n_0 is the refractive index of the condensed medium.

The dependence $k(\omega) = \omega n(\omega)/c_0$ and the parameter $k'_\omega = \partial k / \partial \omega |_{\omega=\omega_0}$, which includes the function $n(\omega)$, approximated by the Sellmeier formula [8], describe the material dispersion in the medium.

The operator $\hat{T} = 1 - (i/\omega_0)\partial/\partial t$ allows reproducing the wave nonstationarity in self-modulation of a pulse, which manifests itself in an increase in the steepness of the trailing edge of the pulse and in the formation of a shock wave of the envelope [6].

The increment of the refractive index $\Delta n_K(r,z,t)$, caused by the Kerr nonlinearity of the medium, is represented in the form of convolution [9]:

$$\Delta n_K(r,z,t) = n_2 \left[(1-g)I(r,z,t) + g \int_{-\infty}^{t} n_2 h(t-\tau) I(r,z,t) d\tau \right], \quad (2.12)$$

where $I = I(r, z, t)$ us the intensity of the light field of the pulse.

The delay function of the nonlinear response, or the so-called Raman response function associated with the combination-active molecular rotations, is given by [8]

$$h(t) = \frac{\tau_1^2 + \tau_2^2}{\tau_1^2 \tau_2} \exp\left(-\frac{t}{\tau_1}\right) \sin\left(\frac{t}{\tau_2}\right). \quad (2.13)$$

For fused quartz: $n_0 \approx 1.45$ is the refractive index; $n_2 \approx 3.54 \cdot 10^{-16}$ cm^2/W is the coefficient of the Kerr nonlinearity at quasistationary radiation; $g = 0.18$ is the partial contribution of the Raman response; $\tau_1 = 32$ fs, $\tau_2 = 12.5$ fs [8].

The nonlinearity of the refractive index Δn_p and the cross section of the inhibitory absorption σ for the field of a light pulse in a plasma are determined by the expressions

$$\Delta n_p(r,z,t) = -\frac{2\pi e^2 N_e(r,z,t)}{n_0 \omega_0^2 m_e} \quad (2.14)$$

and

$$\sigma = \frac{4\pi k_0 e^2}{n_0^2 \omega_0^2 m_e} \frac{v_c}{\omega_0}, \tag{2.15}$$

where $v_c \approx 10^{14}$ s^{-1} is the frequency of electron–ion collisions, m_e and e are the electron mass and charge.

The concentration of the free electrons in the laser plasma $N_e = N_e (r, z, t)$ is described by the kinetic equation

$$\frac{\partial N_e}{\partial t} = W(I)(N_0 - N_e) + v_i N_e, \tag{2.16}$$

where N_0 is the concentration of the neutral atoms (for fused quartz $N_0 = 2 \ 10^{22}$ cm^{-3}); W is the rate of field ionisation determined by the Keldysh equation [10],

$$v_i(r,z,t) = \frac{e^2 |A(r,z,t)|^2}{2U_i m_0 (\omega_0^2 + v_c^2)} v_c \tag{2.17}$$

– frequency of avalanche ionization; U_i is the width of the band gap of the material (for fused quartz $U_i \approx 9$ eV).

For femtosecond radiation at 800 nm with an intensity of $I \approx 10^{14}$ W/cm^2, characteristic for the filamentation process, the avalanche ionization frequency for fused quartz is $v_i \approx 10^{14}$ s^{-1}, and avalanche ionization makes a significant contribution to the increase in the electron concentration during a femtosecond pulse duration. Equation (2.16) does not take into account the recombination of electrons, since its characteristic time is several hundred femtoseconds and significantly exceeds the duration of a femtosecond pulse.

The weakening of the light field of the pulse in the medium is caused by losses in the generation of the laser plasma, so the absorption coefficient of the laser radiation is defined as

$$\alpha(r,z,t) = \frac{K\hbar\omega_0}{I} W(I)(N_0 - N_e), \tag{2.18}$$

where K is the order of multiphotonization of the ionization process:

$$K = \frac{U_i}{\hbar\omega_0} + 1. \tag{2.19}$$

The radiation at the entrance to the medium can be represented as a spectral-limited impulse with a Gaussian distribution of electrical components of the field amplitude in space and time:

$$E(r,0,t) = E_0 \exp\left[-\frac{1}{2}\left(\frac{r^2}{a_0^2} + \frac{t^2}{\tau_0^2}\right)\right]. \tag{2.20}$$

where E_0 is the amplitude of the pulse; a_0 is the radius of the light beam; τ_0 is the parameter that determines the pulse duration.

Equations (2.11)–(2.20) make it possible to numerically simulate the distribution in space and time of the complex amplitude $E(r, z, t)$, of a femtosecond pulse, the electron concentration in a self-induced laser plasma, and calculate the intensity of the spectral components in the plane of the cross section

$$S(r,z,\omega) = \text{const} \cdot \int_{-\infty}^{\infty} E(r,z,t)\exp(-i\Omega t)dt \tag{2.21}$$

the density of the power of the supercontinuum

$$S(z,\omega) = \text{const} \int_0^{\infty} S(r,z,\omega)dr. \tag{2.22}$$

Equation (2.12) includes the following scale scales [6, 8]:
– the diffraction line by analogy with (1.92):

$$L_{df} = \frac{ka^2}{2}, \tag{2.23}$$

where a is the radius of the laser beam, k is the wave number in the medium;
– dispersion length (the phenomena of the dispersion of the medium of the second and third orders):

$$L_d = \tau_0^2 / (2k_2), L_d' = \tau_0^3 / (2k_3), \tag{2.24}$$

where $k_2 = \partial^2 k/\partial\omega^2$, $k_3 = \partial^3 k/\partial\omega^3$, and τ_0 is the initial pulse duration;
– the length of the nonlinear interaction

$$L_{nl} = \frac{c_0}{\omega n_2 n_0 I_0}, \tag{2.25}$$

which is determined by the nonlinear refractive index n_2 and the peak intensity of laser radiation I_0;

– the length of manifestation of the phenomenon of saturation of self-focusing due to the nonlinearity of the fifth order:

$$L_s = \frac{c_0}{\omega n_4 n_0 I_0^2}, \tag{2.26}$$

where n_4 is determined by the fifth-order nonlinear susceptibility χ^5;
– lengths characterizing spatial scales on which phenomena associated with the formation of plasma and multiphoton absorption due to ionization losses are manifested:

$$L_{pl} = \frac{k m_e c_0^2}{2 \pi e \tau_0 N_0 \sigma^{(K)} I_0^K} \tag{2.27}$$

and

$$L_{MPI} = \frac{n \hbar \omega N_0 \sigma^{(K)} I_0^{K-1}}{2}, \tag{2.28}$$

where $\sigma^{(K)}$ is the cross section for multiphoton photon ionization.

Taking into account such a variety of characteristic scales, the numerical simulation of supercontinuum generation in a condensed medium is a multiparameter problem, the analysis of which on the whole set of physical parameters is extremely complicated and unproductive. In connection with this, the diffraction length (2.23) and the ratio of the peak power of the laser pulse are considered as determining parameters in the numerical modelling of the filamentation process in the case of Kerr self-focusing of laser radiation in a medium to the critical power P_c, determined by the relation (1.91).

To establish the connection between the calculated momentum spectrum S_{comp} (λ, z) and the experiment values of S (ω (λ) z) we use the dependence:

$$S_{comp}(\lambda, z) = S(\omega(\lambda), z) \frac{2 \pi c_0 n(\lambda)}{\lambda^2}. \tag{2.29}$$

Figure 1 shows the results of a numerical experiment with the propagation in fused quartz of a femtosecond pulse with a duration of $2\tau_0$ = 80 fs, a central wavelength of λ_0 = 1900 nm, a radius a_0 = 80 μm, an energy of W =3.77 μJ and a peak intensity of I_0 = 2.7 10^{11} W/cm^2.

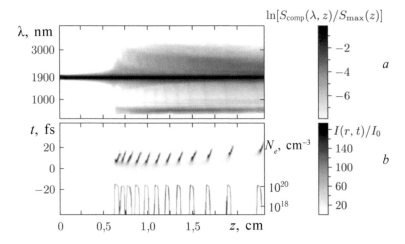

Fig. 1. Tone pictures of the change in the spectrum of a femtosecond pulse $S_{comp}(\lambda, z)/S_{max}(z)$ with a distance z (*a*); the shape of the pulse at the axis I ($r = 0, z, t)/I_0$ and the distribution of the electron density of the laser plasma N_e (z) on the axis (*b*) [6].

For a selected wavelength of laser radiation, the group velocity dispersion in fused quartz is anomalous and the parameter $k_2 \approx 23$ fs²/cm. In the wavelength range from 800 to 2300 nm, the diffraction length of the light beam for femtosecond laser pulses is $L_{df} \approx 3$ cm, and the peak power $P = 5P_c$, which corresponds to the mode of distribution of one filament. In the nonlinear focus, the intensity growth in self-focusing is limited to the generation of a laser plasma, the development of which is determined by the multiphotonic order K, which varies from 6 for $\lambda_0 = 800$ nm to 17 for $\lambda_0 = 2300$ nm.

As can be seen from the curves in Fig. 1 [6], the quasi-periodic localization of the light field in space and time occurs in the process of filamentation of radiation, as well as a significant transformation of the emission spectrum into the Stokes and anti-Stokes regions, as a result of which the supercontinuum spectrum occupies the region from almost 300 to 3000 nm. In this case, the peak value of the light emission intensity on the axis reaches values of $5.2 \cdot 10^{13}$ W/cm², which is almost two orders of magnitude greater than the value of the peak pulse intensity in free space, the beam radius contracts more than an order of magnitude, reaching a size of ~6 μm, and the duration is several periods of the oscillation of the light field. Such a process is cyclically repeated when the pulse propagates in fused quartz.

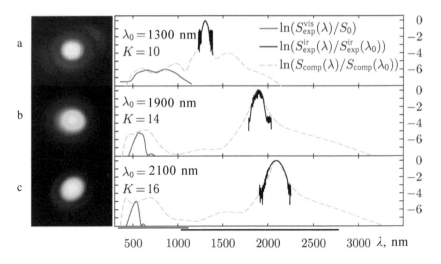

Fig. 2. Spectra of the supercontinuum (right) and its photographs in the far zone (left) during the filamentation of femtosecond radiation under conditions of zero (1300 nm) and anomalous (1900 and 2100 nm) group-velocity dispersion. The spectral sensitivity bands of photodetectors for the visible region of 400–1200 nm and the infrared region of 1100–2700 nm are marked by fat segments on the wavelength axis [11]

Figure 2 [11] shows the calculated (dashed lines) and experimentally measured (solid lines) spectra of the supercontinuum in the filamentation of femtosecond radiation in fused quartz for laser pulses with the following parameters: λ_0 = 1300 nm, W = 2.6 μJ (*a*); λ_0 = 1900 nm,: W = 4.5 μJ (*b*); λ_0 = 2100 nm, W = 8.7 μJ (*s*). For each wavelength, the experimental spectra in the infrared region $S_{exp}^{ir}(\lambda)$ (1100 nm < λ^{ir} < and 2700 nm) are normalized to the maximum value of In the visible area $S_{exp}^{ir}(\lambda_0)$ (400 nm < λ^{vis} < 1100 nm) normalization of experimental spectra is made on the normalization constant S_0 in such a way that the maximum values of $S_{exp}^{vis}(\lambda)/S_0$ coincide with the calculated number of $S_{comp}(\lambda)/S_{comp}(\lambda_0)$ in this spectral band.

For all pulses, the duration was 70 fs, the diameter of the light beam in the constriction was ~100 μm, and the ratio of the peak power to the critical self-focusing power was ~5. The radiating filament region had a length of ~1 mm and was located at a distance ~8 mm from the input plane for radiation with λ_0 = 1300 nm a6.5 mm for radiation with λ_0 = 1900 nm. The wavelength λ_0 =1300 nm corresponded to the zero dispersion of the group velocity in fused quartz, and the wavelengths were λ_0 = 1900 nm and λ_0 = 2100 nm

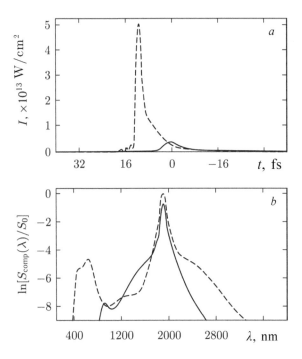

Fig. 3. Calculated dependences of the shapa $I(r = 0/t)$ (a) and the spectrum (i) of a femtosecond-pulse with a central wavelength? $\lambda_0 = 1900$ nm, obtained for medium with an anomalous dispersion of the group velocity. Dashed lines for fused quartz ($U_i = 9$ eV, $K = 14$), solid lines – for the hypothetical medium ($U_{i1} = 4.5$ eV, $K_1 = 7$) [11].

– regions of the strong anomalous dispersion of the group velocity. For radiation at all wavelengths of laser pulses, the experimentally obtained spectra within the dynamic range of spectrometers are close to the calculated spectra. It should also be noted that for pulses, whose wavelength is in the region of the anomalous dispersion of the medium, the intensity of the supercontinuum spectrum in the anti-Stokes region increases nonmonotonically with increasing central wavelength. In this case, an anti-Stokes wing is formed in the form of an isolated maximum in the visible wavelength region. The width of the anti-Stokes wing decreases, and the intensity of the spectral components increases.

This behavior of the dependencies is explained by the fact that the appearance of regions with a high energy density on the axis of propagation of a pulse is inextricably linked with the generation of a laser plasma and a strong phase self-modulation of the light field, which causes broadening of its spectrum. As can be seen

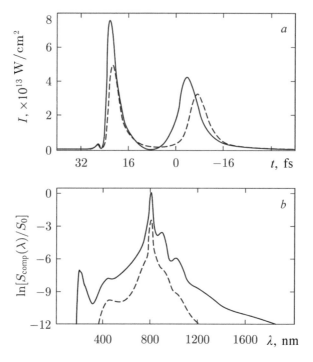

Fig. 4. The calculated dependences of the form I($r = 0$, t) (a) and the femtosecond pulse spectrum obtained for media with normal dispersion of the group velocity at the central wavelength $\lambda_0 = 800$ nm. Dashed line – dependences for fused quartz ($U_i = 9$ eV, $K = 6$) and calculated dependences (solid line) for a hypothetical medium ($U_{i2} = 20$ eV, $K = 13$) (b).

from Fig. 2, the wavelength of the short-wave cutoff of the anti-Stokes components decreases and the anti-Stokes shift increases with increasing multiphoton order, the process of laser plasma generation, which depends both on the band gap and on the energy of the photons used in the radiation (2.19).

The results of the study of the influence of the band gap on the shape of the pulse envelope and the spectrum of the supercontinuum under conditions of anomalous and normal dispersion are shown in Figs. 3 and 4 respectively [11].

Figure 3 shows the calculated pulse profiles on the axis $I(r = 0,t)$ and its spectra obtained for fused quartz with $K = 14$ and a hypothetical medium with $K_1 = 7$ in the region of anomalous dispersion of pulses of 80 fs duration, energy 4 μJ, peak intensity $I_0 = 0.268 \cdot 10^{12}$ W / cm^2, the ratio of peak power to the critical power of self-focusing equal to 5, and the propagation length of 0.7 mm.

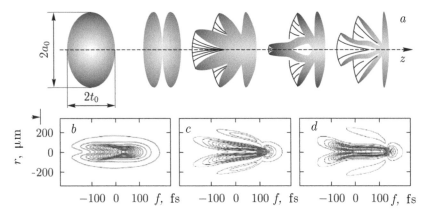

Fig. 5. Illustrating the formation of rings around the filament and the effect refocusing of the pulse: a qualitative picture of the change in the intensity distribution (*a*) and computer-calculated lines of equal intensity in the direction parallel to the direction of propagation of the radiation from the plane pulse (*r,t*) at distances z = 27 m (*b*), 33 m (*c*) and 40 m (*d*). Peak power P_0 = 5P, where P = 6 GW, beam radius 3.5 mm, and the interval between lines of equal intensity in the plane (*r, t*) is equal to $0.25 \cdot 10^{13}$ watts cm^{-2}.

Figure 4 shows the calculated profiles of the pulse on the axis of the light ray – $I(r = 0,t)$ and its spectra obtained for the case of normal dispersion of the group velocity in fused quartz, for radiation with λ_0 = 800 nm, U_i = 9 eV,, K = 6, and for a hypothetical medium with λ_0 = 800 nm, U_{i2} = 20 eV, K_2 = 13. For all radiation pulses, the duration was 80 fs, energy 0.65 mJ, peak intensity 0.111 10^{12} W/cm², the ratio of peak power to critical ~5, and the pulse propagation distance in the medium is 0.91 cm.

As can be seen from Fig. 4, in a medium with normal dispersion of the group velocity with increasing multiphoton order, the momentum, like in the case of an anomalous dispersion of the group velocity, splits into subpulses. However, at a larger value, the trailing edge of the second subpulse becomes steeper, which entails an increase in the anti-Stokes shift, so that $\dfrac{\Delta\omega_{as}}{\Delta\omega_{as2}} \sim \dfrac{K}{K_2}$. With the expansion of the supercontinuum band, a minimum appears in its spectrum, separating the anti-Stokes wing from the central wavelength of the radiation.

Thus, regardless of the nature of the dispersion of the group velocity, the anti-Stokes shift in the spectrum of the supercontinuum upon filamentation of radiation is determined by the multiphoton order of the laser-plasma generation process and increases with increasing K. In multiphoton ionization, which predominates at the

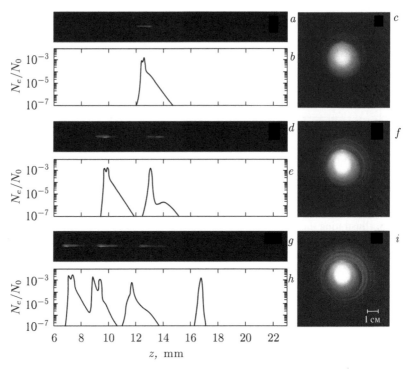

Fig. 6. Splitting of conical emission rings during filamentation of femtosecond pulses with a radiation wavelength of 800 nm and a duration of 35 fs in fused quartz. On the left: photographs of the plasma channels and numerically obtained dependences for the electron concentration on the axis of the plasma channels $N_e(z)$. Right: photographic images of conical emission rings [18].

initial stage of plasma generation, the time gradient of the electron concentration with which the rate of defocusing of the radiation at the tail of the pulse and, consequently, the steepness of the trailing edge are associated increases with increasing K. Phase self-modulation at a large steepness of the trailing edge causes a stronger enrichment of the supercontinuum spectrum with high-frequency harmonics, which is manifested in an increase in the value of the anti-Stokes shift of the frequency.

The anti-Stokes components of the supercontinuum form a conical emission, the divergence of which increases with the growth of their detuning in the short-wave region [1]. The minimum in the spectrum of the supercontinuum separating its anti-Stokes wing from the region in the vicinity of the central wavelength of the radiation is usually explained as the result of the destructive interference of the coherent radiation of the supercontinuum generated by the limited region of the radiating filament [4].

The radiation diverging from the filament axis due to defocusing in the plasma interferes with the light field near the filament, forming around it ring structures. The formation of rings around the filament in the air was first recorded and explained in [12, 13]. The energy losses of laser radiation during filamentation are small, since they are determined only by photoionization of the medium. Therefore, the radiation defocused in the plasma can again focus on the filament axis due to Kerr self-focusing and thereby increase the energy density on its axis (Fig. 5) [15].

The radiation of the supercontinuum is coherent [16], and its interference is the reason for the appearance of a fine structure of the radiation spectrum during the decay of the pulse into subpulses when a distributed source of the supercontinuum is formed in an extended filament [17]. When the pulse is refocused, a sequence of supercontinuum sources is formed in the filament, and as a result of the radiation interference, the continuous spectrum of conical emission splits into a set of discrete rings (Fig. 6) [18].

The effect of refocusing radiation in filaments has a strong influence on the spatiotemporal evolution of radiation during its propagation. As shown in [15], at a peak pulse power slightly exceeding the critical self-focusing power, the process of radiation propagation is close to self-channeling. In this mode, an extended continuous region with a high energy density and electron concentration is formed, whereas in a pulse with a high peak power, the filament, due to multiple plasma defocusing processes and subsequent refocusing, breaks up into a sequence of isolated plasma foci and regions with a high light-field concentration.

As a result of numerical modelling of filamentation of femtosecond laser pulses, the free electron density distributions in the plasma channels on the filament axis were obtained, formed in quartz by femtosecond pulses with an energy of 1.3; 1.9 and 2.4 µJ, respectively (Fig. 6).

As can be seen, as the pulse energy grows a chain of plasma channels forms in the quartz volume arranged in series one after the other. The formation of the chain of several plasma channels is a consequence of repeated refocusing of the laser pulse in a nonlinear medium. It should be noted that the material dispersion of the medium significantly affects the refocusing of the pulse during filamentation. In a medium with high material dispersion, multiple refocusing occurs in a pulse, the peak power of which is hundreds of times greater than the critical power of self-focusing.

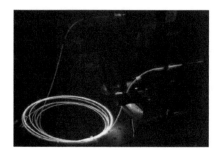 

Fig. 7. Generation of a supercontinuum in optical fibers [6]

It is very important that the conditions for generating a supercontinuum can be realized in optical directing systems, such as optical fibres, photonic crystal fibers and capillaries [1]. The modal dispersion of the radiation and the long length of the nonlinear-optical interaction of the channeled laser pulses make it possible to radically lower the requirements for the power of the laser radiation necessary for generating the supercontinuum. As a result, in such guiding systems it is possible to obtain a high intensity of the light field at a comparatively long length, which ensures a high efficiency of the nonlinear-optical conversion of the pulse spectrum (Fig. 7). Despite the fact that the intensity of the supercontinuum formed in the light-guiding structures is, as a rule, much lower than the intensity of white light generated by powerful focused laser pulses in gases or transparent dielectrics, nevertheless light guides allow to receive radiation supercontinuum with a spectrum overlapping most of the visible range, or even the entire visible and part of the IR ranges, when using pulses with a fairly low peak power of the order of kW [6].

As shown by the results of experiments, in the immunoglobulins, whose peak power is several tens of times higher than the critical self-focusing power in the medium, a number of filaments are formed, the appearance of which is a consequence of the small-scale self-focusing of a high-intensity light field (Fig. 8) [4]. In this case, extended narrow regions with a high energy density are not continuous along the entire length of the filamentation.

Around the filaments originating in the perturbations of the output radiation of the laser, divergent ring structures are formed their interference creates perturbations for the formation of new filaments (Fig. 9) [4]. As a result of the energy competition between the filaments, some of them disappear, and the rest generate the next generation of filaments. As a result, with multiple filamentation,

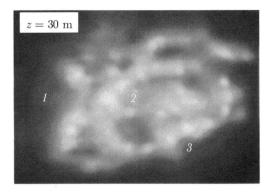

Fig. 8. Distribution of the energy density in the cross section of a beam of pulsed radiation with a wavelength of 800 nm, with a pulse duration of 35 fs, energy 230 mJ and a peak power density of 2.3 TW (~700 P_0) for filamentation in air. 1–3 – 'hot spots' that determine the filaments in the cross section of the beam [4].

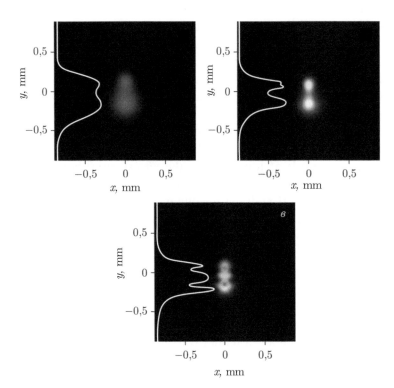

Fig. 9. The picture of the formation of multiple filamentation in a beam of radiation with a duration of 42 fs, a peak power of $P_0/P_c = 4$ when propagating in water. *a* – nucleation, *b* – formation of primary filaments, and *c* – the emergence of new filaments [4].

a dynamic set of extended regions with a high energy density and plasma channels is formed, the location of which in the cross section of the beam changes randomly as the radiation propagates in the medium.

According to the latest data, with increasing pulse power, the density of the number of filaments in the beam section, due to the strong interaction between filaments, tends to saturation. The superposition of the emission of conical emission from such a set of coherent sources in turn leads to the 'washing away' of discrete rings of conical emission and to the formation of a speckle pattern.

2.2. Filamentation of pulsed radiation in gaseous media

An interesting and practically important direction of laser generation of white light is associated with the generation of a supercontinuum when high-intensity femtosecond laser pulses are propagated in gas and liquid media [16–30]. The characteristic peak power of the light pulses of modern lasers used in such experiments lies in the terawatt range, which is many times higher than the critical power of self-focusing in a liquid or in air. The filamentation process is mainly determined by two phenomena: the Kerr nonlinearity of neutral atoms and molecules of the air medium, leading to self-focusing of the light beam, as well as the nonlinearity of the laser plasma that occurs when photoionization of atoms and air molecules in a strong light field leads to defocusing of laser radiation. The free electrons that form as a result of ionization make an essential contribution to the nonlinear response of the gas and have a significant effect on the spectral, temporal and spatial dynamics of the laser pulse. As a result, according to the model of moving foci, the laser pulse focused in the air is transformed into a continuous sequence of points, which creates a luminous filament.

The interest in the phenomenon of filamentation of high-power femtosecond laser pulses in air is largely determined by the prospects for using this phenomenon for environmental monitoring [4, 6].

To study the phenomenon of filamentation and the generation of a supercontinuum in the air, a model was developed for the propagation of high-power femtosecond laser radiation in gases, based, as in section 2.1, using the method of slowly varying amplitudes. In this case, equation (2.11), which describes the propagation of the envelope of a light pulse in a gaseous medium, reduces to the form:

$$2ik_0 \frac{\partial E(r,z,t)}{\partial z} = \hat{T}_{\perp} E(r,z,t) - k_0 k_2 \frac{\partial^2 E}{\partial t^2} + \frac{i}{3} k_0 k_3 \frac{\partial^3 E}{\partial t^3} +$$

$$+ \frac{2k_0^2}{n_0} \hat{T} \left[\Delta n_K (r,z,t) E(r,z,t) \right] + \frac{2k_0^2}{n_0} \hat{T}^{-1} \left[\Delta n_p (r,z,t) E(r,z,t) \right] - \quad (2.30)$$

$$- ik_0 \alpha(r,z,t) E(r,z,t).$$

The first term on the right-hand side of Eq. (2.30) describes the diffraction of the light beam, the second and third the dispersion of the pulse, which is considered in the third approximation of dispersion theory, since a small-scale time structure on the pulse profile is formed in filamentation in air [9].

The Kerr nonlinearity of the air medium is represented by the fourth term. For the considered pulses of femtosecond duration, the contribution of the anharmonicity of the electronic response can be considered instantaneous, whereas the contribution of stimulated Raman scattering to rotational transitions of air medium molecules is nonstationary [20]. As a result, the change in the refractive index caused by the Kerr nonlinearity Δn_K will still have the form (2.12), however, the function a slightly different modified form:

$$h(t) = \theta(t) \Omega^2 \exp\left(-\frac{\Gamma t}{2} \right) \frac{\sin(\Lambda t)}{\Lambda}, \quad (2.31)$$

and

$$\Lambda = \left(\Omega^2 - \frac{\Gamma^2}{4} \right)^{1/2}, \quad (2.32)$$

$\theta(t)$ is the Heaviside function. Parameter g in equation (2.12) for the air medium determines the ratio of the contributions of both mechanisms of the nonlinear response to the overall change in the refractive index of the medium. According to [11], $g = 0.5$; $\Omega = 20$ THz; $\Gamma = 26$ THz.

The fifth term in (2.30) is due to the nonlinearity of the laser plasma, which arises from the multiphoton ionization of air in a strong light field of a femtosecond laser pulse. For air this term also has a different form from (2.14):

$$\Delta n_p = -\frac{2\pi e^2 N_e}{m_e \omega_0^2}, \quad (2.33)$$

where N_e is the concentration of free electrons.

The concentrations of nitrogen photoelectrons (N_e^N) and oxygen (N_e^O) in the air atmosphere ($N_e = N_e^N + N_e^O$) at each point of space obey the ionization equations describing ionization:

$$\frac{\partial N_e^{N,O}}{\partial t} = R^{N,O}(I)(N_0^{N,O} - N_e^{N,O}), \tag{2.34}$$

where N_0^N and N_0^O are the concentrations of nitrogen and oxygen molecules in the Earth atmosphere; $R^N(I)$ and $R^O(I)$ are the ionization rates of nitrogen and oxygen determined by the Perelomov–Popov–Terent'ev model [22], and I is the intensity of the light field.

The last, sixth, term in (2.30) is related to the absorption of the pulse energy during photoionization of the medium. The absorption coefficient in this term is given by the formula:

$$\alpha = \frac{K \hbar \omega_0}{I} \frac{\partial N_e}{\partial t}, \tag{2.35}$$

where ω_0 is the laser radiation frequency; K is the order of multiphotonicity of the ionization process.

The momentum profile and intensity distribution in the transverse section of a femtosecond pulse beam is usually approximated by the Gaussian dependence (2.20).

For the Earth's atmosphere under normal conditions and a central wavelength of 800 nm laser radiation, the main parameters of the medium entering into equation (2.30) have the following values:

$$N_0 \approx 3 \cdot 10^{19} \, \text{cm}^{-3}, \; k_2 \approx 0.2 \, \text{fs}^2 / \text{cm}, \; k_3 \approx 0.1 \, \text{fs}^3 / \text{cm},$$
$$n_2 \approx 4 \cdot 10^{-19} \, \text{cm}^2 / \text{W}, \; \tau_1 \approx 77 \, \text{fs}, \; \tau_2 \approx 62.5 \, \text{fs}.$$

As shown by the results of numerical calculations and experimental studies performed using (2.30), the filament in the air is a continuous sequence of points of nonlinear focusing of the time layers of the pulse. The process begins with a layer in which the power of the self-focused beam begins to exceed the critical power. Following this layer, successive focusing occurs at ever increasing distances from the layers, in which the power decreases with distance from the central layer to the leading edge of the pulse since a significant increase in the intensity of the light pulse occurs in the self-focusing region. This leads to multiphoton ionization of the medium. The

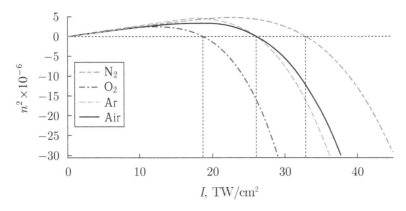

Fig. 10. Dependences of the nonlinear refractive index of gaseous media on the intensity of a femtosecond laser pulse (λ = 800 nm).

electron density in the generated plasma reaches 10^{15}–10^{17} cm^{-3}. The contribution of free electrons to the refractive index of the medium has a negative value:

$$\Delta n_{\text{plasma}} = \frac{\rho}{2\rho_c}, \qquad (2.36)$$

where ρ is the density of the electron plasma produced by the light pulse, and $\rho_c = \varepsilon_0 m_e \omega_0^2 / e^2$ is the critical density at which the plasma becomes opaque (for radiation with a wavelength of 800 nm, the value of ρ_c is ~1.7×10^{21} cm^{-3}). Thus, due to the presence of free electrons in the plasma, a negative addition to the refractive index of the medium leads to the destruction of the Kerr self-focusing process. Figure 10 shows the dependence of the change in the nonlinear refractive index of various gaseous media on the intensity of the femtosecond pulse propagating in them, obtained at room temperature and a gas pressure of 1 atm [23]. As can be seen, the initially observed linear growth of the nonlinear refractive index with increasing pulse intensity passes into saturation, and then begins to decrease and changes the sign to negative. It follows that the process of filamentation in a gaseous medium has the character of a dynamic balance between two effects: Kerr self-focusing and plasma defocusing of the light beam.

Since the laser plasma produced when the intensity of the self-focusing radiation reaches the multiphoton ionization threshold during self-focusing causes defocusing in the momentum layers that follow the focusing layer, this leads to the formation of a supercontinuum intensity in the cross section of the beam behind the nonlinear focus of the rings, the number of which increases toward the trailing edge of the pulse [19].

The dynamic model of moving foci allows one to determine the distance to the beginning of filamentation. Simple arguments are made for this. At the initial stage of the filamentation process, when the intensity of the light beam on the axis does not exceed the photoionization threshold of the medium, a nonlinear focus is formed and the light field changes only as a result of Kerr self-focusing. At the same time, for spectrally limited pulses with a duration of several tens or hundreds of femtoseconds, filamentation in air begins at distances considerably less than the dispersion length, so that the influence of the material dispersion on the nucleation of the filaments can be neglected. If we do not take into account the delay of the nonlinear response of the Kerr medium under the action of a femtosecond pulse, then we can apply a stationary self-focusing model to estimate the distance to the nucleation site of the filament. In this model, the time layer of the pulse, corresponding to its peak power, is focused at the smallest distance, thus setting the position of the beginning of the filamentation. The distance to the beginning of the filament z_{fil} coincides with the self-focusing length of the beam of continuous radiation, the power of which is equal to the peak power of the laser pulse P_0. An empirical formula was obtained in [24] to calculate the length of stationary self-focusing on the basis of a generalization of the results of a numerical study,

$$z_{fil} = \frac{0.367 k_0 a^2}{\left\{ \left[\left(P_0 / P_c \right)^{1/2} - 0.852 \right]^2 - 0.0219 \right\}^{1/2}}, \tag{2.37}$$

where a is the radius of the laser beam in the level e^{-1} from its intensity; P_c is the critical focusing power, determined from (1.90) and depends on the profile of the laser beam. For the air environment $\approx 3.77 \lambda_0^2 / (8\pi n_0 n_2)$ and for laser radiation with a wavelength of 800 nm, the critical power has a value in the range (~1.9–4) GW.

Both numerical analysis and experimental studies show that nonlinear optical phenomena accompanying the propagation of high-intensity pulses in the atmosphere make it possible to obtain radiation from a high-energy supercontinuum. The spectrum of the supercontinuum generated by intense laser pulses in the atmosphere

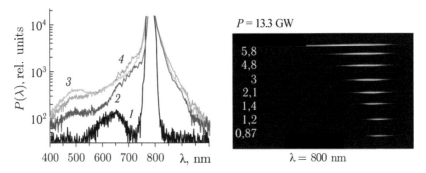

Fig. 11. *a* – dependences of the experimental power spectra of a laser pulse (λ = 800 nm) after the passage of an atmospheric route 85 m long for different initial power in the pulse: P_0 = 5 (1), 13 (2), 45 (3), and 76 (4) GW [25]; *b* – photos of filaments created in the air by pulses of a femtosecond titanium–sapphire laser (λ = 800 nm, τ = 39 fs), focused with a lens with a focal length of 200 mm [26].

(as well as in the liquid) has a pronounced peak at the wavelength of the pump radiation (Fig. 11) [25, 26]. When the supercontinuum has a sufficiently high energy, its spectral density in the visible part of the spectrum decreases rapidly with a characteristic rate of 1–2 orders of magnitude for every 100 nm. The effects of plasma defocusing, the limitation of spatial self-action due to multiphoton absorption, as well as significant phase mismatches of nonlinear optical interactions due to the dispersion profile, prevent the production of more uniform and wider emission spectra.

2.3. Filamentation of laser radiation in the atmosphere

The radiation of the supercontinuum, generated by femtosecond laser radiation during filamentation, is promising for environmental probing [4]. First, this pulsed radiation has a femtosecond duration, which provides a high spatial resolution. Secondly, it has a wide spectrum, in which lie the absorption lines of many polluting impurities, atmospheric gases (ozone, oxygen and nitrogen), as well as greenhouse gases, water vapor, benzene, toluene, etc. (Fig. 12 *a*) [27]. Thus, the supercontinuum generated by the filament makes it possible to provide a broadband spectral-temporal analysis of the multifrequency response of the medium under study with a high spatial resolution.

Compared to a two-wave lidar of differential absorption, when a femtosecond lidar is used, the wide radiation spectrum is generated

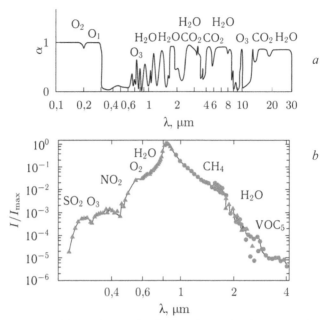

Fig. 12. *a* – Spectral dependence of radiation absorption by the Earth's atmosphere; *b* – spectrum of the supercontinuum.

(Fig. 12 *b*) [28], and there are no big problems in probing narrow absorption lines of substances contained in the atmosphere. The width of the supercontinuum spectrum far exceeds the range of tuning of existing lasers, and the angular distribution of the supercontinuum scattering can provide information on the size and concentration of aerosol particles in the atmosphere. The ability to approximate the source of broadband radiation to the probing object and the existence of large local gradients of the refractive index induced by the filament contribute to an increase in the level of the received signal. It should also be noted that the high-intensity radiation of filaments is capable of causing the dissociation and multiphoton excitation of impurity molecules. The resulting characteristic lines of fluorescence of the medium can also be used for the remote diagnostics of the concentration of pollutants.

To solve the problems of sounding the atmosphere, the problem of the formation of filaments at a remote distance from the source of laser radiation is of great importance. As follows from (2.37), an increase in the diameter of the laser beam *a* leads to an increase in the nonlinear focus, and hence the filament formation region is moved away from the radiation source. Thus, the expansion of the diameter of the laser beam makes it possible to control the location of the localization of the nonlinear focus. However, as practice shows, it is easy to increase the diameter of the laser beam to ~10 cm. Actually, it is necessary to have the possibility of tuning the nonlinear focus within wide limits. For example, if there is a need to adjust the nonlinear focus in the range from 100 m to 1000 m, this entails the need to change the diameter of the laser beam from ~8 cm to ~25 cm, which is a technically difficult task.

As was shown in [24], the introduction of a focusing element into the optical circuit makes it possible to simplify the problem of controlling the nonlinear focus. This technique in the literature is called the scaling of a laser beam. In [24], an expression was obtained for calculating the position of the nonlinear focus in a combined optical system

$$\frac{1}{z_{\text{fil}}^{\text{new}}} = \frac{1}{z_{\text{fil}}} + \frac{1}{F}, \qquad (2.38)$$

where $z_{\text{fil}}^{\text{new}}$ is the new value of the position of the nonlinear focus; F is the right distance of the focusing optical element.

Thus, the combination of the focusing optical element with the process of Kerr self-focusing of the laser beam opens the possibility of controlling the position of the nonlinear focus in the sounding system. The magnitude of the displacement of the nonlinear focus depends on the nonlinear properties of the medium, the power of the laser pulse, and the type of the focusing element. So, if you use a positive lens in the system, the nonlinear focus will approach the radiation source. If the lens is negative, then the divergence of the laser beam provided by it will, on the contrary, move the position of the nonlinear focus away from the radiation source.

The formation of filaments occurs as a result of the redistribution of the radiation intensity both in space (due to Kerr and plasma nonlinearities) and in time (due to phase self-modulation and dispersion). Thus, two types of control of the filamentation process are possible: space control, in which the lateral dimension (spatial scaling) changes, and the laser beam is focused, and also time control, which is performed by changing the initial modulation of the pulse phase [4]. The effect of the initial phase modulation of a laser pulse on filament formation is determined by two factors. The first is to reduce the initial power in the temporal layers of the pulse and does not depend on the sign of the phase modulation. According to (1.100), the pulse duration during phase modulation increases. Due to the pulse stretching in time, at a constant pulse energy, its peak power decreases, as a result, in accordance with expression (2.37), the distance to the filament formation region increases.

The second factor is the preliminary compensation of the dispersion of the group velocity of the pulse, and its result depends on the sign of the phase modulation. In a medium with normal dispersion, a pulse with negative phase modulation contracts in time and its peak power increases. An increase in the peak power of the pulse reduces the distance to the formation region of the filament. On vertical lines, for pulses of terawatt power with phase modulation, it was possible to obtain filaments at heights of up to 2 km.

At the same time, the combination of laser beam scaling, focusing and initial phase modulation is most effective on extended atmospheric routes.

In the vertical sounding of the atmosphere, the features of filamentation of femtosecond laser radiation are observed. These features are due to the dependence of the density of atmospheric air on height, which leads to an increase in the critical power of filament formation with altitude. Therefore, on vertical tracks filaments are formed at a greater distance than on horizontal tracks.

Random fluctuations in the refractive index of the atmosphere cause a random character of the nucleation of filaments in the air and their 'wandering'. In this case, the random displacements of filaments in the plane of the beam cross section are isotropic and their distribution obeys the Rayleigh law [4].

Random fluctuations in the air refractive index, in the case when $P_0 \gg P_c$, initiate the decay of the beam into a multitude of filaments. In this case, the phase fluctuations of the light field caused by turbulence are transformed into intensity fluctuations and play the role of a 'seed' for the formation of a set of filaments [3]. As a result, many filaments are formed which are randomly located in the plane of the cross section of the laser beam and are generated at different distances from the output aperture of the laser system.

Another application of laser filaments is adaptive optics, used in astronomical observations. The latter consists in the creation of 'artificial stars' in the sky (Fig. 13). Created at an altitude of 90 km in the terrestrial atmosphere, a laser filament causes (fluoresce) sodium atoms to glow (fluoresce). Measurements of the radiation flux from this reference 'star' allow to correct the turbulence caused by

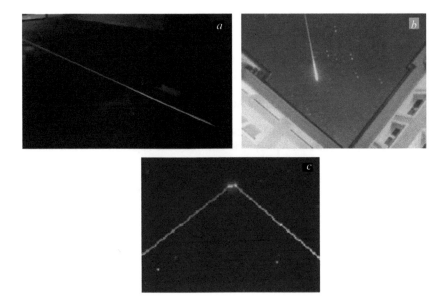

Fig. 13. Photographs illustrating the ignition of 'artificial stars' in the earth's sky using laser filaments [29].

the blurring of the star image. The advantage of such an 'artificial star' is that it can be 'ignited' anywhere in the sky, exactly where it is necessary to correct the image of the nearby object under study.

References

1. Kandidov V.P., et al., Kvant. Elektronika 2009. V. 39, No. 3. P. 205–228.
2. Chin S. L., et al., J. Phys. 2005. V. 83. P. 863–273.
3. Basov N.G.,et al., Zh. Teor. Eksperiment. FIz., 1969. PV 57. P. 1175–1180.
4. Chekalin S.V., Kandidov V.P., Usp. Fiz. Nauk, 2013. V.183, No. 2, P. 133–152.
5. Zheltikov A.M. (Ed.), Special issue of Appl. Phys. 2003. V. B7, No. 2. P. 3–23.
6. Smetanina E.O., et al., Kvant. Elektronika, 2012. Vol. 42, No. 10. P. 913–919.
7. Barbec T., Krausz F., Rev. Lett., 2009. V. 78. P. 545–591.
8. Agrawal G., Non-linear fiber optics, Moscow, Mir, 1996.
9. Stolen R.H., et al., JOSA B. 1989. V. 6, No. 6. P. 1159–1166.
10. Keldysh L.V. Zh. Eksp. Teor. Fz., 1964. P. 49. P. 1945–1957.
11. Smetanina E.O., et al., Kvant. Elektronika, 2012. Vol. 42, No. 10. P. 920–924.
12. Chin S.L., et al., Opt. Commun., 2001. V. 188. P. 181–186.
13. Chin S.L., et al., Opt. Commun., 2002. V. 210, P. 329–341.
14. Kosareva O.G., et. al., J. Nonlin. Opt. Phys. Mater., 1997, No. 2. P. 485–491.
15. Fedotiv V.Y., et al., Appl. Phys. 2010. V. B99. P. 299–306.
16. Chin S.L., et al. Nonlin. Opt. Phys. Mater., 1999. V. 8. P. 121–171.
17. Dormidontov A.E., Kandidov V.P., Laser Physics. 2009. V. 19. P. 1993–2001.
18. Dormidontov A.E., et al., Kvant. Elektronika. 2009. V. 39, No. 7. P. 653–657.

19. Golubtsov I.S., et al., Optika Atmosfery Okeana. 2001. V. 14. P. 335–348.
20. Oleinikov P.A., Platonenko V.T., Laser Physics. 1993. V. 3. P. 618–625.
21. Nibbering, E.T.J., et al., J. Opt. Soc. Am. 1997. V. B14. P. 650–660.
22. Perelomov A.M., et al., Zh. Eksp. Teor. Fiz. V. 50. P. 1393–1409.
23. Loriot V., Opt. express. 2010. V. 17, No. 16. P. 13429–13434.
24. Marburger J.H., Progr. Quant. Electron. 1975. V. 4. P. 35–70.
25. Apeksimov D.V., et al., Optika Atmosfery Okeana. 2009. V. 22, No. 11. P. 1035-1021.
26. Bukin O.A., et al., *ibid.* 2011. V. 24, No. 5. P. 351–358.
27. Seinfeld J., Pandis S., Atmospheric Chemistry and Physics. From Air Pollution to Climat Change. New York: Wiley, 2006.
28. Kasparian J., et al., Science. 2003. V. 301, No. 5629. P. 61–64.
29. Lotshaw W., Ultrashortpulse lasers for space applications. Crosslink, Spring, April, 2011

3

Photonic Crystals

Introduction

Recently, the rapid progress in microelectronics has been facing the problem of the existence of fundamental constraints on the speed of semiconductor devices. In connection with this, an increasing number of studies are devoted to the development of the fundamentals of alternative semiconductor electronics regions: microelectronics of superconductors, spintronics and photonics. The basis of many photonic devices can be photonic crystals – spatially ordered systems with a strictly periodic spatial modulation of the dielectric constant at scales comparable to the wavelengths of radiation in the visible and near infrared ranges. The periodic modulation of the dielectric permittivity of the medium, by analogy with the electronic band structure in the regular crystal lattice of matter, causes the appearance of a photonic band gap, the spectral region within which the propagation of light in a photonic crystal is suppressed in all or in certain definite directions. The presence of a photonic band gap causes the effect of light localization, which makes it possible to use photonic crystal media to create a variety of integrated optical devices, including non-threshold semiconductor lasers, high-Q resonators, optical waveguides, spectral filters, polarizers, etc. The development of this direction started in 1986 [1] and very quickly became fashionable for many leading laboratories in the world.

3.1. Band gaps of photonic crystals

Photonic crystals (PC) are materials that have spatial periodicity of their dielectric permittivity. Under certain conditions, photonic band gaps, that is, spectral regions for which propagation of electromagnetic radiation through a crystal is impossible, can be created in photonic crystals. In this sense, the propagation of light in a photonic crystal is similar to the propagation of electrons and holes in a semiconductor. Electrons, passing through a semiconductor, move in the periodic potential of an ordered lattice of atoms. The interaction between the electron and the periodic potential of the crystal lattice leads to the formation of energy band gaps. As a result, an electron can not propagate in a crystal if its energy falls within the range of the band gap energy. The presence of lattice defects or the violation of its periodicity leads to a local disruption of the band structure and the appearance of new interesting properties of electrons associated with their local states. Therefore, if we replace the electrons with photons and the periodic atomic lattice of the crystal by a periodic change in the dielectric constant of the medium, then it is possible to observe similar phenomena for electromagnetic waves. For photonic crystals, the lattice constant should be in the range 100 nm–1 μm. This range of sizes in photonic crystals can be achieved using both conventional artificial nanofabrication methods and self-assembly methods observed in objects of animate and inanimate nature.

To calculate the optical modes in photonic crystals, it is necessary to solve Maxwell's equations (1.3) for a periodic dielectric medium. Although this problem is much simpler than calculating the propagation of electrons in semiconductors, where it is necessary to take into account multiparticle interactions, it is impossible to obtain an analytical solution of Maxwell's equations for two- and three-dimensional periodic dielectric lattices and the application of numerical methods is required. Nevertheless, many interesting phenomena in photonic crystals can be detected already when considering a simple one-dimensional case of a periodic layered medium [2, 3]. To this end, we consider a one-dimensional photonic crystal made of an infinite number of planar alternating layers of dielectrics of thickness d oriented perpendicular to the direction of z having dielectric permittivities ε_1 and ε_2 (Fig. 1). Optical modes of radiation inside the material are characterized by the wave vector $\mathbf{k} = (k_x, k_y, k_z)$. The medium is also assumed to be nonmagnetic: $\mu_1 = \mu_2 = 1$ and without optical losses.

The radiation propagating in the PC can be represented in the form of the TE- and TM-modes. For TE-modes, the electric field strength vector is parallel to the layer interface, and for the TM-

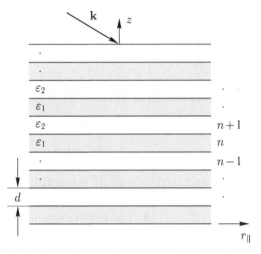

Fig. 1. One-dimensional photonic crystal with an infinite number of planar layers with the spacing d.

modes, on the contrary, the magnetic intensity vector is parallel to the interface. According to this, the complex amplitudes for the TE- and TM-modes are written as:

$$\text{TE-modes}: \mathbf{E}(r) = E(z)e^{i(k_x x + k_y y)}\mathbf{n}_x, \tag{3.1}$$

$$\text{TM-modes}: E_{n,j}(z) = A_{n,j}e^{ik_{zj}(z-nd)} + B_{n,j}e^{-ik_{zj}(z-nd)}, \tag{3.2}$$

where $E(z)$ and $H(z)$ are the corresponding amplitudes of the electric and magnetic fields for the TE- and TM-modes, and $\mathbf{n}_x$ is the unit vector in the x-direction.

When light propagates in a layered medium, it is reflected from the interface of the layers. Therefore, for each n-th layer of the medium, the solution of Maxwell's equations for the corresponding amplitudes of light waves must be a superposition of waves propagating in the forward and backward directions:

$$\text{TE-modes}: E_{n,j}(z) = A_{n,j}e^{ik_{zj}(z-nd)} + B_{n,j}e^{-ik_{zj}(z-nd)}, \tag{3.3}$$

$$\text{TM-modes}: H_{n,j}(z) = A_{n,j}e^{ik_{zj}(z-nd)} + B_{n,j}e^{-ik_{zj}(z-nd)}, \tag{3.4}$$

where $A_{n,j}$ and $B_{n,j}$ are constants that depend on the layer number and dielectric permittivity of the layer medium ε_j, $j = 1,2$. The longitudinal wave number k_{zj} is defined as the projection of the vector **k** on the *x*-axis:

$$k_{zj} = \left(\frac{\omega^2}{c^2} \varepsilon_j - k_{\parallel}^2 \right)^{1/2}$$

(3.5)

where $k_{\parallel}$ is the transverse wave number determined by the following

$$k_{\parallel} = \left(k_x^2 + k_y^2 \right)^{1/2},$$

(3.6)

To find the constants $A_{n,j}$ and $B_{n,j}$, we use the continuity condition for the amplitudes and derivatives of the amplitudes of light waves at the interfaces of regions with different dielectric permittivities: $z = z_n = nd$ and $z = z_{n-1} = (n-1)d$ between *n*-th and $(n + 1)$-th and also $(n-1)$-th and *n*-th layers respectively:

TE-modes:

$$E_{n,1}(z_n) = E_{n+1,2}(z_n),$$

(3.7)

$$E_{n,1}(z_n) = E_{n-1,2}(z_n),$$

(3.8)

$$\frac{d}{dz} E_{n,1}(z_n) = \frac{d}{dz} E_{n+1,2}(z_n),$$

(3.9)

$$\frac{d}{dz} E_{n,1}(z_n) = \frac{d}{dz} E_{n-1,2}(z_n).$$

(3.10)

TM-modes:

$$H_{n,1}(z_n) = H_{n+1,2}(z_n),$$

(3.11)

$$H_{n,1}(z_n) = H_{n-1,2}(z_n),$$

(3.12)

$$\frac{1}{\varepsilon_1} \frac{d}{dz} H_{n,1}(z_n) = \frac{1}{\varepsilon_2} \frac{d}{dz} H_{n+1,2}(z_n).$$

(3.13)

$$\frac{1}{\varepsilon_1} \frac{d}{dz} H_{n,1}(z_n) = \frac{1}{\varepsilon_2} \frac{d}{dz} H_{n-1,2}(z_n).$$

(3.14)

Substituting (3.3) and (3.4) into (3.7)–(3.14), we obtain a system of equations:

$$A_{n,1} + B_{n,1} = A_{n+1,2} e^{-ik_{z2}d} + B_{n+1,2} e^{ik_{z2}d},$$

(3.15)

$$A_{n,1} - B_{n,1} = P_m \left[A_{n+1,2} e^{-ik_{z2}d} - B_{n+1,2} e^{ik_{z2}d} \right], \qquad (3.16)$$

$$A_{n-1,2} + B_{n-1,2} = A_{n,1} e^{-ik_{z1}d} + B_{n,1} e^{ik_{z1}d}, \qquad (3.17)$$

$$A_{n-1,2} + B_{n-1,2} = \frac{1}{P_m} \left[A_{n,1} e^{-ik_{z1}d} - B_{n,1} e^{ik_{z1}d} \right]. \qquad (3.18)$$

In the system of equations (3.15)–(3.18) the factor *pm* depends on the polarization of the radiation propagating in the PC:

$$P_m = \begin{cases} \dfrac{k_{z2}}{k_{z1}}, & \text{for TE-modes,} \\[2mm] \dfrac{\varepsilon_1}{\varepsilon_2} \dfrac{k_{z2}}{k_{z1}}, & \text{for TM-modes.} \end{cases} \qquad (3.19)$$

Thus, we have a system of four equations containing six unknowns, which has no solutions. To solve this system of equations it is necessary to use the Bloch theorem [4], according to which the solution of the system of Schrödinger equations in a periodic medium has the form:

$$\Psi(z) = u(z) e^{ik_B z}, \qquad (3.20)$$

where $u(z)$ is the amplitude of the Bloch function periodic with the period of the lattice 2d, and k_B is the Bloch wave vector.

In our case, the amplitude of the intensity of an electromagnetic wave in a PC with a period 2d will have the form

$$E(z+2d) = E(z) e^{ik_B 2d}. \qquad (3.21)$$

According to the Bloch theorem, the solution of Maxwell's equations for the magnetic field strength $H(z)$ in a periodic environment will have a similar form.

Applying the Bloch theorem to the PC, taking (3.3) into account, leads to the following expression:

$$\{A_{n+1,2} + B_{n+1,2} e^{-2ik_{z2}[z-(n-1)d]}\} = e^{ik_B 2d} \{A_{n-1,2} + B_{n-1,2} e^{-2ik_{z2}[z-(n-1)d]}\}. \qquad (3.22)$$

Since equation (3.22) is valid for any point with coordinate z, we obtain the following recurrence relations for the field amplitudes in the PC:

$$A_{n+1,2} = e^{ik_B 2d} A_{n-1,2}, \tag{3.23}$$

$$B_{n+1,2} = e^{ik_B 2d} B_{n-1,2}. \tag{3.24}$$

The obtained recurrence relations allow us to reduce the system of equations (3.15)–(3.18) to the form:

$$A_{n,1} + B_{n,1} = e^{ik_B 2d}[A_{n-1,2}e^{-ik_{z2}d} + B_{n-1,2}e^{ik_{z2}d}], \tag{3.25}$$

$$A_{n,1} - B_{n,1} = p_m e^{ik_B 2d}[A_{n-1,2}e^{-ik_{z2}d} - B_{n-1,2}e^{ik_{z2}d}], \tag{3.26}$$

$$A_{n-1,2} + B_{n-1,2} = A_{n,1}e^{-ik_{z1}d} + B_{n,1}e^{ik_{z1}d}, \tag{3.27}$$

$$A_{n-1,2} - B_{n-1,2} = \frac{1}{p_m}[A_{n,1}e^{-ik_{z1}d} - B_{n,1}e^{ik_{z1}d}]. \tag{3.28}$$

The resulting system of equations (3.25)–(3.28) is homogeneous and has a nonzero solution if its determinant is zero. Expanding the determinant, we can obtain the characteristic dispersion equation

$$\cos(2k_B d) = \cos(k_{z1}d)\cos(k_{z2}d) - \frac{1}{2}[p_m + \frac{1}{p_m}]\sin(k_{z1}d)\sin(k_{z2}d). \tag{3.29}$$

As $|\cos(2k_B d)| \leq 1$, the solution (3.29) can not exist if the absolute value of its right-hand side is greater than one. The absence of a solution of Eq. (3.29) indicates that *band gaps* appear in the emission spectrum. For example, for the radiation $(k_{z1} = k_1 = [(\omega^2/c^2)\varepsilon_1]^{1/2}$ and $k_{z2} = k_2 = [(\omega^2/c^2)\varepsilon_2]^{1/2}$ normally incident on a one-dimensional PC, for the case when $\varepsilon_1 = 2.25$, and $\varepsilon_2 = 9$, the radiation with a wavelength of $\lambda = 12d$ will be free to pass through the crystal, and the radiation with a wavelength of $\lambda = 9d$ will be reflected.

We note that for the more general case of a one-dimensional PC with a period of the structure $d = a + b$, where a and b are the dimensions of the sections having permittivities ε_1 and ε_2, respectively, the characteristic equation will have the form [5]:

$$\cos(2k_B d) = \cos(k_{z1}a)\cos(k_{z2}b) - \frac{1}{2}[p_m + \frac{1}{p_m}]\sin(k_{z1}a)\sin(k_{z2}b). \tag{3.30}$$

For each Bloch vector k_B equations (3.29), (3.30) define the dispersion relations $\omega(k_\parallel)$. If all possible values of the dispersion relations are shown in the figure, then this figure will be a zone diagram of a one-dimensional PC. As an example, Fig. 2 shows the band diagram for a one-dimensional photonic crystal calculated for the case $\varepsilon_1 = 2.33$ and $\varepsilon_2 = 17.88$, illustrating the existence of band gaps in the PC, i.e., the impossibility of propagation in a crystal of radiation with certain frequencies in certain directions.

We note that the existence of modes propagating in the PC is possible even if the longitudinal wave number k_{zj} is imaginary. The Bloch vector determines the boundaries of the *band gaps*, in particular, for the one-dimensional photonic crystal

$$k_B d = \frac{n\pi}{2}. \tag{3.31}$$

Therefore, for the chosen radiation propagation direction in the PC, given by the transverse projection of the wave vector $k_\parallel$, one can always find the range of values of the change in the radiation frequency for which the PC will be transparent even in the range of the frequencies of the radiation that is incapable of propagating into the PC (Fig. 2).

If a light wave propagating in a vacuum is introduced into a one-dimensional photonic crystal, then only those modes for which the condition $k_\parallel < \omega/c$ is satisfied can propagate in it. The one shown

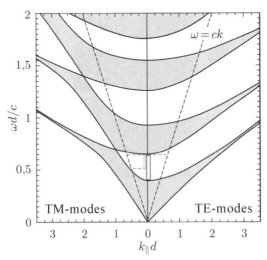

Fig. 2. The band diagram of a one-dimensional photonic crystal. Darkened regions correspond to the band gaps for TE- (the right half-plane) and TM- (the left half-plane) modes of the photonic crystal [2].

in Fig. 2 by the dotted line $\omega = ck$ is called the vacuum line. This line allows one to find the frequency range for which $k_\parallel > k$. In this case, the incident radiation will be effectively reflected, i.e., the photonic crystal will be an ideal mirror. The band gaps will include the frequency range of omnidirectional reflectivity (in Fig. 2 for the main band gap this range is highlighted by coloured rectangles). That is, for light waves with these frequencies in this range there are no propagating waves with $\omega > ck_\parallel$. Therefore, the light of a given frequency incident at any angle to the photonic crystal must be completely reflected.

The concept of a forbidden photon band exists for 2D- and 3D-photonic crystals. Unfortunately, due to the complexity of the task, an analytical solution for photonic bandgap zones of these crystals can not be found in analytical form. Therefore, in practice, as a rule, the result of numerical calculations is used.

Figure 3 shows photographs of photonic crystal semiconductor structures with different types of lattices, as well as the corresponding Brillouin zones. Figure 4 shows examples of photon-crystal zones for two- and three-dimensional PCs, characterizing the dispersion properties of the photonic crystals. As can be seen, in such structures, as well as for one-dimensional PCs, there are forbidden photonic bands for which the propagation of light waves is forbidden. As a result, if the characteristics of the light wave are found to correspond

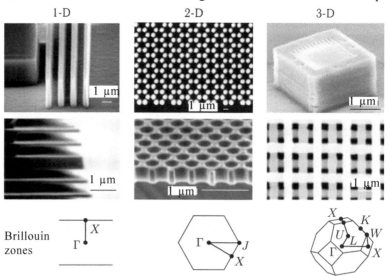

Fig. 3. Different types of photonic crystal structures and corresponding Brillouin zones (the point Γ corresponds to $k = 0$) [6].

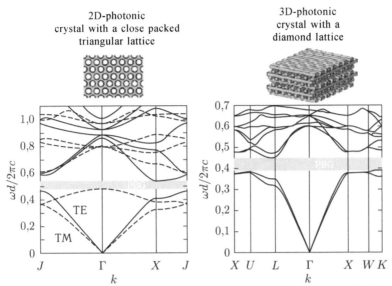

Fig. 4. Photonic band gaps (PBG) for 2D- and 3D-photonic crystals [6].

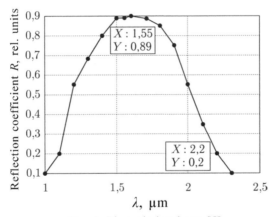

Fig. 5. Photonic band gap [5].

to the photonic band gap, then the light can not enter and propagate in such a photonic crystal.

As an illustration, Figs. 5 and 6 show the results of a study of the normal incidence of a plane light wave with TE-polarization on a two-dimensional photonic crystal [5] with the following parameters: $n_1 = 3.25$ – refractive index of the medium; $n_2 = 1$ is the refractive index of the holes; $r = 0.25$ μm – the radius of the holes; $T_z = 0.6$ μm and $T_x = 1$ μm – the distance between the centres of the holes in the direction of the corresponding axes.

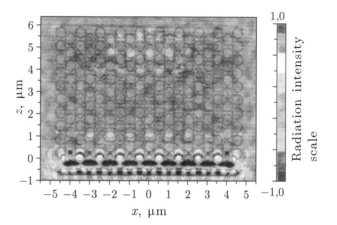

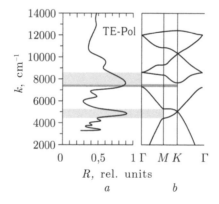

Fig. 6. Illustration of the limitation of the propagation of light (λ = 1.55 μm) in the band gap of the PC in the direction of the z axis (the wave incident on the crystal is completely reflected, without having passed through three nodes of the PC lattice).

Figure 5 shows the calculated spectral dependence of the reflection coefficient for the selected 2D-photonic crystal. As can be seen, for radiation with a wavelength of λ = 1.55 μm, the reflection coefficient is close to unity. This means that this wavelength lies in the photonic band gap of the PC. Figure 6 shows the calculated distribution of the intensity of radiation of a light wave (λ = 1.55 μm) of TE-polarization entering a two-dimensional photonic crystal in the z direction. This figure illustrates that if the wavelength of the incident radiation is in the photonic band gap, then the light in the PC does

Fig. 7. The measured values of the reflection coefficient of TE-polarized waves propagating in the 2D-photonic crystal in the direction Γ–M (a). The gray bands correspond to the position of calculated photonic band gaps (b).

not pass beyond the first three rows of holes and is completely reflected back.

Figure 7 shows the experimental dependence of the reflection coefficient R for TE-polarized radiation in the spectral range from 1.3 to 1.55 μm (Fig. 7 *a*) for the direction $\Gamma-M$ in a two-dimensional photonic crystal whose calculated photonic band structure of which is shown in Fig. 7 *b* [7]. A two-dimensional photonic crystal was formed in a silicon dioxide layer in which a triangular lattice of pores of radius $r = 0.18$ μm spatially separated by a distance $a = 0.5$ μm and a depth of 100 μm. As can be seen, the obtained experimental dependence agrees well with the calculated data and the maxima of the reflection coefficient correspond to the position of photonic forbidden bands.

3.2. Defects in photonic crystals

One of the most interesting properties of photonic crystals is the appearance of localized (defective) modes, which can occur in the photonic band gap when a defect is introduced into the ideal PC.

In such photonic crystals, photons with states within the photonic band gap can not propagate through the PC, but they can localize in the vicinity of the defect region. The simplest way to create a defect in a 2D-photonic crystal is to remove one rod or, conversely, to introduce a defective rod: with a different radius and dielectric constant.

Using the terminology of solid state physics, the states formed by introducing a defect into a periodic structure can be divided into localized and delocalized ones. In electrodynamics, localized states correspond to guiding modes (in the case of extended defects) and resonator (in the case of point defects) modes, and delocalized or emitted modes [9]. The localized states can in turn be divided into modes whose field strengths have the greatest value within the defect (Fig. 8 *a*), and the modes whose field strength has the greatest value at the boundary between the defect and the photonic crystal (Fig. 8 *b*). Delocalized states can also be divided into modes whose field strength decays exponentially in the defect region (Fig. 8 *c*), and the modes whose field strength changes periodically in the defect region (Fig. 8 *d*).

In [10], the mode spectrum was calculated for the one shown in Fig. 8 of a one-dimensional defective photonic crystal based on Si with the following parameters: defect size – L; thickness

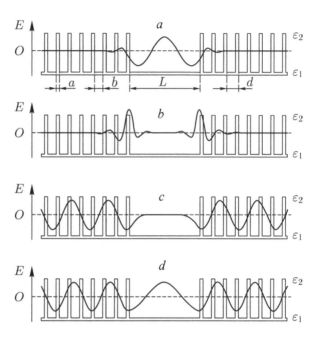

Fig. 8. Schematic representation of defective modes in a photonic crystal [10]. Localized modes: *a* – directional or resonator and *b* – surface. Delocalized modes: *c* – damped in the defect region and *d* – oscillating in the defect region.

of the silicon wafer – *a*; thickness of the air layer – *b*; the period of a photonic crystal – $d = a + b$; $a/b = 0.25$; $b/d = 0.75$; $\varepsilon_1 = 1$; $\varepsilon_2 = 12$; $L = 6d$; the number of layers on each side of the defect $N = 8$. Figure 9 shows the calculated dispersion dependences for the TE- and TM-modes of the defect crystal (localized modes) and gray dispersion dispersion surfaces for the eigenmodes of a homogeneous photonic crystal are shown. It can be seen that the properties of the eigenmodes of a photonic crystal vary greatly depending on the polarization. In particular, as the projection of the wave vector k_z on the *z* axis increases, the band gap for the TE-modes increases, and decreases for the TM-modes. This in turn affects the degree of localization of modes in the region of defects of the PC. Defective modes that fall into the band gap of the PC (Fig. 9) will be localized on the defect the stronger the closer their dispersion curve to the middle of the band gap.

Figure 10 [10] shows the distribution of the transverse components of the light wave fields and the longitudinal component of the Poynting vector for the TE- and TM-modes of the first order

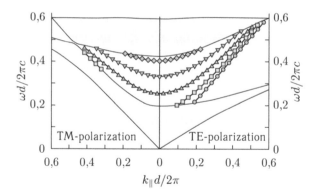

Fig. 9. Dispersion dependences of localized TE- and TM-modes in a one-dimensional PC with a defect. Notation: ○ – mode of zero order; □ – mode of the first order; △ – mode of the second order; ▽ – mode of the third order; ◊ – mode of the fourth order. The shaded regions correspond to the projection of the dispersion of the eigenstates of a defect-free PC on the plane x [10].

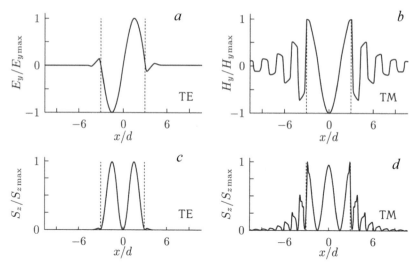

Fig. 10. Distribution of the transverse components of the light wave field for TE- and TM-waves of the first order in the artificial defect region of the PC and the corresponding longitudinal component distributions of the Poynting vector [10].

$(k_x = 0.64\pi/d)$ within the defect localization region. The regions of the defect boundary are indicated by vertical dashed lines. It can be seen from the figure that the field intensity of the TE-mode decreases rapidly with distance from the defect. A different situation is observed for the TM-polarization mode, whose field strength has a sufficiently large value near the boundaries of the defect region.

In [11], the process of formation of defect states in a band gap of the photonic crystal zone of a 1D-photon crystal in the form of a periodic phase lattice was studied by experiments, and the distribution of the refractive index of the crystal varied according to the following law:

$$n(z) = n_a + \frac{1}{2}n_p \sin\left(\frac{4\pi n_a z}{\lambda_0} + \varphi_0\right), \qquad (3.32)$$

where z is the thickness of the film; n_a is the average refractive index of the film; n_p is the modulation amplitude of the refractive index of the film; λ_0 is the wavelength of the light corresponding to the center of the band gap of the PC; φ_0 is the phase term. The period of the PC grating was chosen from the following condition: $\Lambda = \lambda_0/2n_a$. It is well known that such a PC is able to function as a spectral filter. Defects in the PC were created due to the introduction of inhomogeneities in the profile of the refractive index of the lattice. In this work, two phase defects with the size π radian were created in a periodic lattice in the regions $z = (L \pm \Delta z)/2$, where L is the total thickness of the film, and Δz is the distance between the defects.

Figure 11 *a* shows the distribution of the refractive index over the thickness of the PC, in which two phase defects are created, spaced by a distance $\Delta z = 2\Lambda$. Figure 11 *b* shows a photograph of the PC cross section, made in a scanning electron microscope, consisting of

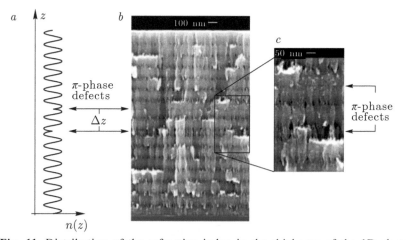

Fig. 11. Distribution of the refractive index in the thickness of the 1D-photonic crystal (*a*); electron microscopic photo of the section of the PC (*b*) and an enlarged image of the PC region containing two phase defects (*c*) [11].

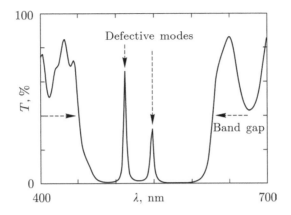

Fig. 12. The transmission spectrum of a PC containing two phase defects [11].

smoothly varying layers of TiO$_2$, successively deposited on a silicon substrate using molecular beam technology. Figure 11 *c* shows an enlarged photograph of the cross section of the PC layers containing two phase lattice defects.

Figure 12 shows the transmission spectrum of a PC containing two phase defects, measured for normal incidence of radiation in the wavelength range from 320 to 800 nm [11]. Apparently, the photonic crystal has a photonic band gap lying in the range from 451 nm to 628 nm with a center near 550 nm. The dependence obtained demonstrates the presence within the photonic band gap of two defect modes localized near the wavelengths 547.3 nm and 512.0 nm.

Thus, the introduction of defects into the photonic band gap opens the possibility of creating additional allowed states in it, which allows light waves initially incapable of propagating into the PC to propagate through the crystal along the lines created by defects [7-9]. In connection with this, one of the most attractive applications of structures with a photonic bandgap is the possibility of creating a new type of optical fibres.

In traditional light guides, light is retained by the phenomenon of total internal reflection occurring at the interface between the core and the shell. The disadvantage of such light guides is the impossibility of realizing their sharp bends, because for large, in comparison with the wavelength, bending radii, a significant part of the radiation directed in them flows out of the core and is lost. This serves as a serious obstacle to the creation of integrated optical

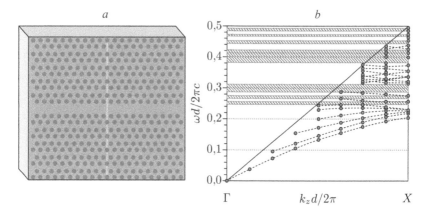

Fig. 13. A planar photonic crystal waveguide, created by bound point defects in a regular photonic crystal structure (*a*), and the mode structure and photonic-crystalline zones (*b*) corresponding to the waveguide [12].

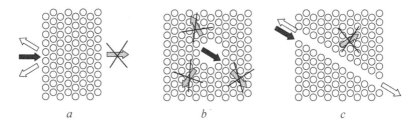

Fig. 14. An illustration of the possible modes of functioning of PC structures.

circuits, since it does not allow bending light guides in the same way as for conductors in microchips.

In waveguides based on materials with a photonic band gap, another physical mechanism for the localization and direction of radiation is used. In such structures, light propagates along the lines of bound defects, each of which supports a localized mode with a frequency lying inside the photonic band gap (Fig. 13).

A waveguide in a 2D-photonic crystal structure is created by removing or adding holes in one of the directions of the main crystallographic axis, as a result of which allowed states are created in the band gap. Due to the introduction of point defects in the crystal, a single- or multimode radiation propagation mode can be realized. The formed modes are classified as evanescent (flowing), standing and travelling waves.

Figure 14 schematically shows the different modes of operation of 2D-photonic crystals. In the case of Fig. 14 *a* the PC acts as a mirror, reflecting the waves incident on it. The creation of a point

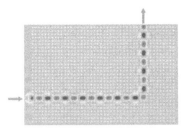

Fig. 15. Photonic-crystal waveguide based on a chain of bound defects.

defect in the PC structure (Fig. 14 *b*) leads to the appearance of bound states in the defective region for an electromagnetic wave, and this region is capable of performing resonator functions, since waves are not able to penetrate deep into the crystal and can exist only in the defective region. The formation of a chain of defects in the PC (Fig. 14 *c*) leads to the appearance of a waveguide regime for waves whose lengths fall into the region of the photonic band gap. Therefore, the electromagnetic wave effectively reflects from the walls of the channel formed by the defects and spreads freely along it.

A chain of coupled point defects is the basis for creating a light-conducting structure in the PC, as shown in Fig. 15.

Instead of a single localized state of an isolated defect, the waveguide supports propagating states (guided modes) with frequencies in a narrow band inside the forbidden band of the perfect PC. The profile of such guided modes along the waveguide is periodic, and they decay exponentially in the transverse direction.

The functioning of defective photonic crystal waveguides is analogous to the effect of resonators, since the propagation of guided waves is forbidden in the surrounding chain of defects of a homogeneous PC medium, as a result of which its boundaries are periodically reflected by radiation, which leads to the propagation of radiation along the defect line, that is, to the formation of a waveguide. As a result, even with the bending of such a waveguide, light is held in it irrespective of the sharpness of the bend (although from the sharp bends some of the energy of the light wave can be reflected back). To illustrate this possibility, Fig. 16 is a micrograph of a waveguide splitter fabricated on the basis of a 2D-defective photonic crystal [13], which allows deflecting the blown radiation perpendicular to the direction of the incident wave without visible losses. This feature of defective photonic crystal waveguides makes it possible to create on their basis a variety of

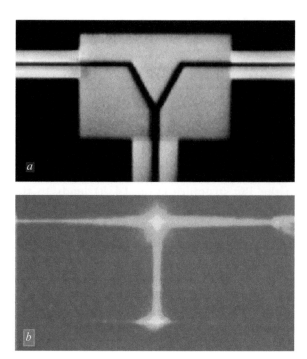

Fig. 16. A micrograph of a rectangular waveguide splitter constructed on the basis of a 2D photonic crystal (*a*); experimental division of the ray of a He–Ne laser (*b*) [13].

functional devices of integrated optics: mirrors, dividers, splitters, modulators, resonators, etc.

3.3. Photonic crystal fibres

Today, the most widely used type of optical fibres is based on the principle of total internal light reflection occurring in two-layered waveguide structures. The technology for the development of this type of fibre was largely determined by the properties of the core materials and the fibre cladding. The main material for the fabrication of optical fibres was fused quartz, which has low losses in the range of visible and near infrared radiation. However, the transition to the wavelength region >2 μm faces the problem of a significant increase in the radiation losses in the quartz core, which requires the search for new optical fibre materials: chalcogenide glasses, fluoride glasses, and polycrystalline materials [14].

Recently, in the technology of creating fibre-optic fibres, another approach has been outlined: instead of searching for special materials

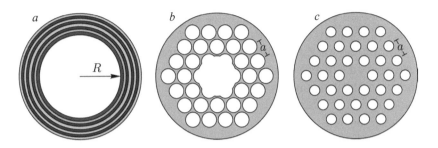

Fig. 17. Various types of optical fibre photonic crystals: a – 1D-Bragg optical fibre with a hollow core; b – 2D holey (D) photonic crystal fibre with a hollow core; in-2D-microstructured photonic crystal fibre.

for the core of light guides, designers began constructing light guides with a hollow core, in which the shell functions as a mirror effectively reflecting the propagating radiation back to the hollow core. Unlike fibre optic lightguides with a solid core, in such fibres, the bulk of the radiation power propagates through the air, the optical properties of which are strikingly different from the optical properties of any solid materials. Since a small part of the radiation passing through the hollow fibre light guides penetrates into the envelope, the optical properties of the shell material, although they affect the characteristics of the transmitted radiation, nevertheless this effect is considerably less than in optical fibres with a solid core. The presence of a hollow core in light guides allows the power of the radiation introduced into them to be increased by several orders, to reduce losses and nonlinear effects, and also to use materials with low optical characteristics for fabricating the optical fibre envelope, but more attractive from the point of view of other properties: mechanical, thermal, etc.

According to the physical mechanism of confinement of light in the hollow core, such fibres can be divided into two classes: optical fibres, in which the air core is surrounded by a surface acting as a Bragg mirror (Fig. 17 a), reflecting the propagating radiation toward the center of the core, and optical fibres in which the localization of light in the hollow core is due to mirror reflection from the shell, which possesses photonic-crystal properties (Fig. 17 b).

In Bragg (BR) optical fibres (Fig. 17 a), the propagation of radiation along the axis of a hollow or filled by a gas or dielectric core is due to the high reflection coefficient that arises from the multipath interference of the incident radiation reflected from the multilayer periodic coating.

Perforated (P) photonic crystal (PC) fibres (P-lightguides) are produced by drawing from a preform made of hollow rods (Fig. 17 *b*). The defect in the microstructure is created by extracting one or more air rods in the centre of the structure. The photonic band gap arising in the transmission spectrum of a two-dimensional periodic lattice provides a high reflection coefficient for radiation propagating along the hollow core.

Microstructured (MS) photonic crystal fibres are another type of photonic crystal fibres in which a quartz or glass core is surrounded by a microstructure with periodically located air holes (Fig. 17 *c*). Such an optical fibre is usually manufactured by drawing at high temperature from a preform made of hollow fibres grouped around a uniform rod. The resulting microstructure defect, corresponding to the absence of one or more air holes in the center of the structure, can perform a function similar to that of the standard fibre-optic waveguide, and provide a waveguide propagation mode. Waveguide modes in MS fibres are formed as a result of interference of reflected and scattered waves (therefore, the rods can be located and not periodically).

Along with the usual waveguide mode of radiation propagation provided by the phenomenon of total internal reflection, under certain conditions the MS fibres provide the formation of conditions for the waveguide propagation of radiation due to the high reflectivity of the envelope in the region of photonic forbidden bands.

To understand the difference between conventional two-layer optical fibres with total internal reflection and photonic-crystal fibres, let us compare their dispersion dependences (Fig. 18) [14]. Figure 18 shows typical dependences of the longitudinal component of the wave vector of the wave propagating in the fibre waveguide (β) on its circular frequency (ω). The transverse components of the wave vector in a medium with a refractive index n are expressed as $k_x^2 + k_y^2 = (n\omega/c)^2 - \beta^2$. Depending on whether the longitudinal component of the wave vector β is greater or less than ($n\omega/c$), the transverse component of the wave vector will assume real or imaginary values, thereby determining whether the light wave propagates in this material in the transverse direction. The line separating these two propagation modes is called the light line of the material or medium and is given by the equation $\beta = n\omega/c$.

In Fig. 18, the light lines of the respective materials are shown in dashed lines, and the solid black curve indicates a schematic representation of the dispersion relation for a typical guided mode in

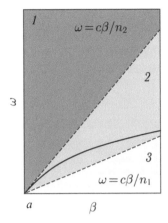 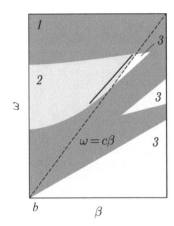

Fig. 18. Schematic dispersion relations for: *a* – two-layer optical fibre with total internal reflection, with refractive indices of the core and shell n_1 and n_2, respectively, and *b* – hole photonic crystal fibre with a hollow core. A solid black curve shows the dispersion relation for the guided mode in each fibre [14].

each fibre. The regions (ω,β) of the diagram denoted by the numeral *1* correspond to modes that can propagate through the envelope, i.e. the modes of radiation. The two areas indicated by the figure correspond to pairs (ω,β), at which light can propagate in the core of the fibre waveguide, but not in the envelope. Thus, in both cases, modes propagating along the fibre core can exist only above the light line corresponding to the refractive index of the core. That is, in a conventional fibre, the guided modes are located between light lines corresponding to the refractive indices of the shell and core materials. And in a fibre lightguide with a photonic-crystal band gap, light can propagate in the core at values of (ω,β), which are above the vacuum line of light and within the photonic band gap of the 1D- and 2D-photonic crystal shell. The regions (ω,β) of the diagram, indicated by the numeral *3*, correspond to the absence of any directional modes.

3.3.1. Bragg fibre optic lightguides

Bragg fibre optic lightguides, which are 1D-photonic crystals, concentrate light in a hollow or dielectric filled core with a mirror interference multilayer shell composed of concentric cylindrical rings with alternating low and high refractive index (Fig. 19) [15].

   In certain frequency ranges, the waves reflected many times from alternating layers of the wave as a result of interference destructively interact with the transmitted waves, which prevents the propagation

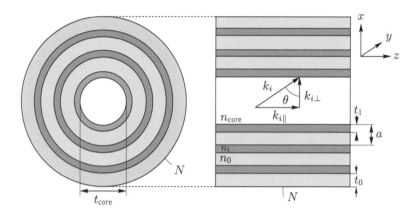

Fig. 19. Schematic representation of a Bragg fibre fibre [15].

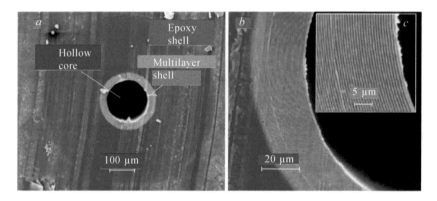

Fig. 20. SEM photograph of the section of a hollow fibre (a) and enlarged photographs of its multilayer (Bragg) shell (*b* and *c*) [16].

of light through the envelope and provides a mechanism for the channelling of radiation in the hollow core of the fibre waveguide. Figure 20 shows a photograph of the cross section of a hollow Bragg fibre optic lightguide obtained using a scanning electron microscope [16]. The light guide consists of a hollow core surrounded by a layered periodic shell, which includes alternating layers of As_2Se_3 with a thickness of 270 nm (light layers in the photograph) and polymeric layers of polyimide with a thickness of 470 nm (in the photo they have a gray color).

Figure 21 shows the calculated dispersion dependences for a given Bragg fibre optic waveguide and shows the experimentally obtained transmission spectrum for an optical fibre having a length of 5 cm [16]. (We draw attention to the fact that the longitudinal components of the wave vector in this figure are expressed in units of $2\pi/a$, where

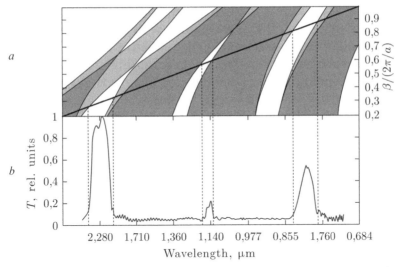

Fig. 21. *a* – dispersion relations for TE- (dark red regions) and TM-modes (brown regions), light line of air (solid dark line); *b* – spectral transmission dependence for a hollow Bragg fibre optic waveguide.

a is the thickness of one pair of layers with high and low refractive indices of the Bragg shell.)

As can be seen from Fig. 21, the photonic band gaps for a given fibre are localized near the spectral regions of 0.8 µm, 1.14 µm and 2.28 µm. It is for wavelengths falling in the region of forbidden photonic bands that a strong reflection is observed at the boundary of the air–multilayer shell in the Bragg fibre, as a result of which the radiation propagates freely along the axis of the core.

Due to the design of the hollow core fibre, most light power is transferred in the region of the hollow core, so the power loss of the guided radiation in the Bragg fibre optic lightguides is negligible. However, these losses are still nonzero due to the exponentially small penetration of the field of guided modes into the mirror layer. Therefore, slight changes in the thickness of the layers, with very little influence on the structure of the field in the core, can make very significant changes to this small exponential tail, which determines the power loss. Even a slight change in the thickness of the first layer (a layer of material with a high refractive index) can 'push' most of the exponential mode tail to the next layer (a layer with a small refractive index) and abruptly increase the amount of losses. In [17], this effect was used as the basis for the development of a method for selecting guided modes of a Bragg fibre waveguide.

A fibre optic fibre that realizes this method of mode selection has been dubbed the 'Bragg waveguide with an intermediate layer'. The scheme of the distribution of the refractive index over the cross section of such an optical fibre is shown in Fig. 22. Apparently, in such a fibre optic fibre between the core and the Bragg mirror shell, an additional layer with a larger thickness and a higher refractive index is introduced with the greater thickness and a higher refractive index than the refractive index of the material of the mirror layers. This intermediate layer in the optical fibre design starts to play the role of an additional resonator, therefore it is quite simple in such a fibre optic fibre, by selecting the parameters of this additional layer, to achieve a low-mode operation mode and make a significant difference (several times) between the nearest modes, thereby their selection.

3.3.2. 2D-photonic crystal fibres

Another form of the periodic shell, which is capable of forming forbidden bands for electromagnetic waves in optical fibres, is the two-dimensional periodic arrangement of materials with sharply differing refractive index values (Fig. 23).

To find the field of an electromagnetic wave in a cylindrically symmetric medium with an inhomogeneous distribution of the refractive index $n = n\,(r)$, we represent the electric and magnetic field strengths in the form $\mathbf{E}(\mathbf{r},t) = \mathrm{Re}\{\mathbf{E}(\mathbf{r})\exp(i\omega t)\}$, $\mathbf{H}(\mathbf{r},t) = \mathrm{Re}\{\mathbf{H}(\mathbf{r})\exp(i\omega t)\}$. Taking this representation into account, the Maxwell equations (see Chapter 1), written for the complex amplitudes of the field of a monochromatic wave, take the form

$$\nabla \times \mathbf{E}(r) = -i\omega\mu\mathbf{H}(r), \tag{3.33}$$

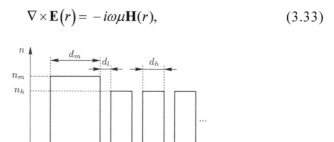

Fig. 22. Distribution of the refractive index over the cross section of a Bragg fibre optic lightguide with an intermediate layer ($d_m > d_l + d_h$).

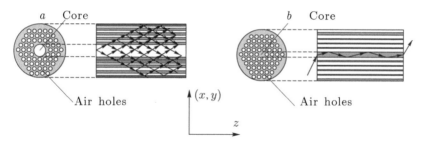

Fig. 23. Scheme of radiation propagation in 2D-photonic-crystal optical fibres: *a* – PC-hole optical fibre with a hollow core (P-lightguide); *b* – PC-microstructured fibre lightguide (MS-lightguide).

$$\nabla \times \mathbf{H}(r) = i\omega n^2(r)\mathbf{E}(r), \tag{3.34}$$

$$\nabla \cdot \left[n^2(r)\mathbf{E}(r) \right] = 0, \tag{3.35}$$

$$\nabla \cdot \mathbf{H}(r) = 0. \tag{3.36}$$

In this case it is assumed that the position of the vector **r** is determined by the coordinates (x, y, z), and ω is the cyclic frequency of the light wave field. Using the system of equations (3.33)–(3.36), we can obtain the following vector equations for the components of the electromagnetic wave field in an inhomogeneous medium:

$$\nabla \times \left[\nabla \times \mathbf{E}(r) \right] - \left(\frac{\omega}{c} \right)^2 n^2(r)\mathbf{E}(r) = 0, \tag{3.37}$$

$$\nabla \times \left[\frac{1}{n^2(r)} \nabla \times \mathbf{H}(r) \right] - \left(\frac{\omega}{c} \right)^2 n^2(r)\mathbf{H}(r) = 0. \tag{3.38}$$

Taking into account the inhomogeneity of the medium of 2D PC fibres in the direction perpendicular to the z axis, we represent the components of the electromagnetic wave vectors in the form of longitudinal ($E_z(r)$) and transverse ($\mathbf{E}_t(r)$) components [18]:

$$\mathbf{E}(r) = \left(\mathbf{E}_t(r) + \mathbf{e}_z E_z(r) \right) e^{-i\beta z}, \tag{3.39}$$

$$\mathbf{H}(r) = \left(\mathbf{H}_t(r) + \mathbf{e}_z H_z(r) \right) e^{-i\beta z}, \tag{3.40}$$

where β is the projection of the wave vector on the z axis (longitudinal component); $\mathbf{k}_\parallel$ is the transverse component of the wave vector lying in the plane (x,y); $n_2(r)(\omega_2/c_2) = (\mathbf{k}_\parallel)2 + \beta_2$.

Similarly, the operator ∇ is also represented in the form

$$\nabla = \nabla_t + \mathbf{e}_z \frac{\partial}{\partial z} = \left(\mathbf{e}_x \frac{\partial}{\partial x} + \mathbf{e}_y \frac{\partial}{\partial y} \right) + \mathbf{e}_z \frac{\partial}{\partial z}, \tag{3.41}$$

where $\mathbf{e}_x$, $\mathbf{e}_y$ and $\mathbf{e}_z$ are the unit vectors in the directions of the axes x, y, and z, respectively.

With allowance for (3.39), (3.40), the wave equations (3.37), (3.38) can be written separately for the longitudinal and transverse components of the field of the PC fibre:

$$[\nabla_t^2 + k^2 n^2 + \nabla_t (\nabla_t ln(n^2))] \mathbf{E}_t = \beta^2 \mathbf{E}_t, \tag{3.42}$$

$$\left[\nabla_t^2 + k^2 n^2 + \left(\nabla_t ln(n^2) \right) \times \left(\nabla_t \times \right) \right] \mathbf{H}_t = \beta^2 \mathbf{H}_t, \tag{3.43}$$

$$\left(\nabla_t^2 + k^2 n^2 \right) E_z - i\beta (\nabla_t ln(n^2) \cdot \mathbf{E}_t) = \beta^2 E_z, \tag{3.44}$$

$$[\nabla_t^2 + k^2 n^2 - \left(\nabla_t ln(n^2) \right) \cdot (\nabla_t)] H_z - i\beta \left(\nabla_t ln(n^2) \cdot \mathbf{H}_t \right) = \beta^2 H_z. \tag{3.45}$$

A system of cylinders with refractive indices n_1 and diameter d infinitely extended along the z-axis is selected as a model of the shell of the 2D-PC lightguide, with the cylinders forming in a medium with a refractive index n_2 in the (x, y) plane a two-dimensional periodic structure with a triangular lattice with spacing Λ (Fig. 24 a).

The nodes of the two-dimensional lattice formed in this way with the vectors of elementary translations $\mathbf{a}_1$ and $\mathbf{a}_2$ are given by the following expression:

$$\mathbf{r}_j = j_1 \mathbf{a}_1 + j_2 \mathbf{a}_2, \tag{3.46}$$

where j_1, j_2 are integers.

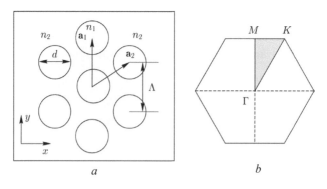

Fig. 24. Elementary cell of 2D-PC (a) and the first Brillouin zone of PC (b).

The existence of the periodicity of the structure of the PC light guide in the (x, y) plane makes it possible to construct its inverse lattice determined by the vectors $\mathbf{B}_1$ and $\mathbf{B}_2$, whose nodes are given by the vector

$$\mathbf{g}_m = m_1\mathbf{b}_1 + m_2\mathbf{b}_2, \tag{3.47}$$

where m_1, m_2 are integers.

The translation vectors of the direct and reciprocal lattices are related by

$$a_j b_m = 2\pi\delta_{jm}, \tag{3.48}$$

where δ_{jm} is the Kronecker symbol, and j, $m = 1$, 2.

Since the refractive index and the dielectric constant of a 2D-photonic-crystal fibre are spatially periodic functions:

$$n\left(\mathbf{r}_\| + \mathbf{r}_j\right) = n(\mathbf{r}_\|), \tag{3.49}$$

$$\varepsilon\left(\mathbf{r}_\| + \mathbf{r}_j\right) = \varepsilon(\mathbf{r}_\|), \tag{3.50}$$

where $\mathbf{r}_\|$ is the vector in the plane (x, y), and they can be represented by the two-dimensional Fourier series, for example

$$n^2\left(r_\|\right) = \varepsilon\left(r_\|\right) = \sum_{g_m}\varepsilon'\left(\mathbf{g}_m\right)\exp(i\mathbf{g}_m\mathbf{r}_\|). \tag{3.51}$$

For both cases of realization of 2D-PC-fibre optic lightguides (Fig. 23), the expansion coefficients in expression (3.51) are in the following form [19]:

$$\varepsilon'\left(\mathbf{g}_m\right) = \begin{cases} \varepsilon_1 + \left(\varepsilon_2 - \varepsilon_1\right)f, & \mathbf{g}_m = 0, \\ \left(\varepsilon_2 - \varepsilon_1\right)f\dfrac{2J_1(|\mathbf{g}_m|d}{|\mathbf{g}_m|d}, & \mathbf{g}_m \neq 0, \end{cases} \tag{3.52}$$

where d is the diameter of the cylinders, J_1 is the Bessel function of the first order; f is the filling coefficient: the ratio of the cross-sectional area of the cylinder (rod) to the area of the unit cell of the lattice, $f = \pi d^2/4(|\mathbf{a}_1 \times \mathbf{a}_2|)$.

The inverse of (3.51) can be represented as a Fourier series:

$$\frac{1}{\varepsilon(\mathbf{r}_\parallel)} = \varepsilon(\mathbf{r}_\parallel)^{-1} = \sum_{\mathbf{g}_m} \alpha'(\mathbf{g}_m) \exp(i\mathbf{g}_m \mathbf{r}_\parallel). \qquad (3.53)$$

The coefficients of the expansion in (3.53) are determined by formulas analogous to (3.52), in which ε_j is replaced by ε_j^{-1}.

Expressions (3.42)–(3.53) are used to find the dispersion relations of radiation in 2D-PC optical fibres. As an example, consider the case of a harmonic light field polarized along the z (E-polarization) axis [19]. In this case, the electric and magnetic field strengths of the wave have the following components:

$$\mathbf{E}(\mathbf{r}_\parallel, t) = \{0, 0, E_z(\mathbf{r}_\parallel)\} \exp(-i\omega t), \qquad (3.54)$$

$$\mathbf{H}(\mathbf{r}_\parallel, t) = \{H_x(\mathbf{r}_\parallel), H_y(\mathbf{r}_\parallel), 0\} \exp(-i\omega t). \qquad (3.55)$$

Using (3.54), (3.55) and (3.42)–(3.45), the following expression is obtained for $E_z(\mathbf{r}_\parallel)$:

$$\frac{1}{\varepsilon(\mathbf{r}_\parallel)}\left[\left(\frac{\partial^2}{\partial x^2} + \frac{\partial^2}{\partial y^2}\right) - \beta^2\right]E_z(\mathbf{r}_\parallel) + \frac{\omega^2}{c^2}E_z(\mathbf{r}_\parallel) = 0. \qquad (3.56)$$

To solve equation (3.56), we use the Bloch theorem. To do this, let us imagine the field strength E_z in a periodic medium in the form

$$E_z(\mathbf{r}_\parallel) = \sum_{\mathbf{g}_m} B_{\mathbf{k}_\parallel}(\mathbf{g}_m) \exp[-i(\mathbf{k}_\parallel + \mathbf{g}_m)\mathbf{r}_\parallel]. \qquad (3.57)$$

Substituting (3.53) and (3.57) into (3.56), we obtain the dispersion relation for E-waves propagating in the 2D-PC optical fibre:

$$\sum_{\mathbf{g}_m}[(\mathbf{k}_\parallel + \mathbf{g}_m)^2 - \beta^2]\alpha'(\mathbf{g}_l - \mathbf{g}_m)B_{\mathbf{k}_\parallel}(\mathbf{g}_m) = \frac{\omega^2}{c^2}B_{\mathbf{k}_\parallel}(g_l), \qquad (3.58)$$

where the indices m and l take integer values in the range from 1 to N, where N is the number of rods (knots) in the PC-fibre.

The resulting system of $2N$ equations (3.58) allow us to find the relationship between β, $k_\parallel$ and ω by a numerical solution of the eigenvalue problem for the coefficient matrix.

The most common realization of a 2D-photonic crystal fibre is a triangular lattice of air channels stretched along the axis z in quartz glass surrounding the core (Fig. 23 b). Such fibres have the advantage that they are made from only one material (usually quartz). The main disadvantage of this kind of photonic crystal fibres lies in the absence of cylindrical symmetry, which allows the guided modes to easily communicate with each other, thereby making them hybrid, without a pronounced polarization state.

Consider a P-fibre lightguide with a hollow core (Fig. 23 *a*). By creating a band gap in the ideal envelope of such a PC fibre, we can expect that it will ensure that the guided modes are retained in the hollow core, which is created in the structure of the PC by the formation along the *z* axis of a hollow cylindrical channel of radius $d/2\Lambda$ in the quartz material.

One of the most important properties of such optical fibres is the ability to support single-mode propagation in a much broader spectral range. An example of the result of numerical simulation of the dispersion relations for such photonic crystals is shown in Fig. 25 [20]. Figure 25 shows the results of calculations of the band diagram for an infinite shell of a P-waveguide with a hollow core, the shell of which is formed by a triangular lattice of air holes ($n_1 = 1$) with a diameter of 0.94Λ in the quartz ($n_2 = 1.45$). In the diagram, a zone called the 'light cone' of a photonic crystal located above the red line (light line) is highlighted in blue. The blue area below the red line corresponds to the radiation which can propagate in the P-waveguide due to total internal reflection. The white colour in the diagram indicates the forbidden regions for which there are

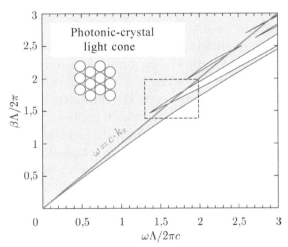

Fig. 25. Band gaps of the shell of the 2D-photonic-crystal hole fibre optic lightguide with a hollow core. The inset shows the scheme of the triangular lattice of air holes ($n_1 = 1$) with a diameter of 0.94Λ in quartz ($n_2 = 1.45$) forming the photonic-crystal shell of the lightguide [20].

no solutions for the propagation of radiation in an infinite shell. It is these areas lying above the light line of air that can be used as a reflecting shell for the propagation of radiation in a quartz core. In Fig. 25 these regions correspond to fingerlike conical gaps that expand with increasing values of β. This property is related to the fact that the air holes in the shell are filled with radiation in different ways for different wavelengths. Consequently, the difference in the refractive indices of the core and the P-waveguide shell turns out to depend on the wavelength, and in such a way that the single-mode condition for the mode of operation is satisfied for a sufficiently wide spectral range.

A criterion for the single-mode radiation propagation regime in ordinary optical fibres satisfies the following condition (for example, [9]):

$$V = \frac{2\pi\rho}{\lambda}\left(n_{co}^2 - n_{cl}^2\right)^{1/2} < 2.405, \qquad (3.59)$$

where ρ is the radius of the core of the optical fibre, and n_{co} and n_{cl} are the refractive indices of the core and the shell of the material, respectively, V is the normalized spatial frequency.

In the limiting case, when the wavelength of optical radiation is much longer than the period of the PC structure Λ, the effective refractive index of the shell n_{cl} can with a good degree of accuracy be represented as a weighted average of the refractive indices of quartz glass and air. When the wavelength of the radiation decreases, the light begins to be displaced in a region with a large refractive index. From this qualitative reasoning it becomes clear that the light with a shorter wavelength senses more solid (quartz) medium than the light with a longer wavelength. Therefore, when the wavelength is reduced, the difference between the refractive indices of the core and the shell in (3.59) decreases, which makes it possible to satisfy the single-mode criterion in a sufficiently large wavelength interval. As noted in [19], for a P-waveguide with a hollow core the parameter of the reduced frequency V in (3.59) can be replaced by an effective parameter

$$V_{eff}^2 = \left(\frac{2\pi\Lambda}{\lambda}\right)^2\left\{f\left(n_2^2 - 1\right) + \frac{\int\nabla_1\Psi dx\,dy}{k\int|\Psi|^2\,dx\,dy}\right\}, \qquad (3.60)$$

where n_2 is the refractive index of quartz glass; Ψ is the function,

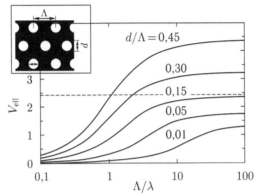

Fig. 26. Dependences of the effective frequency parameter of a P-waveguide with a hollow the core on the ratio of the period of the structure and the wavelength of the radiation, received for various relations d/Λ. The dashed horizontal line corresponds to the value of the cut-off parameter for a conventional quartz fibre waveguide [19].

describing the profile of the field amplitude; f is the coefficient of filling the structure by air holes.

Figure 26 is the dependence of the parameter V_{eff} on the ratio Λ/λ for a P-waveguide with a hollow core made in quartz ($n = 1.45$) for different ratios of the hole diameter d to the period of the structure Λ [19]. As can be seen from the above dependences, for sufficiently small values of d/Λ the PC fiber optic waveguide remains single-mode in a wide range of wavelengths.

For microstructured photonic crystal fibres (Fig. 23 b), the number of guided modes depends on the diameter of the holes d, their period in the shell Λ and the wavelength of radiation λ. Similar to a standard optical fibre, to characterize the waveguide properties of MS fibers, we can also use the concept of the reduced frequency parameter V (3.59). In this case, it is the effective radius of the core, whereas n_{co} and n_{cl} are the refractive indices of the core material and the effective refractive index of the shell, respectively.

The effective radius of the core is usually 0.62 of its radius [21]. The refractive index of the core material is equal to the refractive index of the fused quartz, and the refractive index of the shell is equal to the effective refractive index of the freely propagating mode. For single-mode light-emitting diodes, the mode of single-mode radiation propagation corresponds to condition $V < 4.2$, whereas for conventional telecommunication fibers this mode is observed at $V < 2.405$.

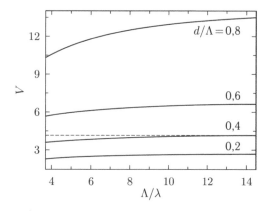

Fig. 27. Calculated dependences of the reduced frequency parameter for the MS-fibre lightguide obtained for different ratios d/Λ, $\Lambda = 7$ µm. The horizontal dashed line denotes the cutoff region for the multimode regime [22].

Figure 27 shows the calculated dependences of the parameter of the reduced frequency (V) of the MS fibres as a function of the normalized wavelength Λ/λ depending on the ratio of the dimensions of the air holes and their period in the shell d/Λ [22]. Unlike standard optical fibers with a stepped profile of the refractive index distribution, the refractive index of the shell for MS-fibre lightguides is dependent on the radiation wavelength.

This fact can be used to reduce the dependence of parameter V on the radiation wavelength and, thereby, to reduce the dependence of the cutoff frequency for the multimode regime of the optical fibre on frequency. As shown in [23], for MS-fibre lightguides with the usual ratio of the diameters of air holes and the period $d/\Lambda \sim 0.48$, parameter V demonstrates a single-mode radiation propagation in the wavelength range from 400 to 1600 nm. This property of the MS-fibre lightguides is extremely useful in generating a supercontinuum, since the generated broadband radiation remains single-mode. At the same time if the ratio d/Λ exceeds 0.48, then the MS-fibre lightguide will function in the multimode mode.

The main drawback of the MS-fibre lightguides is that they have significantly larger losses (more than 1 dB/km) in comparison with traditional quartz fibres (~0.2 dB/km). For MS-fibre lightguides, two mechanisms of the power loss of propagating radiation are characteristic: the intrinsic losses in the material and losses due to micro- and macro-bends. The main radiation losses are due to scattering and absorption in the fibre material. Rayleigh scattering is the determining mechanism of losses in the wavelength range

500–1600 nm. For the radiation with a wavelength of less than 500 nm, absorption in quartz dominates. For wavelengths near 1600 nm, the transmission of the lightguide is limited by strong infrared absorption in the quartz glass, due to the presence of OH-groups in the fibre material.

References

1.   Yablonovich E., Phys. Rev. Lett. 1987. V. 58. P. 2059–2062.
2.   Joannopoulos J.D., et al., Photonic Crystals, Princeton Univ Press, 1995.
3.   Joannopoulos J.D., et al., Nature, 1997. V. 386. P. 143–149.
4.   Bloch F., Z. Phys., 1929. V. 52. P. 555–600.
5.   Diffraction nanophotonics, ed. V.A. Soyfer. Moscow, Fizmatlit, 2011.
6.   Noda S., Baba T., Roadmap on Photonic Crystals. Kluwer Academic, 2003.
7.   Birner A., et al., Adv. Mater. 2001. V. 13, No. 6. P. 377–388.
8.   Sibilia C., Photonic Crystals: Physics and Technology. Springer-Verlag, Italia, 2008..
9.   Kivshar J.S., Agrawal G.P., Optical solitons. From light guides to photonic crystals Russian translation, Moscow, Fizmatlit, 2005.
10.   Spitsyn A.S., Glinsky G.F. Zh. Tekh. Fiz., 2008. V. 78, No. 5. P. 71–77.
11.   Hawkeye M.M., et al., Optics Express. 2010. V. 18, No. 12. P. 13220–13226.
12.   Rajat D., Optical Power Splitting Techniques Using Photonic Crystal Line Defect Waveguides, Ed. University of Western Ontario, 2011.
13.   Ouellette J., Seeing the Future in Photonic Crystals. The Industrial Physicist., American Institute of Physics, 2002. P. 14–17.
14.   Mendez A., Morse T., Handbook of specialized optical fibers. Moscow: Tekhnosfera, 2012..
15.   Rowland K.J.,et al., Optics Express. 2012. V. 20, No. 1. P. 48–62.
16.   Kuriki K., et al., Optic Express. 2004. V. 12, No. 8. P. 1510–1517.
17.   Kulchin Yu.N., et al., Kvant. Elektronika. 2012. Vol. 42, No. 3. P. 235–240.
18.   Shen L.P., et la., J. Lightwave Technol. 2003. V. 21, No. 7. P. 1644–1651.
19.   Zheltikov A.M., Holey wave guides. In: Fundamental optics and spectroscopy. Issue 3. Moscow, FIAN, 2001.
20.   Lee K.K., Proc. of SPIE. 2008. V. 6901. P. 69010K-10.
21.   Koshiba M., Saitoh K., Opt. Express. 2003. V. 11. P. 1746–1756.
22.   Lehtonen M., Application of microstructured fibers (Doctoral dissertation). Espoo, Finland: Helsinki University of Technology. 2005.
23.   Birks T.A., et al., Opt. Lett. 1977. V. 22. P. 961–963.

4

Nonlinear Optics of Fibre Waveguides

Introduction

The development of a technology for the production of optical fibres has made it possible to significantly expand the horizons of nonlinear optics. First of all, this is due to the fact that these optical fibres provide a high degree of radiation concentration in the core material of the fibre, make it possible to realize a long interaction length of the radiation with the medium and open up wide possibilities for controlling the dispersion characteristics of light-conducting structures.

Due to these unique capabilities, PC optic fibres allow achieving high efficiency for nonlinear optical phenomena, including phase self- and cross-modulation of radiation, stimulated Raman scattering and four-wave light interaction, soliton formation, supercontinuum generation, etc. This makes them very promising for use in a variety of practical applications related to metrology, spectroscopy, the transfer of powerful radiation, microprocessing of materials, biomedicine, etc. [1].

4.1. Nonlinear optical processes in optical fibres

The propagation of an electromagnetic wave in an optical fibre (OF) can be described using the scalar wave equation [2]

$$\nabla^2 \mathbf{E}(r,t) - \frac{1}{c^2}\frac{\partial^2 \mathbf{E}(r,t)}{\partial t^2} = -\frac{\partial^2 \mathbf{P}(r,t)}{\partial t^2}, \tag{4.1}$$

where $E(r,t)$ is the strength of the electrical component of the electromagnetic wave; $P(r,t)$ is the polarization of the environment; c is the speed of light in vacuum. As noted in chapter 1, in weak light fields the polarization of the medium has a linear dependence on the magnitude of the field strength. In the field of intense laser radiation, the polarization of the medium includes two components: linear and nonlinear, depending on the radiation intensity:

$$P = P_L + P_{NL} = \varepsilon_0 (\chi^{(1)} E + \chi^{(2)} E \cdot E + \chi^{(3)} E \cdot E \cdot E + ...), \qquad (4.2)$$

where P_L and P_{NL} are the linear and nonlinear components of the polarization of the medium, respectively; ε_0 is the electric constant; $\chi^{(j)}$ is the j-th the order of the dielectric susceptibilities of the medium. Since SiO_2 molecules have a centre of symmetry in quartz glasses – the basic material of traditional fibres and fibre optic fibres, the dielectric susceptibility of even orders makes an insignificant contribution to the nonlinearity of the medium (although impurities inside the fibre or various effects on it can lead to a violation of the inversion symmetry). The largest contribution to the nonlinearity of the medium of optical fibres is made by the frequency-dependent cubic nonlinearity $\chi^{(3)}$ [1].

Since in the general case both linear and nonlinear responses are not instantaneous, but characterized by a delay with respect to the field, then P_L and P_{NL} are related to the electric field by the relations:

$$P_L(r,t) = \varepsilon_0 \int_{-\infty}^{+\infty} \chi^{(1)}(t-t') E(r,t) dt', \qquad (4.3)$$

$$P_{NL}(r,t) = \varepsilon_0 \int_{-\infty}^{+\infty} \int \int \chi^{(3)}(t-t_1, t-t_2, t-t_3) E(r,t_1) \times \qquad (4.4)$$

$$\times E(r,t_2) E(r,t_3) dt_1 dt_2 dt_3.$$

Since the Fourier transform of the convolution is equal to the product of the Fourier transforms, in the Fourier space $P_L(r, \omega)$ and $P_{NL}(r, \omega)$ when self-action have the form

$$P_L = \varepsilon_0 \chi^{(1)}(\omega) E(r,\omega), \qquad (4.5)$$

$$P_{NL} = \varepsilon_0 \chi^{(3)}(\omega) |E(r,\omega)|^2 E(r,\omega). \qquad (4.6)$$

Thus, the dependence of the linear and cubic dielectric susceptibilities of the fibre material on the frequency determine the material dispersion of the linear and nonlinear induced polarizations.

The linearly polarized electric field of a light pulse, propagating in an optical fibre in the direction of the z axis, can be represented in the following form [1, 3]:

$$\mathbf{E}(\mathbf{r},t) = \mathbf{e}_x \left\{ F(x,y) \cdot A(z,t) \left[i \left(\beta_0 z - \omega_0 t \right) \right] + \text{c.c.} \right\}, \qquad (4.7)$$

where $\mathbf{e}_x$ is the unit vector of polarization; ω_0, β_0 is the central frequency in the pulse spectrum and the longitudinal component of its wave vector at this frequency; $F(x, y)$ is the distribution of the mode field over the cross section of the optical fibre; $A(z,t)$ is the envelope of the light pulse field. Further consideration will be carried out under the following assumptions: light guides are considered single-mode and do not have anisotropy, and the field of the light pulse is linearly polarized. This allows us to move from the vector equation (4.1) to the scalar one.

The Kerr nonlinearity of the fibre material leads to the dependence of its refractive index on the intensity of the lightwave field

$$n\left(\omega, |\mathbf{E}|^2\right) = n_0(\omega) + \Delta n = n_0(\omega) + n_2 |\mathbf{E}|^2 = n_0(\omega) + n_2 I, \qquad (4.8)$$

where I is the intensity of the lightwave field; n_0 is the refractive index of the fibre-optic material in the linear mode; Δn is the refractive index induced by light radiation; n_2 is the nonlinear refractive index.

The used intensities of optical fields in an optical fibre create a change in the refractive index Δn induced by the field, which is much smaller than the value of the refractive index in the linear regime i.e. $\Delta n \ll n_0$. This makes it possible to neglect the change in the transverse field distribution under the influence of nonlinear effects.

Proceeding from the foregoing, we can assume that only the slowly changing envelope of the light wave $A(z, t)$ is subject to changes due to dispersion and nonlinearity, which varies slowly with respect to the period of the optical carrier.

Let us also assume that the nonlinear response of the polarization of the medium is instantaneous, and the group velocity of the light pulse does not depend on its intensity. This allows in the expansion of the propagation constant of the pulse $\beta(\omega)$ in degrees $\Delta\omega$ take into account only the terms up to the second order of smallness.

Under these assumptions, from equation (4.1), taking into account (4.7), for the envelope $A(z,t)$ in a coordinate system moving with the group velocity of the pulse $V_g = d\omega/d\beta$ (replacement is made: $t \rightarrow T - z/V_g$), we obtain

$$i\frac{\partial A}{\partial z} = -\frac{i}{2}\alpha A + \frac{1}{2}\frac{d^2\beta}{d\omega^2}\frac{\partial^2 A}{\partial T^2} - \gamma|A|^2 A, \tag{4.9}$$

where α is the radiation attenuation coefficient; $(d^2\beta/d\omega^2)$ is the group velocity dispersion coefficient (GVD);); $\gamma = (n_2\omega_0/cS_{eff})$ is the nonlinear coefficient; S_{eff} is the effective area of the mode depending on the mode profile $F(x,y)$:

$$S_{eff} = \frac{\left[\int\int_{-\infty}^{\infty}|F(x,y)|^2 \, dx \, dy\right]^2}{\int\int_{-\infty}^{\infty}|F(x,y)|^4 \, dx \, dy}. \tag{4.10}$$

In deriving equation (4.9), it was assumed that the envelope of the pulse was slow: $|\partial^2 A/\partial z^2| \ll |\beta\partial A/\partial z)|$, and it was also assumed that the pulse amplitude A was normalized in such a way that $|A|^2$ is the pulse power.

We introduce the concept of the normalized momentum amplitude $U(z,\tau)$:

$$A(z,\tau) = \sqrt{P_0}\exp(-\alpha z)U(z,\tau), \tag{4.11}$$

where P_0 is the initial peak pulse power; $\tau = (t-z/V_g)/T_0$ is the normalized time, T_0 is the initial pulse duration.

As a result of the substitution of (4.11) in (4.10), we obtain

$$i\frac{\partial U}{\partial z} = \frac{\text{sign}\left(\dfrac{d^2\beta}{d\omega^2}\right)}{2L_D}\frac{\partial^2 U}{\partial \tau^2} - \frac{\exp(-\alpha z)}{L_{NL}}|U|^2 U, \tag{4.12}$$

where $L_D = T_0^2/\left|(d^2\beta/d\omega^2)\right|$ is the dispersion length; $L_{NL} = 1/\gamma P_0$ is the nonlinear length.

Dispersion length L_D and nonlinear length L_{NL} characterize the distances at which the dispersion and nonlinear effects become important for the evolution of the pulse along the length of the fibre waveguide L, namely approximately twice the pulse width and the width of the original pulse spectrum, respectively.

Equations (4.9) and (4.12) describe well the propagation of pulses of duration $T_0 \geqslant 1$ ps over quartz optical fibres taking into account the dispersion of group velocities (DGV) and phase self-modulation (PSM). The effects of Mandel'shtam–Brillouin stimulated scattering (SBS) and stimulated Raman scattering (SRS) are not taken into account here.

At a pulse duration of $T_0 < 1$ ps the neglect of certain effects in equations (4.9) and (4.12) becomes incorrect. The broadening of the pulse spectrum leads to an increase in the term $(d^3\beta/d\omega^3)\Delta\omega^3$ in the expansion of $\beta(\omega)$ and they can no longer be neglected. It is also incorrect to neglect the dependence of the group velocity of the light pulse on the intensity and it is also incorrect to neglect Raman self-scattering, since the width the spectrum of the optical pulse approaches the Raman frequency shift. Finally, the nonlinear response of the medium can not be considered instantaneous, since the inertia of the nonlinear response of Raman scattering in quartz fibres is in the range of tens of femtoseconds. As a result, equations (4.9) and (4.12) must be modified to adequately describe the propagation of pulses with a duration shorter than 1 ps along the fibre quartz fibres.

4.2. Waveguide enhancement of the efficiency of nonlinear optical processes in fibre guides

The magnitude of the nonlinear refractive index n_2 for quartz fibres takes different values depending on the wavelength of the light, composition, concentration and transverse distribution of doping impurities, and the pulse duration. This value in quartz, in comparison with other nonlinear media, at least on 2 orders of magnitude less. The same applies to the SRS and SBS amplification coefficients. Nevertheless, in quartz fibres, nonlinear effects can be observed at relatively low powers. This is possible due to two important characteristics of the optical fibre, namely: a small transverse mode size (several microns for a single-mode fibre) and extremely low losses (< 1 dB/km), which ensures long interaction lengths.

The point is that the efficiency of a nonlinear process is determined by the nonlinear coefficients and the product (IL_{eff}), where I is the intensity of the light pulse, and L_{eff} is the effective length of its interaction with the medium. The lightguide makes it possible to substantially increase the effective interaction length. In this section, as an example, we compare the efficiency of nonlinear processes for the cases of bulk (not waveguide) media and optical fibres.

In the case of a bulk medium for the realization of nonlinear effects, the pump radiation is focused by means of a lens into a spot of radius R_0. Because of the diffraction, the spot size increases when light propagates through the medium in accordance with expression

$$r(z) = r_0 \left[1 + \left(\frac{\lambda z}{\pi r_0^2} \right)^2 \right]^{1/2}. \tag{4.13}$$

In a medium with length L for a beam of power P the product of intensity I by the effective interaction length will be:

$$I \cdot L_{\text{eff}} = \int_{-L/2}^{L/2} \frac{P}{\pi r^2(z)} dz < \int_{-\infty}^{\infty} \frac{P}{\pi r^2(z)} dz = \frac{P}{\lambda}. \tag{4.14}$$

As seen from (4.14), the product IL_{eff} is determined by the radiation power and the wavelength of light and, in fact, does not depend either on the degree of focusing of the radiation in the medium or on the length of propagation of light in the medium, since with decreasing r_0 the diffraction divergence of the radiation in the bulk medium increases.

In the case of an optical fibre, diffraction is completely compensated by refraction from the reflecting cladding, so the transverse dimensions of the radiation inside the fibre are unchanged along its length and are determined by the radius of the core a. In this case, the interaction length L_{eff} is limited by the radiation losses in the fibre α. The attenuation of the intensity of the mode directed by the optical fibre can be described equation

$$I(z) = I_0 e^{-\alpha z}, \tag{4.15}$$

where

$$I_0 = \frac{P}{\pi a^2}. \tag{4.16}$$

For this case equation (4.14) has the form

$$I \cdot L_{\text{eff}} = \int_0^L \frac{P}{\pi a^2} e^{-\alpha z} dz = \frac{P}{\pi a^2} \left(\frac{1 - \exp(-\alpha L)}{\alpha} \right). \tag{4.17}$$

In the visible region of the spectrum, due to low losses in optical fibres, the product $\alpha L \gg 1$. Therefore, $L_{\text{eff}} \sim 1/\alpha$. Dividing (4.17) by (4.14), we obtain the ratio of the amplification of nonlinear processes in an optical fibre with respect to the volume medium

$$\frac{(IL_{\text{eff}})_{\text{OF}}}{(IL_{\text{eff}})_{\text{bulk}}} = \frac{\lambda}{\pi a^2 \alpha}. \tag{4.18}$$

Expression (4.18) shows that an increase in the efficiency of nonlinear optical processes in optical fibres is based on the possibility of achieving high intensities at a given radiation power due to the localization of the field in the core region of a fibre with a small radius and achieving large interaction lengths in fibres with low losses. The factor of increasing the efficiency of nonlinear optical processes grows with a decrease in the diameter of the optical fibre. For $\lambda = 0.53$ μm, the absorption coefficient is $\alpha = 2.5 \cdot 10^{-5}$ cm^{-1}, i.e. $\alpha \approx 10$ dB/km and the ratio (4.18) is $\sim 10^7$. For radiation with a wavelength $\lambda = 1.55$ μm, the radiation loss coefficient is $\alpha = 5 \cdot 10^{-7}$ cm^{-1}, i.e. $\alpha \approx 0.2$ dB/km, which makes the ratio (4.18) $\sim 10^9$. Such a large increase in the efficiency of nonlinear processes in optical fibres makes optical waveguides a convenient nonlinear medium for observing numerous nonlinear effects at relatively low powers.

Thus, optical fibres based on quartz glass, characterized by a low nonlinearity with respect to other materials, can provide a significant increase in the efficiency of nonlinear processes due to the large length of interaction of radiation with matter.

4.3. Phase self-modulation of radiation in optical fibres

As reported in section 1.7, the effect of phase self-modulation (PSM) is as follows: when the light pulse propagates through a fibre lightguide, the refractive index depends on the intensity and is described by the expression (1.99) or (4.8). Because of this, different parts of the light pulse with the envelope $A(z,t)$ experience different additional phase advance due to the nonlinearity of the refractive indes. This leads to frequency modulation and to the broadening of the pulse spectrum (in contrast to the linear dispersion, when the width of the pulse spectrum remains unchanged).

We consider the case when the dispersion of the group velocities (DGV) can be neglected: $L_{NL} \ll L_D$. This condition is usually satisfied for relatively long light pulses, $T_0 > 100$ ps, with a peak power of $P_0 > 1$ W. As a result, equation (4.12) takes the form

$$\frac{\partial U}{\partial z} = \frac{i}{L_{NL}} \exp(-\alpha z) |U|^2 U. \tag{4.19}$$

The solution of equation (4.19) has the form

$$U(z,T) = U(0,T) \exp[i\Phi_{NL}(z,T)], \tag{4.20}$$

where $U(0, T)$ is the amplitude of the field at $z = 0$ and the nonlinear phase shift

$$\Phi_{NL}(z,T) = |U(0,T)|^2 \left(\frac{z_{\text{eff}}}{L_{NL}} \right), \tag{4.21}$$

where

$$z_{\text{eff}} = \frac{1}{\alpha}\left[1 - \exp(-\alpha z)\right], \quad L_{NL} = 1/(\gamma P_0). \tag{4.22}$$

It follows from the expressions (4.20)–(4.22) that the PSM causes a phase shift depending on the intensity and z, whereas the shape of the pulse, determined by $|U(z,t)|^2$ and its duration, remain unchanged. The maximum phase shift occurs at the center of the pulse at $T = 0$. If without loss of generality we assume that $U(0,0) = 1$, then from (4.21) we obtain

$$\Phi_{\text{max}} = \frac{z_{\text{eff}}}{L_{NL}} = \gamma P_0 z_{\text{eff}}. \tag{4.23}$$

As a result, it becomes obvious that the nonlinear length L_{NL} is the effective propagation length, on which $\Phi_{\text{max}} = 1$.

It follows from (4.21) that the nonlinear phase shift repeats the shape of the pulse envelope.

The instantaneous frequency is the derivative of the phase with respect to time. The full phase of the pulse Φ taking into account the linear and nonlinear parts is

$$\Phi(z,t) = -\omega_0\left(T + \frac{z}{v_g}\right) + |U(0,T)|^2\left(\frac{z_{\text{eff}}}{L_{NL}}\right). \tag{4.24}$$

Therefore, the instantaneous frequency

$$\omega_{\text{inst}} = -\frac{\partial \Phi}{\partial T} = \omega_0 - \frac{\partial |U(0,T)|^2}{\partial T} \frac{z_{\text{eff}}}{L_{NL}}. \tag{4.25}$$

It follows from (4.25) that the instantaneous frequency due to the PSM changes with time. The change in the instantaneous frequency in time is called frequency modulation, or pulse chirp.

Thus, with PSM the pulse spectrum broadens. This case should be distinguished from the GVD (group velocity dispersion). At GVD also phase and frequency modulation occurs. However, in this case, the pulse width varies (increases) simultaneously. In other words, with GVD there is a redistribution of the spectral components of the pulse as its duration, and with PSM, new frequencies are generated. Generation of new frequency components in the pulse spectrum occurs continuously as it propagates through the fibre, causing broadening of the spectrum. Since $n_2 > 0$, a shift to the Stokes ('red') is realized at the leading edge of the pulse, and in the background to the anti-Stokes ('blue') region of the spectrum (Fig. 1). At the same time, the frequency change will be greater for pulses with steeper fronts.

Let us estimate the width of the spectrum of the frequency-modulated Gaussian pulse due to the PSM. For a Gaussian pulse

$$U(0,t) = \exp\left(-\frac{T^2}{2T_0^2}\right). \tag{4.26}$$

Substituting (4.26) into (4.25) and taking into account that the

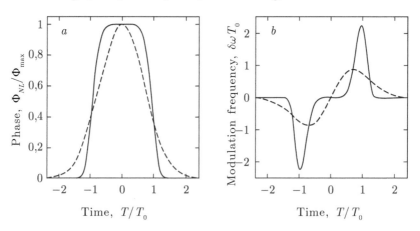

Fig. 1. Change in time of the phase shift (*a*) and frequency modulation (*b*) induced by the PSM in the Gaussian (dashed line) and super-Gaussian (solid line) pulses [2].

variations of the instantaneous radiation frequency $\delta\omega(T)$ are determined by the second term in (4.25), we obtain

$$\delta\omega(T) = \frac{2}{T_0}\left(\frac{T}{T_0}\right)\exp\left(-\frac{T}{T_0}\right)^2\frac{z_{eff}}{L_{NL}}.\qquad(4.27)$$

The width of a spectrum can be estimated on the basis of peak values $\delta\omega(T)$ (see Fig. 1 b). Quantitatively, these peak values are found by calculating the maximum $\delta\omega(T)$ in the equation (4.27). Assuming the first derivative $\delta\omega(T)$ in time equal to zero, we find the maximum value $\delta\omega_{max}$

$$\delta\omega_{max} = 0.86\Delta\omega_0\Phi_{max},\qquad(4.28)$$

where $\Delta\omega_0 = T_0^{-1}$ is the half-width of the spectrum of the initial light pulse without frequency modulation, and Φ_{max} is determined from expression (4.23).

It follows from the expression (4.28) that the broadening of the pulse spectrum is determined by the magnitude of the maximum phase advance Φ_{max}. Hence it follows that the PSM can significantly broaden the spectrum, since Φ_{max} in optical fibres can reach ~100.

The shape of the pulse spectrum $(S(\omega))$ in PSM is described by the expression:

$$S(\omega) = |U(z,\omega)|^2 = \left|\int_{-\infty}^{\infty}U(0,T)\exp[i\Phi_{NL}(z,T)+i(\omega-\omega_0T]dT\right|^2.\qquad(4.29)$$

Figure 2 shows the calculated spectrum of a laser pulse with an envelope in the form of a hyperbolic secant obtained for various values of the initial energies, demonstrating the broadening of the original laser spectrum. The spectrum has a periodic structure with clear maxima and minima. In the framework of the above elementary theory of PSM, which takes into account only the first order of the dispersion of the material, the self-action of the light pulse leads to the symmetric broadening of its spectrum. However, in practice, even at moderate intensities of laser radiation, there are a number of physical mechanisms that lead to asymmetry of the spectral broadening. The three most important of them are related to spatial self-action, the formation of a shock front by the envelope, and the finite time of the nonlinear-optical response of the medium.

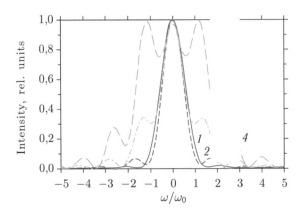

Fig. 2. The calculated dependences of the spectral broadening of the laser pulse, determined by PSM, with the envelope in the form of a hyperbolic secant and the initial duration of 30 fs in an optical fibre with a nonlinear refractive index $n_2 =$ 3.2 10^{-16} cm²/W. Curve 1 shows the initial pulse spectrum. Dependences 2, 3, and 4 were obtained for initial pulse energies of 0.1; 0.2 and 0.3 nJ, respectively [1]

4.4. Effect of dispersion on nonlinear processes in optical fibres

The effects of PSM without GVD taken into account satisfactorily describe the propagation of relatively long pulses ($T_0 > 100$ ps), for which the dispersion length L_D is much greater than the nonlinear length L_{NL} and the length of the optical fibre L. With the shortening of pulses, the dispersion length becomes comparable with the nonlinear length and length of the fibre. In this case, it is necessary to consider the joint effect of the effects of PSM and GVD.

Let us consider the effect of normal dispersion in an optical fibre ($\partial^2\beta/\partial\omega^2 = \beta_2 > 0$). In this case, equation (4.12) takes the form

$$i\frac{\partial U}{\partial \xi} = \frac{1}{2}\frac{\partial^2 U}{\partial \tau^2} - N^2 e^{-\alpha z}|U|^2 U, \qquad (4.30)$$

where $\xi = z/L_D$, $\tau = T/T_0$, and

$$N^2 = \frac{L_D}{L_{NL}} = \frac{\gamma P_0 T_0^2}{|d^2\beta/d\omega^2|} = \frac{\gamma P_0 T_0^2}{|\beta_2|}. \qquad (4.31)$$

Equation (4.30) is solved numerically. The value N determines the relative influence of the effects of the PSM and GVD on the evolution of pulses in an optical fibre.

If $N = 1$, then the PSM and GVD play an equally important role in the process of pulse evolution. Figure 3 shows the evolution of the pulse shape and its spectrum in the case of an initial Gaussian pulse without frequency modulation in the region of normal dispersion for $N = 1$ and $\alpha = 0$.

As can be seen, during the propagation the pulse duration increases, the amplitude value of its intensity decreases, and the spectrum expands. The results of the pulse evolution analysis show that the pulse duration increases much more rapidly than in the case $N = 0$ [2].

This is explained by the fact that the PSM leads to the generation of new frequency components shifted to the long-wave (red) region at the leading edge and to the short-wave (blue) region at the trailing edge of the pulse. Since the red components move faster than the blue ones, in the region of normal dispersion the PSM leads to an increase in the pulse broadening velocity in comparison with the dispersion broadening. This, in turn, leads to a slowing down of the spectral broadening, since the amplitude of the pulse decreases due to the expansion, which leads to a decrease in the effective phase advance Φ_{max} and a decrease in the broadening of the spectrum (see expression (4.28)).

If $N \gg 1$, that is, $L_{\text{D}} \gg L_{\text{NL}}$, then the PSM prevails over GVD However, even a weak effect of dispersion leads to a significant change in the shape of the pulse. The Gaussian pulse becomes close

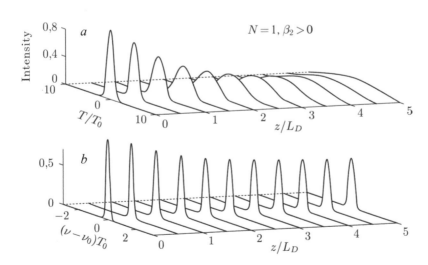

Fig. 3. The temporal (*a*) and spectral (*b*) evolution of a Gaussian pulse without initial frequency modulation under the influence of the PSM in the region of normal dispersion of the fibre at $N = 1$ [2].

to a rectangular one with sharp fronts. The pulse duration increases, and it has a linear frequency modulation on its entire width [3].

In the region of anomalous dispersion of group velocities ($\partial^2 \beta / \partial \omega^2 = \beta_2 < 0$), the nonlinear dynamics of pulse propagation along an optical fibre is described by an equation different from (4.30):

$$i\frac{\partial U}{\partial \xi} = -\frac{1}{2}\frac{\partial^2 U}{\partial \tau^2} - N^2 e^{-\alpha z} |U|^2 U. \tag{4.32}$$

If in (4.32) we ignore the losses in the lightguide and carry out substitution

$$u = NU = \left(\frac{\gamma T_0^2}{|d^2 \beta / d\omega^2|}\right)^{1/2} U, \tag{4.33}$$

we got the equation

$$i\frac{\partial u}{\partial \xi} + \frac{1}{2}\frac{\partial^2 U}{\partial \tau^2} + |u|^2 u = 0. \tag{4.34}$$

When $N = 1$ and the condition $u(0,0) = 1$ is satisfied, the solution of equation (4.34) has the form of a hyperbolic secant:

$$u(\xi,\tau) = \operatorname{sech}(\tau)\exp\left(i\frac{\xi}{2}\right). \tag{4.35}$$

Equation (4.35) is the equation of motion for the fundamental soliton.

In the case of a fundamental soliton, the positive frequency modulation determined by the PSM is compensated for by negative frequency modulation due to the GVD. As a result, as follows from (4.35), when the fundamental soliton propagates through the lightguide, neither the shape of the envelope (hyperbolic secant) nor the pulse duration, nor its frequency spectrum, changes.

If $N > 1$ and N is an integer, then the pulses having the envelope in the form of $u(0,\tau) = N \operatorname{sech}(\tau)$ at the entrance to the optical fibre, are characterized by the periodic dynamics shown in Fig. 4 for $N = 3$. As can be seen, the pulse duration initially decreases. Then, at a distance of $z/L_D = 0.5$, it splits into two. Then both components merge, forming at $z = z_0$ the initial impulse. The distance $z_0 = \pi L_D/2$ is called the soliton period.

In the absence of the dispersion of group velocities, the shape of the pulse would remain unchanged. Since the frequency modulation

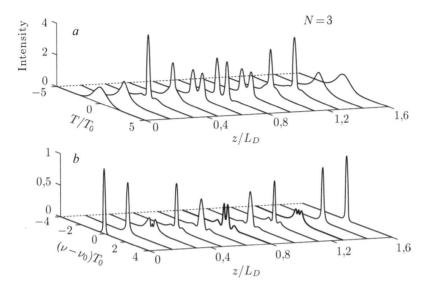

Fig. 4. Dynamics of the time form (*a*) and the spectrum (*b*) of a third-order soliton in one soliton period [3].

due to the GVD in the region of anomalous dispersion has the opposite sign with respect to frequency modulation due to the PSM, the pulse is compressed. This behavior of the pulse at the initial stage can be used to create devices for compressing the duration of laser pulses (see Chapter 5).

4.5. Phase cross-modulation of pulses in fibre guides

Phase cross modulation (PCM) is a nonlinear process, which consists in the fact that if two or more optical waves propagate through a fibre waveguide, then the effective refractive index of a wave depends not only on the intensity of the wave itself, but also on the intensity of other waves propagating with it together. The PCM is always accompanied by a PSM.

To illustrate the PCM process, let us consider the propagation of two waves with different frequencies along the fibre, but with the same polarizations. In the quasi-monochromatic approximation, the total electric field of the lightwave can be represented in the form:

$$\mathbf{E}(x,y,z,t) = \frac{1}{2}\mathbf{e}_x[E_1 \exp(-i\omega_1 t) + E_2 \exp(-i\omega_2 t)] + \text{c.c.,} \qquad (4.36)$$

where e_x is the unit polarization vector, and

$$E_1(x,y,z,t) = F_1(x,y) A_1(z,t) \exp(i\beta_1 z),$$ (4.37)

$$E_2(x,y,z,t) = F_2(x,y) A_2(z,t) \exp(i\beta_2 z),$$ (4.38)

where $F_1(x,y)$, $F_2(x,y)$ – distribution of the mode fields in the cross-section $A_1(z,t)$, $A_2(z,t)$ – the slowly varying amplitudes of the modes; β_1 and β_2 are the propagation modes at the frequencies ω_1 and ω_2, respectively.

Substituting (4.37) and (4.38) into (4.5), we obtain the following expression for the nonlinear component of the polarization:

$$\mathbf{P}_{NL} = \frac{1}{2} e_x \left[P_{NL}(\omega_1) \exp(-i\omega_1 t) + P_{NL}(\omega_2) \exp(-i\omega_2 t) + \dots \right],$$ (4.39)

where

$$P_{NL}(\omega_1) = \chi_{eff} \left(|E_1|^2 + 2|E_2|^2 \right) E_1.$$ (4.40)

$$P_{NL}(\omega_2) = \chi_{eff} \left(|E_2|^2 + 2|E_1|^2 \right) E_1,$$ (4.41)

χ_{eff} is the effective susceptibility of the fibre material.

The nonlinear polarization induced by the total light field (4.39) has terms separated as 'dots', which denotes the presence of terms oscillating with new total and difference frequencies. The effective generation of new frequency components occurs only when the phase-matching condition is satisfied, in the absence of which we neglect these terms.

The remaining two terms in (4.40) and (4.41) contribute to the nonlinear addition to the refractive index, which for the first wave with frequency ω_1 is

$$\Delta n_1 = n_2 (|E_1|^2 + 2|E_2|^2),$$ (4.42)

and for a wave with frequency ω_2

$$\Delta n_2 = n_2 \left(|E_2|^2 + 2|E_1|^2 \right).$$ (4.43)

It follows from (4.42) and (4.43) that the nonlinear addition to the refractive index of an optical wave depends both on the intensity of the wave itself and on the intensity of another wave propagating along the fibre waveguide.

Propagating through an optical fibre, the modes acquire a nonlinear phase shift:

$$\Phi_1^{NL} = \frac{\omega_1 z}{c} \Delta n_2 \left(|E_1|^2 + 2|E_2|^2 \right),$$ (4.44

$$\Phi_2^{NL} = \frac{\omega_2 z}{c} \Delta n_2 \left(|E_2|^2 + 2|E_1|^2 \right).$$ (4.45)

As follows from (4.44) and (4.45), the first term in them is responsible for the effect of the PSM, the second arises from the phase modulation of one wave by another wave propagating with it, and is responsible for the PCM. Because of the interference of the two fields on the right-hand side of (4.44) and (4.45), before the second summands there is a coefficient 2 which shows that the PCM is 2 times more effective than the PSM at the same intensity for waves with the same polarizations, but with different frequencies. The reason for this lies in the number of terms of nonlinear polarization, obtained on the basis of expression (4.39). When the optical frequencies of the two waves are different, the number of terms is doubled in comparison with the degenerate case. It follows from (4.44) and (4.45) that if two waves simultaneously propagate in an optical fibre, one of which is 'weak' and the other 'strong', then a larger nonlinear phase shift is realized for the 'weak' wave due to the PCM, while the 'strong' wave is experienced only by the PSM, the effect of which is 2 times less than that of the PCM. In this case, for both waves, as in the case of the PSM, the broadening of the spectrum of the laser radiation propagating in the fibre waveguide is observed.

4.6. Four-wave mixing of waves

Four-wave mixing (FWM) is a nonlinear process determined by the Kerr nonlinearity due to the dependence of the refractive index on the intensity. In FWM, in the general case, four different optical fields participate. This process is determined by the real, non-resonance terms of the cubic nonlinear susceptibility: $\chi^{(3)}(\omega_4 = \omega_1 + \omega_2 + \omega_3)$, $\chi^{(3)}(\omega_4 = \omega_1 + \omega_2 - \omega_3)$, etc.

In the FWM process, as in the PSM and PCM processes, new frequencies are generated in the laser emission spectrum. The peculiarity of FWM in comparison with PSM and PCM is that the frequencies generated in FWM differ substantially from the frequencies of the initial fields. Therefore, because of the variance, it is important for the FWM to fulfill both the law of conservation of energy

$$\omega_1 + \omega_2 = \omega_3 + \omega_4,$$

and phase-matching conditions.

Let us consider a four-wave mixing of optical waves with frequencies ω_1, ω_2, ω_3, ω_4 linearly polarized along the X axis. The total electric field in an optical fibre is

$$
\begin{aligned}
\mathbf{E} &= \sum_1^4 \mathbf{E}_j = \mathbf{e}_x \frac{1}{2} \sum_1^4 \mathcal{E}_j \exp\left[i\left(\beta_j z - \omega_j t\right)\right] + \text{c.c.} = \\
&= \mathbf{e}_x \frac{1}{2} \sum_1^4 F_j(x,y) A_j(z) \exp\left[i\left(\beta_j z - \omega_j t\right)\right] + \text{c.c.},
\end{aligned}
\tag{4.46}
$$

where $\mathcal{E}_j = F_j(x,y) A_j(z)$, β_j are the propagation constants of the interacting modes of an optical fibre with frequencies ω_j.

The induced nonlinear polarization can be represented in the form

$$\mathbf{P}_{NL} = \mathbf{e}_x \frac{1}{2} \sum_1^4 P_j \exp\left[i\left(\beta_j z - \omega_j t\right)\right] + \text{c.c.} \tag{4.47}$$

Nonlinear polarization P_j for $j = 1...4$ consists of a large umber of terms including the product of three electric field strengths. For example, P_4 is expressed as

$$
\begin{aligned}
P_4 &= \left\{\left[|\mathcal{E}_4|^2 + 2\left(|\mathcal{E}_1|^2 + |\mathcal{E}_2|^2 + |\mathcal{E}_3|^2\right)\right] \mathcal{E}_4 + 2\mathcal{E}_1\mathcal{E}_2\mathcal{E}_3 \exp(i\theta_+) \right. \\
&\quad \left. + 2\mathcal{E}_1\mathcal{E}_2\mathcal{E}_3^* \exp(i\theta_+) + +2\mathcal{E}_1\mathcal{E}_2\mathcal{E}_3^* \exp(i\theta\_) + ...\right\},
\end{aligned}
\tag{4.48}
$$

where

$$\theta_+ = \left(\beta_1 + \beta_2 + v_3 - \beta_4\right)z - \left(\omega_1 + \omega_2 + \omega_3 - \omega_4\right)t, \tag{4.49}$$

$$\theta_- = \left(\beta_1 + \beta_2 - \beta_3 - \beta_4\right)z - \left(\omega_1 + \omega_2 + \omega_3 - \omega_4\right)t. \tag{4.50}$$

The term proportional to $\mathcal{E}_4$ in (4.48) is responsible for the effects of the PSM and PCM. The remaining terms are due to the four-wave mixing.

It is relatively easy to ensure that the phase-matching condition is fulfilled in the case of partially degenerate four-wave mixing, when $\omega_1 = \omega_2$. In this case, a powerful pump wave with a frequency ω_1 generates two symmetrically located sidebands with frequencies ω_3 and ω_4, shifted from the pumping frequency by an amount Ω_S:

$$\Omega_S = \omega_1 - \omega_3 = \omega_4 - \omega_1, \tag{4.51}$$

where $\omega_3 < \omega_4$.

Low-frequency (ω_3) and high-frequency (ω_4) spectral components are called Stokes and anti-Stokes components, respectively. The Stokes and anti-Stokes waves are also called signal and idler waves. If only the pump radiation is introduced into the optical fibre and phase matching is performed, then the generation of Stokes and anti-Stokes waves with frequencies ω_3 and ω_4 can be initiated by noise. On the other hand, if a weak signal of frequency ω_3 is injected into the optical fibre along with the pumping, then it amplifies, at the same time a new frequency wave is generated ω_4. This process is called parametric amplification.

Four-wave mixing and parametric amplification can be mathematically described on the basis of the coupled-wave method [3].

In single-mode optical fibres, when the pump wavelength lies in the region of anomalous dispersion of group velocities and is significantly separated from (zero-dispersion wavelength), lateral spectral bands with frequencies $\omega_1 \pm \Omega_S$ are generated, where

$$\Omega_S = \left(\frac{2\gamma P_0}{d^2\beta/d\omega^2} \right)^{1/2}. \tag{4.52}$$

The width of the parametric amplification band ($\Delta\Omega$) of the FWM process at high pumping powers (P_0) is approximately equal to:

$$\Delta\Omega = \left(\frac{d^2\beta}{d\omega^2} \Omega_S L_{NL} \right)^{-1}, \tag{4.53}$$

where $L_{NL} = (\gamma P_0)^{-1}$ and $L_{NL} \ll L$.

Using equation (4.53) we can estimate the width of the band for the FWM process in an optical fibre. Usually, the parameter of the dispersion of group velocity in quartz optical fibres is $|\partial^2\beta/d\omega^2| = 20{-}60$ ps²/km. For $L_{NL} \sim 1$ m we get $\Delta\Omega \sim 10{-}100$

GHz. In this case, detuning of the amplifying frequencies will be $\Omega_S/2\pi \sim$ 10–100 THz, and the efficiency of conversion of the pump energy into the energy of the Stokes–anti-Stokes components with four-wave mixing reaches tens of percent.

4.7. Stimulated Raman scattering (SRS) radiation in optical fibres

The nonlinear effects considered above (PSM, PCM, FWM) can be called the elastic scattering of light in the sense that there is no exchange of energy between the electromagnetic field and the dielectric medium. Another class of nonlinear effects is associated with stimulated inelastic scattering, in which an optical field transfers part of its energy to a nonlinear medium.

This category includes two important nonlinear effects, which are often combined by the term Raman scattering of radiation. Both of them are associated with the appearance of vibrational modes of quartz, the main material of fibre light guides. These are effects of stimulated Raman scattering (SRS) and stimulated Mandel'shtam-Brillouin scattering (SBS). The main difference between these effects is the fact that optical phonons participate in SRS, whereas in SBS acoustic phonons [2].

SRS in quartz optical fibres is a nonlinear process of generation or amplification of a Stokes wave with a frequency ω_c shifted by 13 THz with respect to the frequency ω_p of the pumping field of a light wave propagating along an optical fibre. The intensity of the Stokes wave increases in such a way that most of the pump energy passes into it. This occurs when the pump power exceeds the threshold level. Excess of energy, determined by the difference in frequencies $\omega_p - \omega_S > 0$, is transferred to the optical vibrational mode of quartz.

From the point of view of quantum theory, Raman scattering is a two-photon process in which one photon $\omega_p(k_p)$ is absorbed, and one photon $\omega_S(k_S)$ is emitted. In this case, the medium receives a portion of energy in the form of a quantum of molecular vibrations, equal to $h(\omega_p - \omega_S) = h\Omega$, where Ω is the frequency of molecular vibrations.

Since the vibrational modes of quartz molecules do not have dispersion, the phase matching condition for stimulated Raman scattering is automatically satisfied. From the classical point of view, the SRS process, like the one considered above, can be considered on the basis of the theory of coupled waves. This allows one to obtain an expression for the power of the Stokes wave [3]

$$P_S(L) = P_S(0)\exp\left(g_R P_0 \frac{L_{\text{eff}}}{S_{\text{eff}}} - \alpha_S L\right), \qquad (4.54)$$

where P_0 is the pump power at the entrance to the optical fibre $(z = 0)$; $P_S(0)$ is the Stokes wave power at $(z = 0)$; $P_S(L)$ is the power of the Stokes wave at a distance $z = L$, g_R is the coefficient of Raman amplification, S_{eff} is the effective cross-sectional area of the fibre, and the effective length

$$L_{\text{eff}} = \frac{1}{\alpha_p}\left[1 - \exp(-\alpha_p L)\right], \qquad (4.55)$$

where α_S and α_p are the loss coefficients for the Stokes wave and pump wave, respectively.

As can be seen from (4.54), the power of the Stokes wave increases exponentially if the gain exceeds the loss. The enhancement of the Stokes wave, determined by expression (4.54), is not related to the inversion of populations, but is due to the transfer of energy from the pump wave as a result of energy exchange, determined by the coupling coefficient, in which the cubic nonlinear susceptibility $\chi_R^{(3)}$ appears. In this case, the Stokes component of stimulated Raman scattering has all the characteristics of a laser wave: coherence, directivity, and high intensity.

An important feature of Raman amplification in fibre optic fibres from fused quartz is a large frequency range (up to 40 THz) with a wide gain maximum at frequency tuning of 13 THz. This behavior is due to the non-crystalline nature of quartz glass. In amorphous materials, such as fused quartz, the frequency bands of molecular vibrations overlap and create a continuum. As a result, Raman amplification in quartz fibres exists over a wide range of frequencies. Due to this property, optical fibres can act as broadband amplifiers.

The SRS is a threshold process. The SRS threshold is defined as such the pumping power at the beginning (at the input) of the optical fibre at which the power of the Stokes wave at the output of the optical fibre becomes equal to the pumping power:

$$P_S(L) = P_p(L) = P_0 \exp(-\alpha_p L), \qquad (4.56)$$

where P_0 is the power of the pump radiation at the entrance to the optical fibre.

Power $P_S(L)$ is determined from the expression (4.54), into which the Stokes wave power enters $P_S(0)$ at the entrance to the optical fibre. If a wave with a Stokes frequency is not fed to the input of a fibre waveguide, then $P_S(0) = 0$. Then the SRS process can evolve from spontaneous Raman scattering, which arises on the entire length of the fibre waveguide. This is equivalent to having one photon per mode at the entrance to the optical fibre. Consequently, (4.54) can be represented in the form

$$P_S(L) = \int_{-\infty}^{\infty} \hbar\omega \exp\left[g_R(\omega) \cdot P_0 \frac{L_{\text{eff}}}{S_{\text{eff}}} - \alpha_S L \right] d\omega. \qquad (4.57)$$

The integral (4.57) can be estimated as

$$P_S(L) = \hbar\omega_S \Delta\omega_{\text{eff}} \exp\left[g_R(\omega) \cdot P_0 \frac{L_{\text{eff}}}{S_{\text{eff}}} - \alpha_S L \right], \qquad (4.58)$$

where $\Delta\omega_{\text{eff}}$ is the effective bandwidth of the Stokes radiation with the centre at the peak of amplification at $\omega = \omega_c$.

From the expressions (4.56) and (4.58), in the $\alpha_p \approx \alpha_S$ approximation, the threshold condition is written in the form

$$\hbar\omega_S \Delta\omega_{\text{eff}} \exp\left[g_R(\omega) \cdot P_0 \frac{L_{\text{eff}}}{S_{\text{eff}}} \right] = P_0. \qquad (4.59)$$

The threshold pump power P_{th}, when the pumping radiation and the Stokes radiation are co-directed, must satisfy the relation [2]

$$g_R(\omega) \cdot P_{\text{th}} \frac{L_{\text{eff}}}{S_{\text{eff}}} \approx 15. \qquad (4.60)$$

It has been shown experimentally that SRS in optical fibres is also observed in the opposite direction of the pump and Stokes waves, but with less amplification. In this case, for the agreement of theory and experiment on the right-hand side of expression (4.60) there must be a number 20.

As noted in [3], when pumping at a wavelength of 1 μm, the maximum value of the gain of the Stokes wave g_R in a quartz optical fibre is capable of reaching a maximum value of 10^{-13} m/W.

In practice, when the power of the Stokes wave becomes sufficient to satisfy (4.60), it serves as a pump to generate a second-order Stokes wave. Such a cascade SRS process can lead to the generation of many orders of Stokes waves, the number of which depends on the input pump power.

Considering stimulated Raman scattering, so far we have neglected the phenomenon of self-action of light waves. If propagation of ultrashort pulses of duration <1 ps is considered, the width of the spectrum of such pulses becomes so large that the neglect of the effect of self-action in stimulated Raman scattering becomes unjustified. In pulses with a large spectral width (>1 THz) under the action of stimulated Raman scattering, the low-frequency components of the spectrum can be amplified, obtaining energy from the high-frequency components of the spectrum of the same pulse. This phenomenon is called intrapulse Raman scattering. As a result, the spectrum of a short pulse shifts to the long-wavelength region when propagating in an optical fibre. This phenomenon is called the frequency shift caused by SRS. Physically, the SRS effect is explained by the retarded (oscillatory) nonlinear response (4.1) and (4.9), but we need to use the general form of the nonlinear polarization (1.43) and (4.4).

Nonlinear effects that depend on the intensity can be taken into account by introducing a third-order nonlinear susceptibility (response) in the form [2]:

$$\chi^{(3)}\left(t-t_1, t-t_2, t-t_3\right) = \chi^{(3)} R(t-t_1)\delta(t_1-t_2)\delta(t_1-t_3), \qquad (4.61)$$

where $R(t)$ is the nonlinear response function normalized in such a way that $\int_{-\infty}^{\infty} R(t)dt = 1.$.

Assuming that the electron response is practically instantaneous, the function $R(t)$ can be written in the following form [3]:

$$R(t) = \left(1 - f_R\right)\delta\left(t - t_e\right) + f_R h_R(t), \qquad (4.62)$$

where t_e corresponds to a negligibly short delay of the electronic response ($t_e < 1$ fs), and f_R corresponds to the relative fraction of the contribution of the inertial (delayed) molecular (vibrational) response to the magnitude of the nonlinear polarization. The form of the molecular response function $h_R(t)$ is determined by the nature of the vibrations of the quartz molecules excited by the optical field. Due to the amorphous nature of the quartz fibre, calculation of $h_R(t)$ is extremely difficult. Therefore, in practical calculations, the interrelation of this function with the experimentally measured spectrum of stimulated Raman amplification is used. Indeed, the imaginary part of the Fourier transform of the function $h_R(t)$ is related to the SRS-gain spectrum by the following expression [2]:

$$g_R(\Delta\omega) = \frac{\omega_0}{cn(\omega_0)} f_R \chi^{(3)} \mathrm{Im}[\tilde{h}_R(\Delta\omega)], \qquad (4.63)$$

where $\Delta\omega = \omega - \omega_0$, and Im is the imaginary part.

The real part of the function $\tilde{h}_R(\Delta\omega)$ can be obtained from imaginary part in terms of the Kramers–Kronig relations. Performing the inverse Fourier transform of the function $\tilde{h}_R(\Delta\omega)$, we obtain the molecular response function $h_R(t)$. Figure 5 shows the time dependence of the function $h_R(t)$ obtained from the SRS gain spectrum in a quartz fibre, measured experimentally.

As can be seen from Fig. 5, the function $h_R(t)$ has the form of rapidly damped oscillations, which allows us to use the following approximate expression for its description [3]:

$$h_R(t) = \frac{\tau_1^2 + \tau_2^2}{\tau_1 \tau_2^2} \exp\left(-\frac{t}{\tau_2}\right) \sin\left(\frac{t}{\tau_1}\right). \qquad (4.64)$$

The values of time parameters τ_1 and τ_2 are obtained by the fitting method, which ensures the best correspondence to the SRS gain

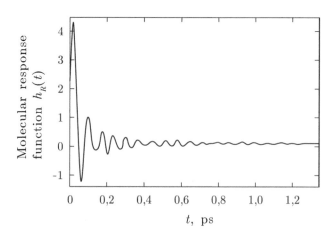

Fig. 5. Time dependence of the molecular response function $h_R(t)$ [3].

spectrum. For optical fibres, the following values were obtained: $\tau_1 = 12.2$ fs and $\tau_2 = 32$ fs. The value of coefficient f_R can be estimated from the experimental values of the maximum SRS amplification, and it amounts to $f_R \sim 0.18$.

4.8. Stimulated Mandel'shtam–Brillouin scattering in optical fibres

Stimulated Mandel'shtam–Brillouin scattering (SBS) is a nonlinear process that can occur in fibres with a radiation power much lower than required for stimulated Raman scattering. SBS is manifested in the form of generation of a Stokes wave propagating in the direction opposite to the pump wave and containing a significant part of the initial energy. Both the SBS and SRS processes occurring in optical fibres are associated with the excitation of vibrational modes in quartz. However, if an optical phonon participates in the stimulated Raman scattering, then in the SBS it is an acoustic phonon.

In quantum mechanics, the SBS process is described as the annihilation of a pump photon having a frequency ω_p and a wave vector $\mathbf{k}_p$, and a simultaneous appearance of a Stokes photon having a frequency ω_S and a wave vector $\mathbf{k}_S$ and an acoustic phonon with a frequency ω_a and a wave vector $\mathbf{k}_a$. Therefore, the following relations must be satisfied from the laws of conservation of energy and momentum for the SBS process in optical fibres:

$$\omega_a = \omega_p - \omega_S. \tag{4.65}$$

$$\mathbf{k}_a = \mathbf{k}_p - \mathbf{k}_S. \tag{4.66}$$

In the framework of the classical theory of SBS, the process of interaction of three waves is based on the phenomenon of electrostriction in quartz glass material, due to which an acoustic wave is generated at the frequency of beating of the pump wave and the Stokes wave, leading to periodic modulation of the refractive index of the medium. The lattice of the refractive index induced by the acoustic wave dissipates the pump radiation into the Stokes wave as a result of Bragg diffraction. A scattered Stokes wave will be amplified if the power that enters it as a result of scattering of the pump wave exceeds the attenuation in the fibre, i.e., if the pump wave power exceeds a certain critical value.

Assuming that in the SBS process $|k_p| \approx |k_{sl}|$, the phase-matching conditions (4.65) and (4.66) allow one to obtain the following dispersion relation:

$$\omega_a = |k_a| V_a = 2V_a |k_p| \sin \frac{\theta}{2}. \tag{4.67}$$

where θ is the angle between the propagation directions of the pump wave and the Stokes scattering wave; V_a is the velocity of the acoustic wave in the material of the optical fibre.

As follows from (4.65) and (4.67), the change in the frequency of the Stokes wave depends on the scattering angle θ. In particular, for scattering in the opposite direction ($\theta = \pi$) it will be maximum, and for incident scattering ($\theta = 0$) it will be equal to zero, i.e., in the passing direction the SBS process is absent.

In a single-mode optical fibre, only the inverse direction of propagation of the SBS ($\theta = \pi$) is possible, so the frequency shift of the Stokes wave will be determined by the expression

$$v_B = \frac{\omega_a}{2\pi} = \frac{2nV_a}{\lambda_p}, \tag{4.68}$$

where n is the refractive index of the core material of the fibre.

As can be seen from (4.68), for single-mode optical fibres, the frequency detuning of the SBS depends only on the pump wavelength. For example, for a pumping wavelength of $\lambda_p = 1.55$ µm and typical parameters of optical fibres made of quartz glass $n = 1.45$ and $V_a = 5.96$ km/s, the frequency tuning is equal to 9 $V_B = 11.1$ GHz.

From the classical point of view, the SBS process, as well as the processes of FWM and SRS considered above, can be considered on the basis of the theory of coupled waves [3]. In the description of this process, the resonance part of the Brillouin susceptibility $\chi_B^{(3)}$ is responsible for the SBS phenomenon and is purely imaginary.

In the approximation of the inexhaustibility of continuous pumping, the dependence of the Stokes radiation power of scattering has the form

$$P_S(0) = P_S(L) \exp\left[g_B(\omega) P_p(0) \frac{L_{\text{eff}}}{S_{\text{eff}}} - \alpha_S L \right], \tag{4.69}$$

where $P_p(0)$ is the pump power at the entrance to the optical fibre; $P_S(0)$ is the Stokes wave power at $z = 0$; $P_S(L)$ is the power of the

Stokes wave at $z = L$; S_{eff} is the effective cross-sectional area of the optical fibre; L_{eff} is the effective interaction length; g_B is the SBS gain; α_s is the decay coefficient of the Stokes wave

The maximum gain at $\omega = 2\pi v_B$ is equal to

$$g_B(v_B) = \frac{\omega_S^2}{\varepsilon_0 c^3 |k_S| n_c} \text{Im } \chi_B^{(3)} = \frac{4\pi\gamma_e^2}{n_c \lambda_p^2 \rho_0 c V_a \Delta v_B} = \frac{8\pi^2 \gamma_e^2}{n_c \lambda_p^2 \rho_0 c V_a \Gamma_B}, \qquad (4.70)$$

where ε_0 is the dielectric constant; c is the speed of light in vacuum; ρ_0 is the density of the material; γ_e is the electrostriction coefficient; Δv_B is the width of the SBS-gain spectrum, $\Delta v_B = \Gamma_B/(2\pi)$; Γ_B is the temporal increment of the attenuation of the acoustic wave, $\Gamma_B = 1/T_B$, where T_B is the phonon lifetime.

For typical values of fused quartz parameters: $\gamma_e = 0.902$; $\rho_0 = 2210$ kg/m³; $\Delta v_B = 10$ MHz for $\lambda = 1.5$ μm, for the gain factor of SBS scattering is $g_B \approx (3-5) \cdot 10^{-11}$ m/W. This value is almost three orders of magnitude greater than the gain for stimulated Raman scattering ($g_R \approx (3-5) \cdot 10^{-11}$ m/W).

It follows from equation (4.69) that the amplification of SBS radiation is possible if radiation with this frequency is introduced into the optical fibre from the opposite end at the point $z = L$. But in practice such a signal is absent, and the Stokes wave of SBS arises from the noise of Mandel'shtam–Brillouin spontaneous scattering. As in the case of stimulated Raman scattering, this is equivalent to the presence of one photon per mode.

SBS is a threshold process. With continuous pumping, the threshold value of the pumping power (P_0^{cr}) satisfies relation

$$g_B P_0^{cr} \frac{L_{eff}}{S_{eff}} \approx 20, \qquad (4.71)$$

where g_B is the peak value of the SBS gain factor, determined from (4.70).

For $\lambda_p = 1.55$ μm, $S_{eff} = 50$ μm², $L_{eff} = 20$ kg, $g_B = 5 \cdot 10^{-11}$ m/W we obtain the quantity $P_0^{cr} = 1$ mW. This corresponds to a power density of $2 \cdot 10^3$ W/cm² and the electric field strength of the wave is 10^3 V/cm. For comparison, the intensity of the intra-atomic field is of the order of $3 \cdot 10^8$ V/cm. Such a low threshold makes SBS the dominant nonlinear process in optical fibres when they are continuously pumped.

The bandwidth of the SBS gain spectrum Δv_B is related to the decay time of the acoustic wave (the phonon lifetime) T_B, $\Gamma_B = 1/T_B$, and has a very small value of $\Delta v_B \approx 10\text{--}100$ MHz, which is much smaller than the width of the gain spectrum for the SRS process $\Delta v_R \sim 5$ THz. Comparison of the gain and optical lifetime of the optical (in stimulated Raman scattering) and acoustic (for SBS) phonons shows that the greater inertia of the process corresponds to a larger value of the gain. The large inertia (narrow band) of SBS, as compared with stimulated Raman scattering, leads to the fact that if the width of the spectrum Δv_p of the pump radiation introduced into the optical fibre is much higher than Δv_B, the SBS amplification decreases substantially. This is true for both continuous pumping and pulse pumping with pulse duration $T_0 \ll T_B$, since only that part of the signal energy whose spectrum coincides with the SBS spectrum has a Lorentz profile with a width at half-height Δv_p, the peak value of the SBS-gain is given by the expression

$$g_B = \frac{\Delta v_B}{\Delta v_B + \Delta v_p} g_B(v_B), \tag{4.72}$$

where $g_B(v_B)$ is found from expression (4.70).

So, if $\Delta v_p \gg \Delta v_B$, the SBS-gain decreases $\Delta v_p/\Delta v_B$ times.

In the case of SBS, just as in the case of stimulated Raman scattering, a cascade process is possible: the generation of higher-order Stokes components after the power of the component of the lowest order reaches the SBS threshold.

4.9. The propagation of ultrashort laser pulses in optical fibres

As noted in section 4.1 of this chapter, in order to describe the process of propagation of ultrashort laser pulses (USP) ($T_0 < 1$ ps) in optical fibres, it is necessary to modify equation (4.9). Indeed, the shortening of the pulse duration leads to a substantial expansion of its spectrum. Therefore, in the decomposition $\beta(\omega)$ in degrees, with growth $\Delta\omega = (\omega - \omega_0)$, we can not neglect members of a higher order, in the first place $(d^3\beta/d\omega^3)(\omega-\omega_0)^3$. This is especially important when the wavelength of the pulse radiation is close to the region of zero dispersion, since in this case the value of the derivative is close to zero ($d^2\beta/d\omega^2 \sim 0$), as a result of which the role of the cubic dispersion ($d^3\beta/d\omega^3$) significantly increases.

Along with this, the dependence of the nonlinear additive on the refractive index of the optical fibre on the intensity (maximum at the vertex of the pulse and decreases on the leading and trailing fronts) leads to deformation of the shape of the pulse envelope. This leads to a decrease in the velocity of propagation of the central part of the pulse in comparison with its fronts, which causes a steepening of the trailing edge of the pulse. This effect is known as the formation of an envelope shock wave. Therefore, for its description, in equation (4.9), in addition to the zeroth term $\sim|A|^2A$, it is necessary to add a term of the first order of smallness $\sim(\partial|A|^2/\partial T)A$.

For the ultrashort pulse, the nonlinear polarization response of the environment is instantaneous. As a result, the cubic polarization of the material of the fibre lightguide is expressed by the following integral relation (4.4):

$$\mathbf{P}_{NL}\left(\mathbf{r},t\right)=$$
$$=\varepsilon_0\oint_{\infty}\chi^{(3)}\left(t-t_1,t-t_2,t-t_3\right)(\mathbf{E}(\mathbf{r},t_1)\times\mathbf{E}(\mathbf{r},t_2))\cdot\mathbf{E}(\mathbf{r},t_3)dt_1dt_2dt_3. \tag{4.73}$$

In this expression, the real, non-resonance part $\chi^{(3)}$ contributes to the change in the real part of the refractive index of the optical fibre material, and the imaginary, resonant $\chi^{(3)}$ is connected with the SRS amplification in the fibre lightguide.

Taking into account the foregoing, in the small perturbation approximation and the slowness of the change in the pulse envelope, the generalized equation (4.9) for the propagation of an ultrashort pulses with a central frequency ω_0 in a fibre waveguide takes the form [3]:

$$\frac{\partial A}{\partial z}+\frac{\alpha}{2}A+\frac{i}{2}\frac{d^2\beta}{d\omega^2}\frac{\partial^2 A}{\partial T^2}-\frac{1}{6}\frac{d^3\beta}{d\omega^3}\frac{\partial^3 A}{\partial T^3}=$$
$$=i\gamma\left[|A|^2 A+\frac{2i}{\omega_0}\frac{\partial(|A|^2 A)}{\partial T}-T_R A\frac{\partial|A|^2}{\partial T}\right], \tag{4.74}$$

where

$$T_R=\int_0^{\infty}tR(t)dt, \tag{4.75}$$

and $R(t)$ is the nonlinear response function defined by expressions (4.62), (4.63).

Using the concept of the normalized amplitude (4.11), under the condition that there is no loss in the optical fibre, equation (4.74) can be reduced to the form

$$
i\frac{\partial U}{\partial z} - \frac{\text{sign}\left(d^2\beta/d\omega^2\right)}{2L_D}\frac{\partial^2 U}{\partial T^2} + i\frac{\text{sign}\left(d^3\beta/d\omega^3\right)}{6L_D'}\frac{\partial^3 U}{\partial T^3} =
$$
$$
= -\frac{1}{L_{NL}}\left[|U|^2 U + \frac{2i}{\omega_0}\frac{\partial}{\partial T}\left(|U|^2 U\right) - T_R U\frac{\partial}{\partial T}\left(|U|^2\right)\right],
$$

(4.76)

where

$$
L_D = \frac{T_0^2}{\left|d^2\beta/d\omega^2\right|}; \quad L_D' = \frac{T_0^3}{\left|d^3\beta/d\omega^3\right|}; \quad L_{NL} = \frac{1}{\gamma P_0}.
$$

Using the normalized variables for time

$$
\tau = \frac{(t - z/V_g)}{T_0}, \quad \tau_R = \frac{T_R}{T_0},
$$

and the length

$$
\xi = \frac{z}{L_D},
$$

we transform (4.76) to the form

$$
i\frac{\partial U}{\partial \xi} - \frac{\text{sign}\left(d^2\beta/d\omega^2\right)}{2}\frac{\partial^2 U}{\partial \tau^2} + i\,\text{sign}\left(\frac{d^3\beta}{d\omega^3}\right)\frac{\partial^3 U}{\partial \tau^3} =
$$
$$
= -\frac{L_D}{L_{NL}}\left[|U|^2 U + is\frac{\partial}{\partial \tau}\left(|U|^2 U\right) - \tau_R U\frac{\partial}{\partial \tau}\left(|U|^2\right)\right].
$$

(4.77)

In the correct equation (4.77) for the pulses of duration less than 1 ps, the following parameters are introduced:

$$
\delta = \frac{d^3\beta/d\omega^3}{\partial\left|d^3\beta/d\omega^2\right|T_0}; \quad s = \frac{2}{\omega_0 T_0}; \quad \tau_R = \frac{T_R}{T_0},
$$

(4.78)

which, respectively, determine the effects of linear dispersion of higher order, the dependence of the group velocity on the intensity and the delay of the nonlinear response.

In order to understand the role of the contribution of various effects to the propagation of ultrashort pulses in an optical fibre, we consider the cases when the carrier wavelength lies in the regions of normal and anomalous dispersion of the fibre.

4.9.1. Pumping in the region of normal dispersion

1. Let us consider the structure of the pulse in the case when there is no influence of nonlinear terms, when the propagation of ultrashort pulses in an optical fibre is mainly influenced by the dispersion of the group velocity ($N \gg 1$). As a result, equation (4.76) takes the form

$$i\frac{\partial U}{\partial z} - \frac{\text{sign}\left(d^2\beta/d\omega^2\right)}{2L_D}\frac{\partial^2 U}{\partial T^2} + i\frac{\text{sign}\left(d^3\beta/d\omega^3\right)}{6L'_D}\frac{\partial^3 U}{\partial T^3} = 0. \quad (4.79)$$

As shown in [4], taking into account the cubic dispersion term in (4.79) in the case when the pulse wavelength is in the region of normal dispersion leads to the appearance of asymmetry in the spectrum of the transmitted pulse and the appearance of oscillations on its leading and trailing fronts, the character of which depends on of the sign of $d^3\beta/d\omega^3$.

Figure 6 shows the calculated dependences of the evolution of the time dependence of the form of ultrashort pulses for cases with a negative value of the cubic dispersion term: $d^3\beta/d\omega^3 < 0$. As can be seen, when the wavelength of the carrier pulse is in the region of positive dispersion, the presence of a negative additive $d^3\beta/d\omega^3$ leads to the appearance of oscillations of considerable amplitude at the leading edge of the pulse transmitted through the medium.

The influence of the cubic dispersion term becomes significant when $L_D \geq L'_D$ or when the wavelength of the carrier pulse is $\lambda_0 = 2\pi c/\omega_0$ coincides with the wavelength of zero dispersion (λ_D) and $d^2\beta/d\omega^2 = 0$. The time dependence of the shape of the ultrashort pulse for the case $d^2\beta/d\omega^2 = 0$ clearly demonstrates an increase in the influence of the cubic dispersion term $d^3\beta/d\omega^3$, which manifests itself in the mixing of the oscillations of the temporal shape of the pulse to its trailing edge and the increase in the amplitude of the oscillation modulation of the pulse intensity.

2. In the case when the influence of nonlinear processes in the optical fibre can not be neglected ($N \geq 1$), the joint effect of the GVD (group velocity dispersion) and PSM (phase self-modulation)

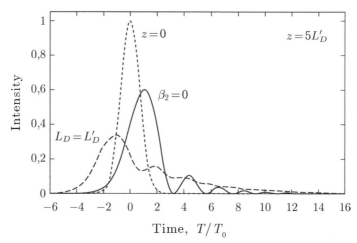

Fig. 6. Calculated time dependences of the ultrashort pulse shape in the presence of cubic dispersion. The initial Gaussian pulse shape (dotted curve) at the input to the optical fibre ($z = 0$); The shape of the pulses at $z = 5L'_D$ (solid line) when $\lambda_0 = 2\pi c/\omega_0 = \lambda_D$ and when $\lambda_0 < \lambda_D$, $d^3\beta/d\omega^3 < 0$ and $L_D = L'_D$ (dotted line) [3].

processes on the shape and spectrum of the ultrashort pulses should be considered. In this case, equation (4.76), if the nonlinear terms of higher order are neglected, takes the form

$$i\frac{\partial U}{\partial z} - \frac{\text{sign}(d^2\beta/d\omega^2)}{2L_D}\frac{\partial^2 U}{\partial T^2} + i\frac{\text{sign}(d^3\beta/d\omega^3)}{6L'_D}\frac{\partial^3 U}{\partial T^3} = -\frac{1}{L_{NL}}|U|^2 U. \tag{4.80}$$

In the absence of losses ($\alpha = 0$) and the influence of the cubic term, equation (4.80) coincides with equation (4.9) describing the process of phase self-modulation of a light pulse in an optical fibre. As shown in section 4.3, for the case of normal dispersion ($d^2\beta/d\omega^2 > 0$), taking into account only the first order of the dispersion of the material ($d^3\beta/d\omega^3 = 0$), the temporary self-action of the light pulse (PSM) leads to a symmetric broadening of its spectrum (Fig. 2). However, in practice, even at moderate intensities of laser radiation, the influence of cubic and higher terms of material dispersion begins to affect, leading to the asymmetry of the spectral broadening.

Under conditions where the wavelength of the carrier pulse (λ_0) is close to the wavelength of zero dispersion (λ_D), since $d^2\beta/d\omega^2 \sim 0$ the dispersion effects in (4.80) will be determined by the cubic dispersion term $d^3\beta/d\omega^3$. In this case, the efficiency of the effect of dispersion and nonlinear processes in the optical fibre should be estimated by the parameter

$$\tilde{N}^2 = \frac{L'_D}{L_{NL}} = \frac{\gamma P_0 T_0^3}{\left| d^3 \beta / d\omega^3 \right|}. \tag{4.81}$$

Figure 7 shows the calculated dependences of the time shape and spectrum of a Gaussian ultrashort pulse as the passage through an optical fibre of length z = $5L'_D$ for $\tilde{N}$ = 1 and λ_0 = λ_D ($d^2\beta/d\omega^2$ = 0), when $d^3\beta/d\omega^3$ > 0, obtained in [2].

Comparison of Fig. 6 and Fig. 7 *a* shows that the contribution of the PSM is to increase the number of oscillations at the trailing edge of the pulse, but with a smaller depth of modulation. The contribution of the cubic term of the GVD consists in the transformation of the symmetrical pulse spectrum arising from the PSM from two maxima (Fig. 2) into a spectrum consisting of two asymmetric maxima (Fig. 7 *b*).

An increase in the efficiency of nonlinear processes in optical fibres is shown in Fig. 8 [3]. As can be seen, under similar conditions, a decrease in the length L_{NL} by an order of magnitude: $\tilde{N}$ = 10 already at smaller fibre lengths z = 0.1 L'_D leads to significant modulation of the trailing edge of the pulse with a more significant depth and redistribution of the pulse energy along two asymmetric spectral maxima. One maximum is in the region of normal dispersion, the energy of which is dissipated during the propagation process. And the other maximum is in the region of anomalous dispersion,

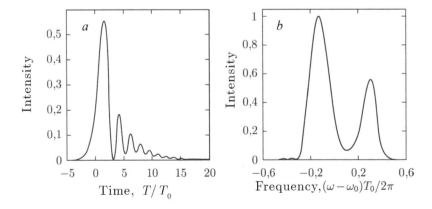

Fig. 7. The shape (*a*) and spectrum of the pulse (*b*) formed after passage of the Gaussian ultrashort pulse through an optical fibre with length z = $5L'_D$ at $\tilde{N}$ = 1 and λ_0 = λ_D [2].

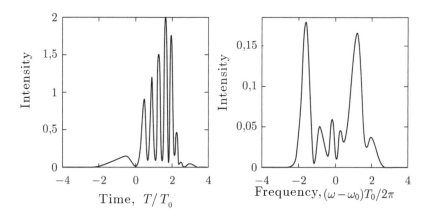

Fig. 8. The shape (*a*) and spectrum of the pulse (*b*) formed after passage of the Gaussian ultrashort pulse through an optical fibre with length $z = 0.1L'_D$ at $\tilde{N} = 10$ and $\lambda_0 = \lambda_D$ [2].

which opens the possibility of the formation of solitons in optical fibres (see Section 3).

3. Taking into account the subsequent nonlinear terms on the right-hand side of equation (4.75), we can distinguish a nonlinear effect associated with the formation of a shock wave of the pulse envelope of the ultrashort pulse. As noted above, this nonlinear effect is due to the appearance of the dependence of the group velocity on the intensity of the pulse. The formation of a shock wave, in turn, leads to an asymmetry in the broadening of the pulse spectrum due to the PSM effect.

To understand the physical essence of the phenomenon, we first consider the case of the dispersion of the group velocity in fibres and the lack of inertness of the process, that is, we set

$$\frac{d^2\beta}{d\omega} = 0, \quad \frac{d^3\beta}{d\omega^3} = 0, \quad T_R U \frac{\partial(|U|^2)}{\partial T} = 0.$$

Then in the moving coordinate system, when $\tau = (t - z/V_g)/T_0$, and $Z = z/L_{NL}$, equation (4.76) leads to the form:

$$i\frac{\partial U}{\partial Z} = -\left[|U|^2 U + s\frac{\partial}{\partial \tau}\left(|U|^2 U\right)\right]. \tag{4.82}$$

Equation (4.82) is an equation for describing the front of a shock

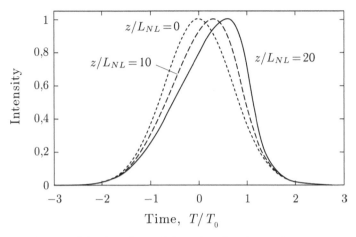

Fig. 9. Dependences of the time form of Gaussian ultrashort pulses upon propagation through the optical fibre obtained for $s = 0.01$ [3]. The initial shape of the pulse at the entrance to the optical fibre ($z = 0$) is the dotted line and the shape of the pulse after passing through them to the optical fibre of distances $z = 10L_{NL}$ is the dashed line and $z = 20L_{NL}$ the solid line.

wave and has an analytical solution. An analysis of the solution obtained shows that the dependence of the group velocity of the pulse on intensity leads to the fact that when the pulse propagates through optical fibres, the pulse becomes asymmetric. With increasing distance from the entrance to the optical fibre (z) the trailing edge of the pulse becomes steeper, and its maximum shifts to the trailing edge. The process of formation of a shock light wave is clearly illustrated in Fig. 9, which shows the calculated dependences of the change in the shape of a Gaussian ultrashort pulse as it passes through different distances along the fibre lightguide, obtained for $s = 2/(\omega_0\tau_0)= 0.01$ [3].

It is the increase in the steepness of the trailing edge of the ultrashort pulse that leads to the formation of a shock wave. The critical length of the fibre light guide on which the shock wave is formed, can be found from the condition $\partial U/\partial\tau \rightarrow \infty$. This condition allows us to obtain an estimate formula for determining the critical length [3]:

$$z_c \approx 0.4\frac{L_{NL}}{s}. \tag{4.83}$$

For the ultrashort pulse with parameters $T_0 \sim 100$ fs and $P_0 \sim 1$ kW, the value of z_c is ~ 1 m.

The foregoing assumption made in the derivation of equation (4.81) that the quadratic material dispersion in equation (4.76) is zero is never satisfied in practice for the following reasons:

1. Due to the finite duration of the ultrashort pulse, the pulse is always characterized by a certain spectral width $\Delta\omega \sim 2\pi/T_0$. Therefore, even if $d^2\beta/d\omega^2 = 0$ even for the central frequency, the quadratic dispersion will be nonzero for the spectral wings of the pulse.

2. Due to the spectral broadening of the pulse due to the PSM, the pulse always produces an eigenvalue of the quadratic dispersion, which can be estimated as:

$$\frac{d^2\beta}{d\omega^2} = \frac{d^3\beta}{d\omega^3} \cdot \left|\frac{\delta\omega_{max}}{2\pi}\right|, \qquad (4.84)$$

where $\delta\omega_{max}$ is the maximum frequency shift in the spectrum of the ultrashort pulse due to the PSM.

3. The steeper the wave front of the pulse, the shorter is its effective duration ($T_{eff} < T_0$). As a result, the pulse spectrum will be broadened ($\Delta\omega \sim 2\pi/T_{eff}$), and hence dispersion will be greater.

On the basis of the above reasons, an optical shock wave corresponding to an extremely steep rear edge of the pulse, due to the GVD, is never formed in practice. The GVD weakens the shock wave, broadening its steep rear edge. In addition, radiation losses in the fibre also delay the formation of an optical shock wave.

Thus, in the region of normal dispersion, the combined effect of the GVD and PSM processes, as well as the consideration of dispersion effects of higher order, leads to asymmetry of the USP envelope and its spectrum, to the appearance of oscillations in the envelope of the pulse and its spectrum, which leads to a change in the shape of the envelope and broadening of the ultrashort spectrum pulses in the Stokes and anti-Stokes regions.

4.9.2. Pumping in the region of anomalous dispersion

In section 4.3, it was shown that if the carrier wavelength is in the region of anomalous dispersion ($\partial^2\beta/\partial\omega^2 = \beta_2 < 0$), this leads to the formation of solitons in the fibre lightguide.

Let us consider the influence of the dependence of the group velocity on the intensity and the PSM on the behaviour of the ultrashort pulses for the case when the wavelength of the carrier pulse is in the region of anomalous quadratic dispersion. In this case, taking into account the neglect of the influence of cubic dispersion ($d^3\beta/d\omega^3 = 0$, $\delta = 0$) and the instantaneous nonlinear response ($\tau_R = 0$), equation (4.77) for the pulse envelope has the form:

$$i\frac{\partial U}{\partial \xi} + \frac{1}{2}\frac{\partial^2 U}{\partial \tau^2} = -N^2\left[|U|^2 U + is\frac{\partial}{\partial \tau}\left(|U|^2 U\right)\right], \qquad (4.85)$$

where $N^2 = L_D/L_{NL}$.

Equation (4.85) has analytic solutions in the form of solitons.

Figure 10 shows the calculated dependences illustrating the evolution process of the ultrashort pulse, which corresponds to the propagation in the optical fibre of the fundamental soliton ($N = 1$) [3]. Apparently, due to the dependence of the material dispersion of the group velocity on the intensity, the vertex of the pulse moves in the fibre more slowly than its edges and this leads to its shift to the trailing edge of the pulse. The dispersion of group velocities also weakens the steepness of the front of the ultrashort pulse.

For higher-order solitons ($N \geq 2$), the effect of the group velocity dependence on the intensity leads to the breakdown of such solitons into constituent parts. This phenomenon is called the decay of the bound state of a soliton. In particular, in the absence of the dependence of the group velocity on the intensity, two solitons of the second order ($N = 2$) form a bound state and propagate with the same velocity (they are degenerate). The phenomenon of the dependence of the group velocity on the intensity removes this degeneracy. As

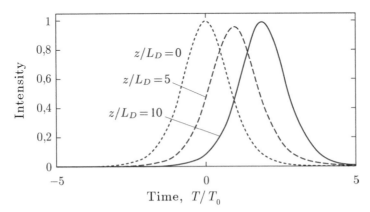

Fig. 10. Dependences of the time form of Gaussian ultrashort pulses when propagating through the optical fibre, obtained for $N = 1$ and $s = 0.2$ [3]. The dotted curve corresponds to the initial pulse shape, the dashed and solid curves are obtained for z/L_D values equal to 5 and 10, respectively.

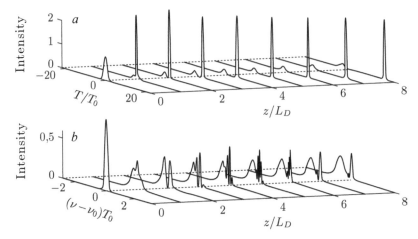

Fig. 11. Calculated dependences illustrating the dynamics of the pulse shape (*a*) and its spectrum (*b*) upon decay in the optical fibre of a soliton of the second order ($N = 2$ and $\tau_R = 0.01$) under the influence of intrapulse Raman scattering of light [3].

a result, these two solitons begin to propagate at different rates, which leads to their increasing spatial separation as they propagate into the fibre. This effect is well traced from the discussion in Fig. 11 *a*, which shows the dependence of the dynamics of the shape of the pulse in the optical fibre [3].

The delay effect of the nonlinear response in the fibre lightguide, taken into account on the right-hand side of equation (4.84) proportional to the τ_R term, dominates over the dependence of the group velocity on intensity, also causing the decay of solitons of higher order. The delay of the nonlinear response is equivalent to the dispersion of the nonlinear medium in the frequency representation. A short light pulse propagating in a medium with a delayed nonlinearity undergoes a low-frequency shift. Therefore, the spectral broadening induced by the retarded nonlinearity is equivalent to Raman scattering.

Another important feature of the retardation process is that when taking into account the dependence of the group velocity on the intensity, both solitons are delayed. In this case, a low-intensity soliton is accelerated and appears on the leading edge of the initial pulse, while a soliton with a higher intensity slows down and turns out to be shifted to the trailing edge.

Splitting the initial pulse into two or more solitons leads to a redistribution of energy between them. As a result, the energy of the initial pulse is concentrated in a sequence of short pulses, which includes new mechanisms for nonlinear broadening of the laser radiation spectrum. As noted in [1], already the earliest experiments on the propagation of pulses of pico- and nanosecond duration in the OF showed that the laser pulse spectrum is enriched due to cascade SRS and parametric four-wave interaction (FWI). Therefore, the new spectral components arising as a result of stimulated Raman scattering and FWI transformations of the ultrashort spectrum are then broadened by phase self- and cross-modulation, merge and lead to the generation of radiation with a broad continuous emission spectrum at the output from OF.

Figure 11 shows the dynamics of the shape and the spectrum of the ultrashort pulse, which illustrates the decay of a soliton under the action of intrapulse Raman scattering of radiation [3]. An extremely important feature of the emission spectrum shown in Fig. 11 *b* is its significant shift to the long-wave region. This effect in the experiments is called the self-shift of the soliton frequency and is interpreted as stimulated Raman scattering of the pulse, which arises even if the SRS (stimulated Raman scattering) threshold is not attained. The input pulse, which is a higher-order soliton, is shortened in the initial phase of propagation with simultaneous broadening of the spectrum. The spectral broadening of the 'red' spectrum wing provides a seed for subsequent Raman amplification. Thus, the 'blue' components of the pulse spectrum through the SRS interaction serve as pumping for the 'red' components. In the temporal representation of the pulse shape, SRS, the transfer of energy into the region of the 'red' components of the pulse is manifested in the appearance of a more powerful pulse that lags behind the initial one.

Thus, in the region of anomalous dispersion, accounting for dispersion and nonlinear effects of higher order in equations (4.75) and (4.76) leads to asymmetry of the USP envelope and its spectrum, to the appearance of oscillations in the envelope of the pulse and its spectrum, as well as to the decay of the soliton and self-shift of its frequency in the Stokes region.

4.10. Generation of a supercontinuum in optical fibres

An interesting and practically important direction of laser generation of white light is associated with the generation of a supercontinuum in the propagation of high-intensity ultrashort pulses in gas, liquid, and condensed media [4]. The characteristic peak power of the light pulses used in such problems lies in the terawatt range. As shown in chapter 2, high-intensity laser radiation intensity leads to ionization of the medium. The free electrons formed as a result of ionization make an essential contribution to the nonlinear response of the medium and have a significant effect on the spectral, temporal and spatial dynamics of the laser pulse.

Along with the effects of ionization in the generation of a supercontinuum by means of ultrashort pulses, an important role is played by the effects of spatial self-action, the formation of the shock front of the USP envelope, the delayed nonlinearity of the medium, and the effects of high-order dispersion.

For a quantitative description of the process of supercontinuum generation in the OF, when the power of the USP injected into them is insufficient to start the ionization process, the following equation should be used [2,3]:

$$
\frac{\partial A}{\partial z} + \frac{\alpha}{2} A + \sum_{k \geq 2} \frac{i^{k+1}}{k!} \frac{\partial^k A}{\partial T^k} =
$$
$$
= i\gamma \left(1 + \frac{i}{\omega_0} \frac{\partial}{\partial T} \right) \left[A(z,T) \int_0^\infty R(t) |A(z,T-t)|^2 \, dt \right],
$$

(4.86)

where α is the absorption coefficient for radiation in the OF, and the expression $R(t)$ corresponds to (4.62).

Equation (4.86) does not have an analytical solution and requires the use of numerical calculation algorithms.

For laser pulses of duration more than 10 fs, equation (4.86), taking into account equation (4.74), can be represented in a more simplified form

$$
\frac{\partial A}{\partial z} + \underbrace{\frac{\alpha}{2} A}_{\text{I}} + \underbrace{\sum_{k \geq 2} \frac{i^{k+1}}{k!} \frac{\partial^k A}{\partial T^k}}_{\text{II}} =
$$
$$
= i\gamma \left[\underbrace{|A|^2 A}_{\text{III}} + \underbrace{\frac{2i}{\omega_0} \frac{\partial}{\partial T} \left(|A|^2 A \right)}_{\text{IV}} - \underbrace{T_R A \frac{\partial |A|^2}{\partial T}}_{\text{V}} \right],
$$

(4.87)

where T_R is determined according to (4.75).

Equation (4.86) allows us to explicitly distinguish the action of all the main nonlinear optical processes affecting the shape and spectrum of laser pulses propagating in the OF: I – power loss of radiation; II – dispersion effects; III – phase self-modulation and phase cross-modulation; II + III + IV – formation of solitons and light shock waves; V – effects of inelastic scattering of photons by phonons, caused by the phenomena of stimulated Raman scattering and SBS in the material of the OF. When it comes to bulk environments, one of the above effects can dominate the others and be isolated and studied independently. At the same time, fibre-optic fibres are able to provide long lengths of nonlinear-optical interaction of laser pulses with the medium and, thereby, significantly reduce the requirements for the power of the laser pulse necessary for generating the supercontinuum.

However, the use of optical fibres makes it possible to obtain supercontinuum radiation with a spectrum overlapping most of the visible and a significant part of the near infrared radiation ranges when using pulses with a kilowatt peak power.

The creation of fibre-optic supercontinuum sources capable of blocking the visible and near-IR regions of the emission spectrum was made possible by the creation of photonic-crystal optical fibres.

4.11. Nonlinear properties of photonic crystal fibres

In the sections 4.2 to 4.10, when analyzing nonlinear optical processes in optical fibres, standard optical fibres based on quartz glass, which are characterized by low nonlinearity in comparison with other materials, were used in optical communication systems. To develop nonlinear effects in them, as a rule, fairly long sections of the interaction length of the radiation with the fibre material are required. At the same time, lightguides with high nonlinearity are of considerable interest for a number of scientific and practical applications. For example, the use of highly nonlinear optical fibres in optical amplifiers and SRS-based lasers facilitates the problem of obtaining a homogeneous fibre region by substantially reducing the interaction length.

Highly nonlinear fibres are also effective for obtaining powerful extremely short optical pulses. With the help of highly nonlinear fibres, it is possible to obtain pulses with a wide spectrum (supercontinuum) exceeding an octave. Highly nonlinear fibres provide a significant increase in the speed of the soliton self-shift of the frequency, which is of no small importance for effectively tuning the frequency of ultrashort laser pulses.

Thus, undoubtedly, the problem of creating optical fibres with a high nonlinearity becomes urgent.

A measure of the nonlinear properties of optical fibres is the nonlinearity parameter γ in equation (4.9)

$$\gamma = \frac{n_2 \omega}{c S_{\text{eff}}},$$

(4.88)

which determines the length of the nonlinear interaction and which depends on the nonlinear optical properties of the optical fibre material (n_2) and on the effective area of the emission mode in the optical fibre (S_{eff}).

The non-linear parameter γ for a standard quartz fibre waveguide with a core area of 30–70 μm² is (1–2) W⁻¹ km⁻¹. The value of γ for quartz fibres can be increased by decreasing the diameter of the core, which leads to a decrease in the effective mode area. In particular, a decrease in the effective mode area to ~10 μm² makes it possible to increase the value of the nonlinear parameter by an order of magnitude.

Since the effective radius of the waveguide mode is defined as

$$r \approx a + \frac{1}{\sqrt{k^2 n_c^2 - \beta^2}},$$

(4.89)

then a decrease in the radius to a size comparable or smaller than the wavelength of the radiation propagating in the fibre becomes ineffective, due to the penetration of the mode field beyond the core of the fibre.

Further reduction of the effective mode area within the use of one material can be achieved by increasing the difference in the refractive indices of the core (n_c) and the cladding (n_{cl}) of the optical fibre. For a standard quartz fibre waveguide, the maximum $\Delta n = n_c - n_{cl} = 0.45$ is reached only if the cladding material is air. Such fibres are called fibres with constriction and are made by simple melting and subsequent stretching of a standard fibre lightguide (Fig. 12).

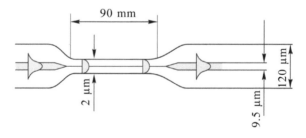

Fig. 12. Scheme of an optical fibre with a constriction [6].

If the diameter of the core in the waist region is 2 μm, then the nonlinear parameter has a value of $\gamma \sim 100$ W^{-1} km^{-1}. Theoretically, the diameter of the core of the fibre in the waist region can be brought to 0.74λ, which corresponds directly to the propagation in the optical fibre of ~75% of the radiation power. As a result, for a radiation wavelength $\lambda = 0.8$ μm and $n_2 = 2.6 \; 10^{-20}$ m^2/W, the value of the nonlinear parameter will be $\gamma \sim 660$ W^{-1} km^{-1}.

However, due to the extreme fragility of the design, the problems associated with the manufacture and subsequent operation have not yet made it possible to achieve such a high value of the nonlinear parameter.

With the advent of MS- and PC optical fibres, nonlinear waveguide optics entered a new phase. The latter is due to the fact that, being much less brittle, these optical fibres provide a high concentration of radiation in the core of the fibre and open up wide possibilities for controlling the spectral profile of the dispersion characteristics Due to these unique properties, the MS- (Fig. 13 *a*) and P- (Fig. 13 *b*) photonic–crystal fibres made on the basis of fused quartz allow to achieve high efficiency for all basic nonlinear optical phenomena when using non-amplified ultrashort laser pulses and open up new possibilities for frequency conversion, spectrum transformation and the control of laser pulse parameters.

4.11.1. Dispersive properties of microstructured (MS) optical fibres

MS-photonic crystal fibres are one of the most widely used in practice. As seen from Fig. 13 *a*, the MS optic fibres consist of a small diameter core of quartz glass surrounded by a periodic array of air holes forming a cladding. The air holes lower the effective refractive index of the cladding, and also provide a large jump in the refractive index (since the refractive index of the core is $n_c = 1.45$, and the effective index of refraction of the cladding, depending on the structure parameters, is in the range from 1 to 1.45), which allows the radiation to propagate practically along the quartz core. The parameters of the microstructure of the cladding can be changed, which opens the possibility for a change in the light-guiding properties of the MS fibres in a wide range. Figure 14 *a* shows the calculated dependences for the diameter of the main guided mode in the MS optical fibre on the period of the triangular lattice of the array of holes in the surrounding cladding, obtained

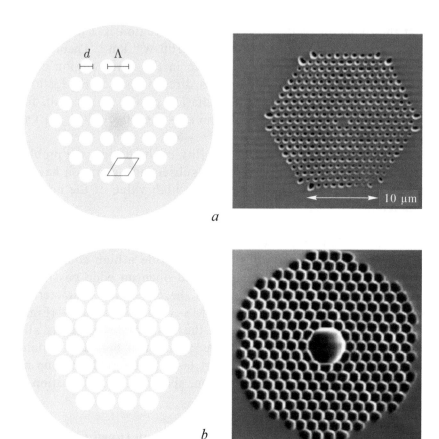

a

b

Fig. 13. Schematic representation of the microstructured (MS) and holey (H)optical fibres and scanning electron microscope photograph (*a*) of the MS and (*b*) P-optical fibres (diameter of the central air channel 9.3 μm) [7].

for radiation with a wavelength of 1.5 μm [8] and the calculated dependence of the effective area of the fundamental mode on the wavelength of the radiation (Fig. 14 *b*).

These dependences show a tendency for the increase the effective mode size in the MS optical fibre with decreasing ratio *d*/Λ for small values of Λ. As a result, this should lead to delocalization of the fashion and its infiltration into the surrounding core. On the contrary, the use of large *d*/Λ ratios in MS lightguides, at small sizes of Λ, makes it possible to approximate the effective refractive index of the microstructured cladding to 1 and, as for the drawn fibres, to localize the mode in a very small core of the MS fibre optic for sufficient a large range of wavelengths of visible and near IR radiation.

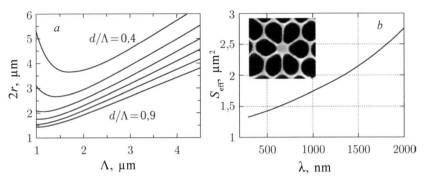

Fig. 14. *a* – The calculated dependences of the diameter of the main mode (2*r*) of the MS optical fibre (photo is shown in Fig. 13 *a*) from the period of holes in the triangular cladding array (Λ) for various ratios of hole diameters and their placement period (*d*/Λ) [7]. *b* – Dependence of the effective area of the fundamental mode on the wavelength of the radiation in the MS optical fibre (the photograph of the section is shown in the inset), having a core diameter of 1.7 μm [8].

As a result of this, the choice of the ratio *d*/Λ makes it possible to fabricate optic fibres with a high nonlinearity, providing for them the value of the nonlinear parameter γ close to the drawn fibres ~100 W⁻¹ km⁻¹.

MS lightguides, like drawn fibres, have a number of unique properties. In particular, by changing the structure and geometry of the holes in the microstructured cladding of such fibres, it is possible to form the desired dispersion profile of the guided modes. This feature of the MS optic fibres is illustrated by the dispersion dependences of the main mode direction on the emission wavelength for MS-fibre waveguides of different geometric sizes, shown in Fig. 15.

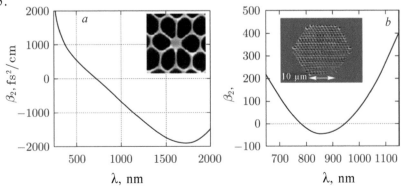

Fig.15. Experimentally measured dispersion dependences (2) for the main guided mode on the emission wavelength for MS-fibres, the photographs of which are shown in inserts [8].

The group velocity dispersion (GVD) of the MS-fibres was measured by the method of white-light interferometry. As follows from the above dependences, the dispersion properties of MS-fibre lightguides can have significant differences from the dispersion characteristics of conventional quartz fibres. In particular, by selecting the cladding structure, the wavelength of the zero dispersion for the MS-fibre lightguides can be shifted to the visible range of the spectrum. It is the presence of a microstructured cladding that makes the dispersion characteristics for MS-fibre lightguides unlike the characteristics of standard quartz fibres. The results shown in Fig. 15 show that the dispersion dependence of the fundamental mode of an MS optical fibre can have both one (Fig. 15 *a*) and two zero-dispersion wavelengths with wavelengths of 780 nm and 945 nm, between which there is a narrow region of anomalous dispersion (Fig. 15 *b*).

In MS optical fibres, the effective refractive index of the cladding and the guided mode depends on the period of the air-hole placement Λ and on the ratio of the sizes of air holes (d) to their period d/Λ in the structure. By varying these characteristic dimensions, it turns out to be possible to vary the dispersion characteristics of the MS fibres in a wide range. Figure 16 shows for comparison the calculated dependences of the group delay dispersion (GDD) of an ordinary telecommunication quartz fibre waveguide and several MS-fibres with different ratios in the cladding. Because for large values d/Λ the effective refractive index of the cladding is close to 1 and this

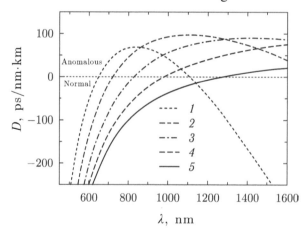

Fig. 16. Calculated profiles of the dispersion of group delay (*D*) for the MS fibre lightguides [7]: 1 – Λ = 1 μm, d/Λ = 0.95; 2 – Λ = 1.5 μm, d/Λ = 0.9; 3 – Λ = 2 μm, d/Λ = 0.8; 4 – Λ = 3 μm, d/Λ = 0.7; 5 – conventional quartz fibre lightguide.

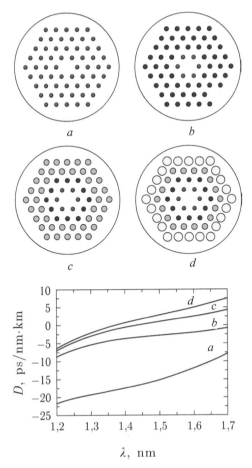

Fig. 17. Calculated dependences of group delay dispersion (D) for MS-fibres with air holes. The period of the hole structure is $\Lambda = 2$ μm; diameters of air holes: (*a*) $d_1 = d_2 = d_3 = d_4 = 0.5$ μm; (*b*) $d_1 = 0.5$ μm, $d_2 = d_3 = d_4 = 0.6$ μm; (*c*) $d_1 = 0.5$ μm, $d_2 = 0.6$ μm, $d_3 = d_4 = 0.7$ μm; (*d*) $d_1 = 0.5$ μm, $d_2 = 0.6$ μm, $d_3 = 0.7$ μm, $d_4 = 1.8$ μm.

leads to the appearance of large values of the waveguide dispersion in comparison with traditional optical fibres. As a result, the wavelength for the first zero of the group dispersion of the MS fibre is shifted to the visible region, while for the second one it remains in the infrared region of the spectrum. It is this type of dispersion that is most preferable for the effective nonlinear-optical generation of the supercontinuum in MS fibres. As shown in Fig. 17, the variation of the dimensions and arrangement of the holes in the cladding of the MS optic fibre leads to a considerable potential in the formation of the dispersion profile of the optical fibre.

In connection with this, the change in the structure of such fibres can form the required dispersion profile, and a significant difference in the refractive indices of the core and the cladding causes a high degree of localization of the light radiation in the fibre core, which allows achieving high efficiencies of nonlinear optical interactions for ultrashort laser pulses of low power.

The nonlinear optical properties of MS lightguides can be increased by using materials having a nonlinear refractive index 2 time greater than that of fused quartz. Such materials include lead–silicate, chalcogenide glasses and glasses with an admixture of tellurite and bismuth oxides.

An example of lead–silicate glasses can be glass grade SF57 [10]. The refractive index of SF57 glass at a wavelength of 1.55 μm is equal to 1.81. The dispersion of the group velocity in a bulk glass of the SF57 grade passes through zero at a wavelength of 1970 nm (Fig. 18). The nonlinear refractive index of these glasses is n_2 ~4.1 10^{-19} m²/W, which is approximately an order of magnitude greater than that of fused quartz n_2 ~ $(2-3)\cdot10^{-20}$ m²/W. In connection with this, it is possible to achieve values of the nonlinearity index γ ~ 640 W⁻¹ km⁻¹ for MS-fibres based on such glasses. Variations in the size and geometry of the holes in the microstructured cladding make it possible to vary the dispersion properties of the lightguides within a wide range (Fig. 17) and achieve values of β_2 ~ 40 ps²/km for MS fibres in the vicinity of λ = 1.33 μm, in contrast to β_2 ~ −30 ps²/km for the fibres made from fused quartz.

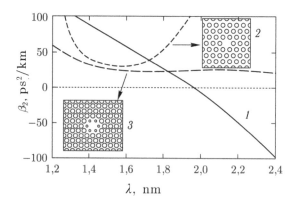

Fig.18. Calculated spectral dependences of the group velocity dispersion for a bulk sample of SF57 glass (1) and for MS fibres based on SF57 glass with the following parameters: Λ = 1.7 μm, d = 0.29 μm (2) and Λ = 1.07 μm, d_1 = 0.2 μm, d_2 = 0.37 μm (3) [10].

Another type of SF6 lead-silicate glass is also known. The nonlinear index of such glasses is $n_2 \sim 2.2 \cdot 10^{-19}$ m²/W. The dispersion for these glasses in the vicinity of a wavelength of 1550 nm is essentially positive. As a result, for MS-fibres with core diameters of 2.6 and 4 μm, the dispersion of the group velocity passes through zero in the region of 1.3 and 1.4 μm wavelengths. This made it possible to ensure in such MS fibres a high efficiency of supercontinuum generation when pumped with femtosecond pulses of duration ~60 fs with a central emission wavelength of 1560 nm [10].

Of particular interest are the chalcogenide glasses [1, 11, 12] for the creation of light-emitting diodes with an increased nonlinearity. The refractive index of these glasses is significantly (1.5–2 times) higher than the refractive index of quartz glass, and the nonlinear refractive index of such materials reaches values of $\sim 10^{-17}$ m²/W, which is almost 40 times higher than the corresponding value for quartz glass. All this allows us to create MS optical fibres from chalcogenide glass, for which the nonlinearity index can reach values $\gamma > 1000$ W⁻¹ km⁻¹.

Thus, the high contrast of the refractive index of the core and the cladding allows, by varying the size of the holes, their number and relative location, to substantially control the dispersion properties of the MS fibres: to shift the wavelength of the zero value of its dispersion, to change the slope of the dispersion curve, to achieve repeated intersection of the dispersion curve with the wavelength axis, and to control the size of the cross-sectional area of the mode.

4.11.2. Generation of a supercontinuum in MS optical fibres, for which the pump pulse wavelength lies in the region of anomalous dispersion

The generation of a supercontinuum is not a manifestation of any single nonlinear optical effect. Generally, the generation of a supercontinuum is the result of the addition of actions of a variety of nonlinear optical phenomena, such as self-modulation of the phase (PSM), decay of high-order solitons, modulation instability, stimulated Raman scattering, shock wave formation, four-wave mixing (FWM) which, if the pump pulse wavelength is close to the zero-dispersion wave length and is in the region of anomalous dispersion, lead to the appearance of an extremely broad spectrum of radiation at the exit of the optical fibres [13].

In special cases of large pump wavelengths and anomalous dispersion, the formation of solitons begins to play a decisive role in the formation of the supercontinuum [13, 14]. At the initial stage, their effect on the broadening of the spectrum is manifested through the decay of solitons of high order, and at a later stage through the resonance interaction of solitons with dispersion waves based on the inclusion of Kerr and Raman nonlinearities [15, 16, 17].

This section is devoted to the process of supercontinuum generation in MS optical fibres, when the wavelength of the pump radiation lies in the region of anomalous dispersion, which is based on the result of the interaction of solitons with dispersion waves and solitons. To describe the process of supercontinuum generation in MS optical fibres by means of ultrashort pulses, one should use equations (4.86) and (4.87), which are capable of taking into account higher-order dispersion phenomena and the finite time of the nonlinear response of medium polarization.

Dispersion parameters of higher orders for fibre light guides depend very much on the diameter of the fibre core. For illustration, Fig. 19 shows the calculated spectral dependences of the dispersion parameters $\beta_2(a)$ and $\beta_3(b)$ obtained for quartz filaments of various diameters surrounded by an air envelope. The dashed lines in Fig. 19 show the dependences for β_2 and β_3 corresponding to a quartz filament of radius $a = 5$ μm. In practice, this size is characteristic of the core size of standard quartz fibres. As can be seen from Fig. 19 a, for a given type of a light-guide quartz filament, the wavelength of zero dispersion ($\beta_2 = 0$) is $\lambda_D = 1.25$ μm. Reducing the filament radius to 1 μm results in the zero-dispersion wavelength shifting to

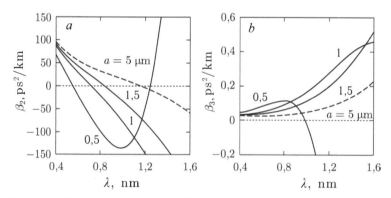

Fig. 19. Spectral dependences of the dispersion parameters β_2 and β_3 for a quartz filament surrounded by an air envelope, with a change in its radius from 0.5 to 5 μm [12].

a region of 0.8 μm. When the filament radius approaches 0.5 μm, two zero-dispersion regions appear on the dispersion relation, and the third-order dispersion relation (β_3) in the wavelength range over 1 μm rushes to the negative range (Fig. 19 *b*).

Since the diameter of the core of the MS fibres made of fused quartz is usually 0.5–2 μm, and for a large value d/Λ the effective index of their holey cladding is close to the refractive index of air, the character of the dispersion dependences shows in Fig. 19 for a quartz filament very well illustrates the behaviour of the dispersion parameters for MS fibres. In connection with this, MS fibre lightguides, while preserving the single-mode radiation propagation regime in a wide spectral range, in contrast to conventional optical fibres, have an anomalous dispersion region in the shorter wavelength range up to wavelengths of $\lambda \sim 800$ nm. In this case, the absolute value of the dispersion of the group delay in the MS optical fibres can reach 103 ps/(nm·km), which is almost an order of magnitude higher than in the conventional optical fibres.

As follows from expression (4.87), various nonlinear optical effects simultaneously contribute to the process of propagation of ultrashort laser pulses in MS fibres. In this case, it is of fundamental importance in what dispersion region for the MS fibre lightguide is the frequency (wavelength) of the carrier of the exciting USP.

In the case when the carrier frequency of the exciting pulse is in the region of anomalous dispersion, when the supercontinuum of USP is generated with a duration of <100 fs, the dominant process leading to a broadening of the pulse spectrum is the decay of solitons [18].

Figure 20 shows the scheme of the supercontinuum generation in an MS fibre optic due to the decay of high-order solitons. The main condition for the effective generation of the supercontinuum is that the carrier frequency of the laser pulse (ω_0) introduced into the MS fibre is near the zero-dispersion frequency (ω_d) and lies in the region of anomalous dispersion. In the figure, the dispersion dependence of the MS fibre is shown by the gray line. Since the wavelength of the carrier pulse is close to the wavelength of the zero dispersion, the value is $\beta_2 \sim 0$. As a result, the dispersion effects in the optic fibre will be determined to a large extent by the cubic dispersion term β_3. The dispersion relation for β_3 in the figure is indicated by the green line. Under these conditions, a high-order soliton will form in the fibre (see section 4.9.1): $N \gg 1$, the order of which increases with increasing power of the pulse introduced into the waveguide. The existence of this soliton is the result of

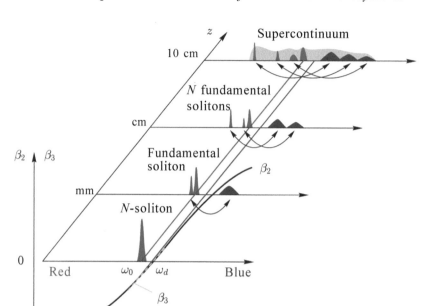

Fig. 20. Scheme of the process of supercontinuum generation in an MS fibre optic, based on the decay of solitons.

a balance between the processes of self-modulation of the phase and the dispersion of the radiation of the pulse in the optical fibre, leading to a periodic expansion of its spectrum and an increase in its activity. The perturbations of the soliton due to the processes of intrasoliton SRS scattering, the self-action of the pulse (the formation of a shock light wave) and the dispersion of higher orders, lead to the fact that in the process of propagation along the MS optical fibre the soliton of high order will decay into several fundamental solitons ($N = 1$) whenever its amplitude reaches its maximum in the frequency domain. The characteristic length at which the soliton decay (L_{break}) will be observed is determined by the following expression [18]:

$$L_{break} = \frac{L_D}{N} = \sqrt{\frac{T_0^2}{\gamma P_0 |\beta_2|}}. \tag{4.90}$$

The earlier the fundamental soliton separates from a high-order soliton, the greater will its amplitude and group velocity be and shorter the duration.

In the course of further propagation along the MS fibre lightguide, the spectrum of the separated fundamental soliton will shift to the 'red' region of the spectrum due to the SRS phenomenon. The SRS is capable of substantially mixing the spectrum of the fundamental soliton. For example, the front of the soliton corresponding to the 'blue' part of the spectrum, due to the SRS shift, can act as a pump for the waves of the trailing edge of the soliton corresponding to the 'red' region of the spectrum. As a result, the central (carrier) wavelength will shift to the region of 'red' (long) waves. This effect is called the soliton self-shift of the frequency.

In the process of decay of a high-order soliton, each soliton formed, in order to retain its shape, radiates a frequency shifted to the 'blue' region of non-soliton radiation whose wavelength depends on the phase-matching condition [8]:

$$\Delta\phi = \phi_S - \phi_{NS} = 0, \qquad (4.91)$$

where ϕ_S is the soliton phase, and ϕ_{NS} is the phase of the non-soliton wave

$$\phi_S = \left(\left(\beta(\omega_S) + \frac{n_2 I \omega_S}{c} \right) z - \omega_S t, \right. \qquad (4.92)$$

$$\phi_{NS} = \beta(\omega_{NS}) z - \omega_{NS} t. \qquad (4.93)$$

Thus, each process of division of a high-order soliton turns out to be resonantly connected with a linearly dispersive wave whose spectrum is shifted to the 'blue' region of the radiation spectrum and located in the region of normal dispersion. The magnitude of the displacement of the frequency of the dispersive wave with respect to the carrier frequency of the soliton is [12]

$$\Omega_d \sim \frac{3\beta_2}{\beta_3} + \frac{\gamma\beta_3 P_S}{3\beta_2^2}, \qquad (4.94)$$

where P_s is the power of the soliton.

In fact, this process is due to the effect of PSM and FWI, arising from the mixing of fundamental and high-order solitons.

The appearance of dispersion non-soliton radiation occurs when the phase-matching ratio is satisfied.

Since the spectrum of the formed non-soliton dispersion radiation lies in the region of normal dispersion, then this radiation propagates in the form of ordinary dispersive waves.

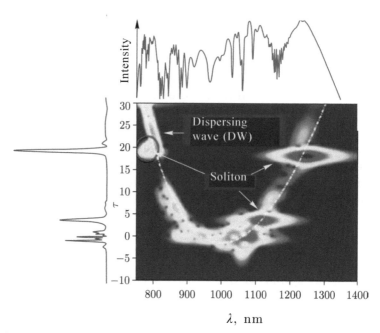

Fig. 21. Spectrogram of radiation of a supercontinuum formed by a pump pulse with a wavelength λ_p = 1060 nm and a duration of 110 fs after passing a distance of 30 cm in the MS optical fibre, the wavelength of zero dispersion for which is equal to 987 nm; τ – normalized group delay [19].

Since part of the soliton spectrum is resonantly associated with the dispersive wave, the energy of this spectral part of the soliton is redistributed into the energy of the dispersive wave. Since in this case the soliton loses energy from the blue part of the spectrum, the energy remaining inside the soliton will be redistributed between its other spectral components and the final result will be the red shift of its spectrum.

As a result, around the wavelength of zero dispersion there is a spectral gap. After this, the initial broadening of the emission spectrum includes mechanisms of phase self-modulation, four-wave mixing, stimulated Raman scattering and dispersion, which lead to further broadening of the pulse spectrum and filling this spectral gap.

Thus, the result of the addition of all the above-listed nonlinear optical processes is the formation of a supercontinuum of broadband radiation at the output of the MS-fibre lightguide.

The process of generating a supercontinuum is traced in detail in the spectrogram shown in Fig. 21. Initially, due to the self-modulation effect of the pulse phase, a symmetrical broadening of the spectrum occurs. Then a soliton is formed, whose self-shifting frequency and

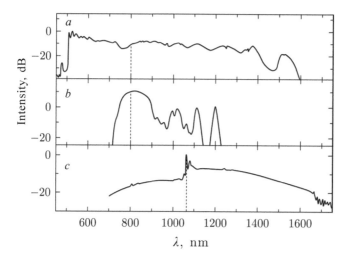

Fig. 22. Experimentally measured spectral dependences of the supercontinuum generated in optic fibres for laser pulses of femtosecond duration, for which the carrier wavelength of the ultrashort pulse is located in the region of anomalous (*a*) and normal (*b*) dispersion, and also for laser pulses of nanosecond duration with a central wavelength in the region of anomalous dispersion (*c*). The dashed vertical line indicates the position of the carrier wavelength of the pulse [10].

stimulated Raman scattering lead to a shift in the emission spectrum to the 'red' region. The subsequent resonant coupling of the soliton with the dispersive wave (it is clearly seen from Figure 21 that the group velocities of the soliton and the dispersive wave are the same) leads to the formation of a 'blue' wing in the radiation spectrum and pumping some of the radiation power into it.

Generation of a supercontinuum in MS fibres is also possible if the central frequency of the ultrashort pulse of the input pulse lies in the region of normal dispersion [20]. In this case, the dominant nonlinear optical process is the process of self-modulation of the radiation phase (see Section 4.9.1).

Thus, the parameters of the laser pulse (wavelength, duration and power) and the characteristics of the MS fibre optic (γ, dispersion, mode composition, length) are the determining factors in the generation of the supercontinuum. The most significant parameter for generating a supercontinuum is the wavelength of the pump pulse, or rather, its location in the region of normal or anomalous dispersion for an MS optical fibre, and also the duration of the pump pulse: femtosecond or pico-/nanosecond.

Figure 22 shows the experimentally obtained spectra of supercontinuum generation in various MS fibres.

The spectral dependence (Fig. 22 *a*) of the supercontinuum was obtained using a titanium–sapphire laser emitting pulses with a wavelength $\lambda_p = 800$ nm, duration $T_0 = 100$ s and repetition frequency 80 MHz. The experiments used a quartz MS fibre lightguide with the following characteristics: length 75 cm, core diameter 2 μm, wavelength of zero dispersion $\lambda_D = 760$ nm. The wavelength of the pump pulse was in the region of the anomalous dispersion of the optical fibre. As a result, with a mean pump pulse power of 100 mW, a supercontinuum was generated in the optic fibre, whose spectrum

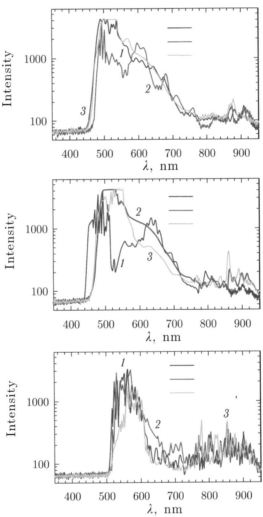

Fig. 23. Spectra of supercontinuum in an MS fibre optical fibre ($\lambda_D = 800$ nm), obtained for different values of the duration of a laser pumping pulse with a wavelength of 826 nm [10].

ranged from 500 to 1600 nm. Figure 22 *b* shows the spectrum of the supercontinuum obtained in an MS optic fibre with a zero-dispersion wavelength $\lambda_D = 940$ nm, when the wavelength of the pump pulse lies in the region of normal dispersion. All other parameters of the laser and optical fibre are similar to the previous case.

Figure 22 *c* characterizes the spectrum of the supercontinuum generated by the emission of a pulsed nanosecond Nd:YAG laser with a wavelength $\lambda_p = 1064$ nm, pulse repetition frequency 30 kHz and average power 180 mW. In the experiments we used a quartz fibre-optic of length 20 m, core diameter 4.8 μm and zero-dispersion wavelength $\lambda_D = 1040$ nm. The wavelength of laser radiation was in the region of the anomalous dispersion of the optical fibre.

As follows from the above dependences, the generation of a supercontinuum in MS light guides is most effective when anomalous dispersion and femtosecond pulses are used for the pump radiation.

Figure 23 shows the experimental spectra of the supercontinuum generated by laser pulses of various durations in an MS optic fibre. It can be seen from these dependences that if the pulse duration lies in the range from 66 to 106 fs, then the energy momentum is concentrated in the vicinity of certain spectral regions. The highest intensity corresponds to the region from 500 to 600 nm. This phenomenon is usually associated with the destructive interference of spectral components arising as a result of the process of self-modulation of the phase of radiation of a pulse.

With a further increase in the pulse duration, an expansion of its spectrum is observed and a decrease in the intensity of the spectral components due to the redistribution of energy. At a pulse duration of more than 177 fs, 202 and 247 fs, the supercontinuum spectrum in the long-wave region is unstable. The widest spectrum with the lowest fluctuations of the spectral components is observed at pulse durations of 83 and 106 fs.

Thus, the maximum spectral expansion of the supercontinuum spectrum into the 'red' spectral region is observed for long pulses. In fact, short-duration pulses create high-order solitons that are rapidly destroyed due to the inclusion of higher-order dispersion and transmit their energy to the emission of the 'blue' region of the spectrum. Therefore, the longer the pulse duration becomes, the greater the role of stimulated Raman scattering. This leads to a shift in the spectrum of the supercontinuum to the long-wavelength region.

In [21], the process of supercontinuum generation in MS optic fibres with a core diameter of 2.5 μm, a length of 38 cm and a zero-dispersion wave length of λ_D =790 nm, was studied theoretically and experimentally, depending on the pump radiation power. The pump wavelength corresponded to the pulse wavelength of a titanium-sapphire laser of λ_p = 800 nm, the pulse duration was 60 fs. For these radiation parameters, the values of the nonlinearity coefficient were determined γ = 80 W^{-1} km^{-1} and the corresponding values of the dispersion coefficients: β_2 = –2.1 fs^2/mm; β_3 = 69.83 fs^3/mm; β_4 = –73.25 fs^4/mm; β_5 = 191.05 fs^5/mm; β_6 = –727.14 fs^6/mm; β_7 = 1549.4 fs^7/mm. Figure 24 presents the experimental and calculated

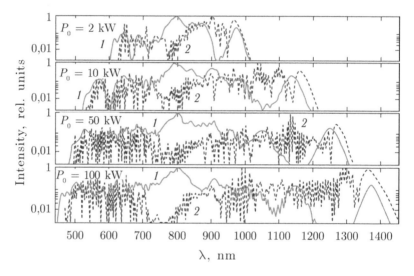

Fig. 24. Comparative dependences of the experimental (1) and calculated (2) dependences of the efficiency of supercontinuum generation in MS optical fibres on the power of pulsed pump radiation [21].

dependences for the efficiency of supercontinuum generation, obtained with a change in pump radiation power from 2 to 100 kW. Comparison of experimental data and numerical calculations shows quite satisfactory agreement between them in determining the width and nature of the supercontinuum spectrum.

More detailed studies of the process of supercontinuum generation were carried out in a work in which the influence of the energy and time characteristics of laser radiation on the generation process was simultaneously studied [14]. In the numerical simulation of the process, the authors based themselves on the description of the supercontinuum generation process as a result of the decay of a high-order soliton and non-soliton scattering of radiation (NSR), which

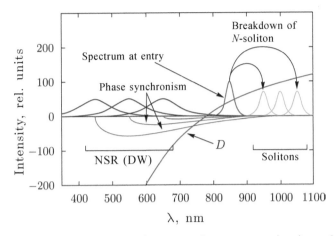

Fig. 25. Scheme of the process of supercontinuum generation in an MS optical fibre [14].

is the result of the resonant coupling of fundamental solitons with dispersion waves (DW).

The scheme for modeling the process of supercontinuum generation is shown in Fig. 25.

As the object of the study, we selected MS fibres from fused quartz with a core diameter of 2.5 μm, a length of 40 cm and a wavelength of zero dispersion $\lambda_D = 780$ nm. In experiments, pulsed emission of a titanium–sapphire laser with a wavelength of 850 nm and a pulse duration of 29 and 100 fs was used.

Figure 26 shows the experimental and calculated dependences of the intensity of the supercontinuum in an MS optical fibre obtained by using pulses of laser radiation with a duration of 29 and 100 fs of different power. Figure 26 *a* shows the measured spectra of the supercontinuum obtained with a pulse duration of 29 fs and a peak intensity of 27 GW/cm², 0.24 TW/cm² and 0.7 TW/cm², corresponding to pulse energies of 0.024, 0.21 and 0.62 nJ, respectively. Figure 26 *b* shows analogous dependences obtained with the use of laser pulses of 100 fs duration. As can be seen from Fig. 26 *a*, in the case when the pulse duration is 29 fs, and its intensity is 27 GW/cm², there is an insignificant broadening of the laser pulse spectrum, mainly to the 'blue' region. At the same time, an increase in the pulse duration to 100 fs (Fig. 26 *b*) leads to the generation of a supercontinuum in a wide spectral range from 500 to 1100 nm. With the intensity of the pumping pulse increasing by a factor of 100 ($I_0 = 0.24$ TW/cm²), the spectra of the supercontinuum generated by pulses of 29 fs duration expand predominantly into the 'blue' region of the spectrum

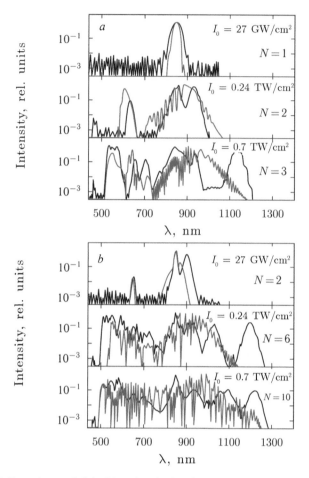

Fig. 26. Experimental (black) and calculated (red) dependences of the intensity of the supercontinuum in an MS optical fibre obtained using pulses of laser radiation of duration 29 (*a*) and 100 fs (*b*) of different power [14].

and practically do not expand into 'red', although the intensity of radiation in these regions increases. For pulses of 100 fs duration, a strong increase in the intensity of the spectral components in the 'blue' region of the spectrum and the appearance of intense spectral peaks in the 'blue' are characteristic.

The subsequent increase in the pump intensity to 0.7 TW/cm² leads to a significant broadening of the supercontinuum spectrum, and in both cases it begins to occupy the region from 350 to 1300 nm. A comparison of the experimental dependences shown in Figs. 26 *a* and *b* demonstrates an explicit dependence of the process of broadening of the supercontinuum spectrum on the pulse duration: for an equal intensity, longer pulses undergo a larger spectral broadening.

In [14], the observed supercontinuum generation in MS fibres was numerically simulated using the schemes shown in Fig. 25. The results of numerical calculations are shown in Figs. 26 *a* and 26 *b* with red lines. A good agreement between the calculated and experimental dependences confirms the legitimacy of the supercontinuum generation mechanism proposed in Ref. [14] for the generation of supercontinuum in MS fibres, when their USP are excited, whose emission wavelength lies in the region of the anomalous dispersion of the optical fibre.

Indeed, when the wavelength of the pump radiation lies in the region of anomalous dispersion, high-order solitons are formed at the entrance to the fibre optic fibre. The order of these solitons is determined according to the expression (4.32) as $N = \sqrt{\gamma T_0^2 P_0 / |\beta_2|}$ where T_0 is the duration and P_0 is the amplitude of the pump pulse power. Since the wavelength of the pump pulse is near the zero-dispersion wavelength, a non-zero third-order dispersion causes the rapid decay of a high-order soliton (N) into fundamental solitons ($N = 1$), followed by a 'red' shift of their frequency and the transformation of some of their energy into the 'blue' region of the spectrum through the establishment of phase synchronism with non-soliton (dispersive) radiation waves, whose spectrum lies in the region of normal dispersion, (see Figs. 21 and 25). Figures 26 *a* and 26 *b* show the calculated values of the order of the solitons N for all these cases. As can be seen, the increase in the power and duration of the pump pulses leads to an increase in the number of fundamental solitons formed in the MS fibre optic fibres and, accordingly, to the expansion of the supercontinuum spectrum. Since the increase in the pulse duration leads to an increase in the number of fundamental solitons at the same power density of the pump pulses, the generation efficiency and broadening of the supercontinuum spectrum for longer pulses turn out to be higher. In the case when the number of excited solitons coincides in both cases: $N = 2$ (Figs. 26 *a* and 26 *b*), regardless of the duration of the pump pulses, the spectral distributions for the generated supercontinuum practically coincide. All this confirms the soliton character of supercontinuum generation in MS optical fibres, when the length of the pump wave is in the region of anomalous dispersion.

4.11.3. Generation of supercontinuum in MS fibre lightguides in pumping in the region of normal dispersion

The determining factors in the generation of supercontinuum in MS fibres are: the location of the wavelength of laser pumping radiation with respect to the wavelength of the zero dispersion of the fibre, the duration of the pump pulse and its peak power. The dispersion of the fibre and, especially, its sign determine the type of nonlinear optical processes involved in the formation of the supercontinuum. When the wavelength of the pump pulse lies in the region of normal dispersion of the MS fibre, the self-modulation of the phase of the pulse and stimulated Raman scattering that determine the spreading of the spectrum into the long-wavelength region are the determining processes in the generation of the supercontinuum. Figure 27 shows the spectrum of the supercontinuum obtained in an MS fibre with a core diameter of 2.5 μm and a length of 12.5 cm when it is pumped by pulsed radiation of 25 fs duration and wavelength of the wave $\lambda_p = 800$ nm with an average power of 53 mW and a repetition rate of 76 MHz. The pump wavelength is in the region of normal dispersion of the MS lightguide having a zero-dispersion wavelength of 875 nm.

The process of supercontinuum generation in the region of the normal dispersion of the light guide changes greatly when the pump

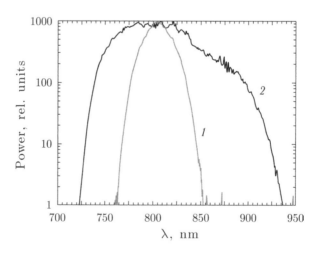

Fig. 27. Spectra of the pump pulse (1) of a titanium–sapphire laser with a duration of 25 fs at a wavelength of 800 nm and a supercontinuum (2) generated in an MS fibre with a core diameter of 2.5 μm, a length of 12.5 cm and a zero-dispersion wavelength of 875 nm, when the pump wavelength is in the region of normal dispersion of the fibre [22].

wavelength approaches the zero-dispersion wavelength of the MS fibre, since new nonlinear optical effects begin to be included.

Figure 28 shows the experimentally measured spectra of a supercontinuum formed in an MS fibre (core diameter 2.5 µm, dispersion of group delay at a pump wavelength $\lambda_p = 800$ nm: $D = -50$ ps/nm · km, zero-dispersion wavelength: $\lambda_D = 900$ nm) with a titanium–sapphire laser with an average pulse power of 50 mW, for four different wavelengths.

Pumping at a wavelength of 800 nm leads to a broadening of the emission spectrum in the same way as shown in Fig. 22 *b*. But in this case the pump wavelength is closer to the wavelength of zero dispersion. As a result of PSM and SRS processes, the emission spectrum is expanded into the region of the anomalous dispersion of the optical fibre.

In turn, this leads to the formation of a soliton with a carrier wavelength of about 940 nm, which as a result of the self-shift of the frequency broadens the supercontinuum spectrum into the long-wave region as the pump pulse power increases [23].

Figure 29 shows the dependences of the spectrum of the supercontinuum in the MS optical fibre on the power of the pulses of laser radiation with a duration of 100 fs and a repetition rate of 76 MHz at a wavelength of $\lambda_p = 875$ nm, which demonstrate the cooperative effect of the effects of PSM and soliton shift of

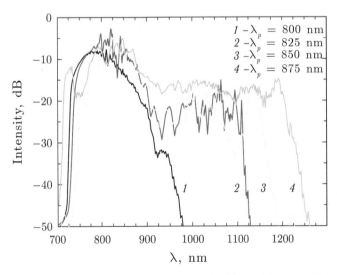

Fig. 28. Supercontinuum formed in an MS optic fibre with a wavelength of zero dispersion $\lambda_D = 900$ nm by pulses of a titanium–sapphire laser with different wavelengths (λ_p) of 100 fs duration and a repetition rate of 76 MHz [22].

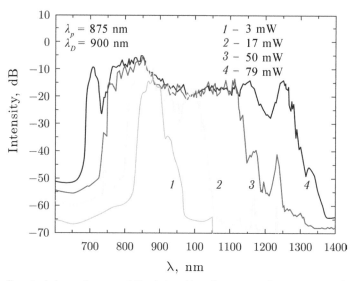

Fig. 29. Spectral dependences of the intensity of supercontinuum generation in an MS lightguide on the power of laser pulses of 100 fs duration and the repetition frequency of 76 MHz at a wavelength of $\lambda_p = 875$ nm [22].

the radiation frequency. As can be seen, in order to maximize the expansion of the spectrum of the supercontinuum generated by ultrashort pulses by laser pulses, it is necessary to increase the power of the radiation introduced into the lightguide and to maximally approximate λ_D to λ_p.

If the wavelength of the pump pulse is substantially separated from the wavelength of the zero dispersion of the MS fibre, this makes the supercontinuum generation process extremely ineffective, since in this case only two nonlinear optical processes are involved: PSM and SRS. For illustration, Fig. 30 (dependence *1*) shows the result of supercontinuum generation in an MS optic fibre with a core diameter of 2.5 μm with a wavelength of zero dispersion of 740 nm with laser radiation with a wavelength $\lambda_p = 532$ nm and a power of ~3 kW. Such a large difference in the respective wavelengths makes the supercontinuum generation process extremely heterogeneous and narrow-band. If, simultaneously with radiation $\lambda_p = 532$ nm, radiation is introduced into the MS fibre with a wavelength $\lambda_p = 1064$ nm, the situation changes significantly. Since radiation with a wavelength of $\lambda_p = 1064$ nm lies in the region of the anomalous dispersion of the optical fibre, the mechanisms of the dispersion–soliton formation of the supercontinuum in the ranges 560–1600 nm are included. Since pulses with wavelengths $\lambda_p = 532$ nm and $\lambda_p = 1064$ nm are

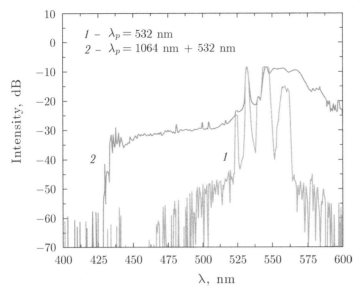

Fig. 30. Generation of short-wave radiation from a supercontinuum using the result of mixing pulses with different wavelengths [24].

introduced into the light guide simultaneously, the mechanism of the four-wave mixing of the waves generated by these pulses occurs. This leads to a significant expansion of the emission spectrum of the supercontinuum into the short-wavelength region. As can be seen from Fig. 30 of the dependence *2*, such a simultaneous action of radiation pulses with different wavelengths makes it possible to provide an extremely homogeneous emission spectrum from a 100 nm plateau shifted to the short-wavelength region.

4.11.4. Generation of supercontinuum in MS optical fibres having two zero-dispersion wavelengths

As follows from the previous consideration (see sections 4.11.2 and 4.11.3), broadband generation of the supercontinuum in MS fibres is observed in the propagation of femtosecond pulses whose wavelength lies near the zero-dispersion wavelength. In this case, the most efficient transformation of radiation occurs when the wavelength of the pump pulse is in the region of anomalous dispersion of the MS fibre optical waveguide. In this case, the formation of the supercontinuum is mainly due to the enhancement of short-wave dispersion waves (DW) resonantly associated with the pump pulse and the fundamental solitons arising from the decay of high-order solitons whose spectrum is gradually shifted to the infrared region due to SRS. As a result, two processes depending on the power and duration of the pump pulse are observed: the expansion (shift) of the supercontinuum emission spectrum into the 'blue' region (the normal dispersion region) and the expansion of the supercontinuum spectrum into the 'infrared' region (located in the region of anomalous dispersion). The main factor, which limits the short-wave boundary of the supercontinuum spectrum, is the position of the zero-dispersion wave length on the dispersion curve and the relationship between the values of λ_D and λ_p [25].

The more the values of the pump wavelength and the zero-dispersion wavelength lying in the region of the anomalous dispersion are separated, the shorter are the wavelengths generated in the 'blue' region of the spectrum. However, the intensity of the waves in the 'blue' part of the spectrum decreases strongly with increasing difference between λ_D and λ_p. For this reason, the pump wavelength can not be greatly spaced with respect to the wavelength of the zero dispersion to provide any appreciable optical intensity in the 'blue' region. But for the same reason, the efficiency of spectrum broadening will be limited supercontinuum in the infrared region [26], which is extremely important when using supercontinuum radiation for spectroscopy.

In order to increase the expansion of the supercontinuum spectrum into the infrared region, it is proposed in [25] to use the same mechanism that is used to efficiently generate the 'blue' components. For this purpose, it is proposed to use MS fibre lightguides having two zero-dispersion wavelengths (see Fig. 29). In this case, the wavelength of the pump pulse (λ_p) lying in the region of the anomalous dispersion of the light guide is chosen so close to the value of the zero-dispersion wavelength (λ_{D1}), so as to ensure a partial penetration of the pulse spectrum into the normal dispersion region of the optic fibre. Then, due to the resonant coupling of the fundamental solitons arising from the decay of a high-order soliton to the freely dispersive waves in the region of normal dispersion, the wave of zero dispersion wave (λ_{D1}) localized near the short-wave (or visible) radiation region (denoted in Figure 31 as BDW), the spectrum of which expands due to the PSM in the 'blue' region.

Since the high-order soliton generated by the high-power pump pulse propagates in the MS lightguide due to the SRS process, it will shift its wavelength to the infrared region, its wavelength will gradually approach the second value of the zero-dispersion

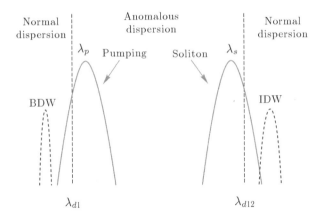

Fig. 31. Scheme of the process of amplification of dispersion waves that extend the spectrum of the supercontinuum to the 'blue' (BDW) and infrared (IDW) regions, in an MS optical fibre having two zero-dispersion wavelengths in the visible and near infrared regions of the spectrum λ_{D1} and λ_{D2}, respectively [25].

wavelength (λ_{D2}). As a result, the soliton spectrum will partially penetrate into the region of normal dispersion lying to the right of the zero-dispersion wavelength (λ_{D2}), which is in the infrared region of the spectrum. In this case, the soliton will act as a pump pulse for the resonance amplification of dispersion waves formed in the infrared region of the spectrum, which are designated as (IDW), which will expand the spectrum of the supercontinuum created in the infrared region. Thus, the presence of two zero-dispersion wavelengths in the MS lightguide makes it possible to form an expanded supercontinuum spectrum that is effectively amplified both in the visible ('blue') and in the IR regions of the spectrum.

Table 1 shows the properties of four types of MS fibre lightguides with two zero-dispersion wavelengths in which the supercontinuum generation was generated by radiation of a titanium–sapphire laser (λ_p = 800 nm) with a pulse duration of 200 fs.

Figure 32 shows the calculated dependences of the group delay dispersion for these optical fibres [25].

The spectra of the supercontinuum generated in these MS fibres with a length of 1.5 m, obtained for different values of the power of the pump pulses introduced in them, are shown in Fig. 33.

The visible part of the supercontinuum (Fig. 33 *a–d*) is practically identical for all four types of fibres. The latter is due to the fact that for short wavelengths these fibres have comparable dispersion profiles. When low-energy pump pulses are used, the supercontinuum

Table 1. Characteristic parameters of MS fibre lightguides (MSF): d – the diameter of holes; Λ – period; MD – mode diameter

	MSF1	MSF2	MSF3	MSF4
Λ, μm	1.4	1.37	1.33	1.22
d/Λ	0.67	0.65	0.62	0.63
MD, μm ($\lambda_p \approx 800$ nm)	1.3	1.3	1.2	1.1
λ_{D1}, nm	740	730	750	690
λ_{D2}, nm	1700	1610	1515	1390

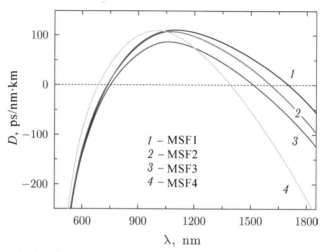

Fig. 32. Calculated group delay dispersion profiles for MS fibres the parameters of which are given in Table 1.

spectrum does not reach wavelengths close to the wavelength of the zero-dispersion IR band (λ_{D2}). As the power of the pump pulses increases, the supercontinuum spectrum begins to expand and attains wavelengths comparable to the wavelength of the zero-dispersion IR range. After this, as the power of the pump pulses increases, a wide peak begins to appear in the spectra of the supercontinuum of light-emitting diodes with numbers 2–4, the amplitude and width of which increase as the input into the power fibre, while the corresponding wavelengths exceed the values of λ_{D2}.

Thus, the generation of a supercontinuum in MS optical fibres, the dispersion dependence for which has two values of the dispersion zero-wave, seems extremely interesting, as it opens up new possibilities for generating broadband radiation. Of particular importance here are the conditions when the zero-dispersion points

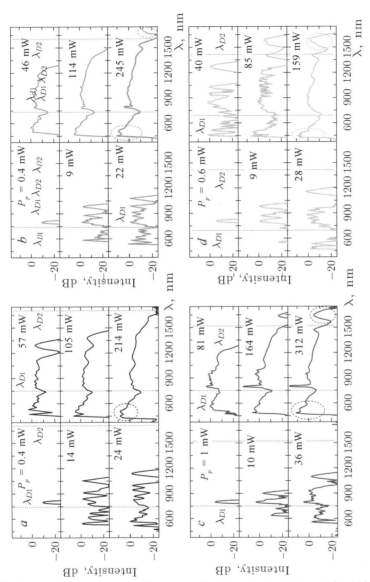

Fig. 33. Experimentally measured supercontinuum spectra obtained for MS fibre lightguides of the MSF1 (*a*), MSF2 (*b*), MSF3 (*c*) and MSF4 (*d*) type at different values of the power of the pumping pulse P_p [25]. The lines show the position of the zero dispersion wavelengths (λ_{D1} and λ_{D2}), the ellipses show the maxima of the dispersion in the visible (BDW) and infrared (IDW) regions of the spectrum.

are not far apart from each other in the spectrum, for example, as in Figs. 15 and 16.

Figure 34 *a* shows the spectral dependence for a supercontinuum obtained using a 5 cm long MS lightguide having two zero-dispersion wavelengths of 779 nm and 945 nm, using pulsed radiation with a duration of 40 fs and a wavelength of 790 nm. As can be seen, in this case there is a significant difference from the results obtained in the generation of the supercontinuum, when the wavelength of the pump

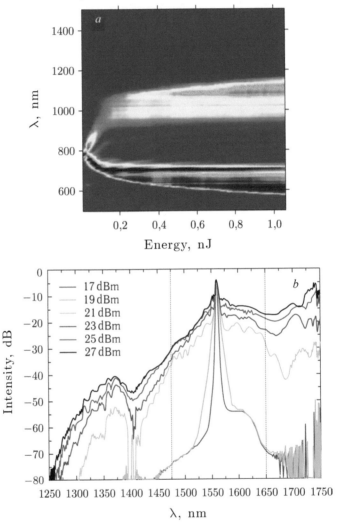

Fig. 34. *a* – supercontinuum obtained using a 5 cm fibre lightguide having two zero-dispersion wavelengths of 779 nm and 945 nm, using pulsed radiation with a duration of 40 fs and a wavelength of 790 nm; *b* – spectrum of the supercontinuum in an MS fibre with two zero-dispersion wavelengths (shown by dashed vertical lines), obtained using laser pulses with a wavelength of 1555 nm with a duration of 2 ps at different pump power levels [27].

radiation is simply in the region of anomalous dispersion. As can be seen from Fig. 34 *a*, the entire pulse energy whose wavelength is in the region between the wavelengths of zero dispersion is redistributed between two spectral peaks, each of which lies in the region of normal dispersion of the fibre. As shown in [22], the efficiency of converting the pump pulse power into supercontinuum radiation reaches up to 99%.

Both these intense lateral peaks are very flat and, very importantly, they are practically insensitive to fluctuations in the spectrum and power of the pump pulse, ensuring high stability of their spectrum. In connection with this, when the pump pulse wavelength changes, only the energy relations between the two peaks change, then the position of the peaks remains unchanged.

Since the position of the centres of these peaks is uniquely determined by the position of the wavelengths of zero dispersion, it is possible to achieve a specific, pre-determined spectrum of supercontinuum radiation by changing the dispersion properties of the MS fibre and thereby create optical fibres that can generate radiation in any visible and near IR-spectrum regions.

In this case, it is not necessary to use femtosecond radiation to generate a supercontinuum in MS lightguides with two zero-dispersion wavelengths. Figure 34 *b* demonstrates the possibility of generating a supercontinuum in an MS fibre with two zero-dispersion wavelengths (shown by the dotted vertical lines: 1475 and 1650 nm) using laser pulses of a wavelength of 1555 nm with a duration of 2 ps.

4.11.5. Nonlinear optical properties of holey PC optical fibres

The technology of fabricating PC fibre-optic lightguides having a hollow light-conducting core practically does not differ from the technology of analogous MS-fibre lightguides, having a continuous light-guiding core. The main difference of this fibre is that the light-conducting core is not a quartz rod, but an air cavity with a diameter exceeding the diameter d of the regular air channels in the cladding (Fig. 13 *b*). Such a structure can direct visible and near IR radiation when $d/\Lambda \sim 0.3$. In this case, the waveguide regime is provided exclusively by the band structure of the photonic crystal. In contrast to the standard fibre lightguides, radiation in holey PC lightguides with a hollow, light-guided core, due to photonic-crystal reflection from a structured cladding, extends predominantly in the hollow core, rather than in the quartz. In this connection, it would seem that the losses in such optical fibres should be very low, since material absorption and Rayleigh scattering in air is negligible compared to quartz glass. However, the experimental losses are large: 13 dB/km at $\lambda = 1.5$ μm. This is due to the fact that the losses in such optical fibres are determined not by scattering and absorption, but by the final reflecting ability of the periodic cladding structure in the transverse direction, which in turn is determined by the perfection of the periodic structure, the extent in the radial direction and the strict observance of the condition of equality of the period of the structure to an integer number of half-waves radiation.

In contrast to highly nonlinear quartz MS fibres, which have a small-diameter core, photonic-crystal fibre fibres with a hollow core have a very small overlap between the guided mode and the photonic-crystalline quartz cladding, which causes extremely low nonlinear optical properties of these fibres. Nevertheless, it is very important to have an idea of their nonlinear optical characteristics, since holey PC-fibre lightguides are widely used in the design of a variety of sensors [29].

The nonlinear optical coefficient of holey PC optical fibres are 100 times smaller than in ordinary quartz optical fibres [30]. Since the radiation mainly propagates in holey fibres in the hollow core and partly in the photonic-crystal cladding, the expression for the nonlinear coefficient can be represented in the form [28]

$$\gamma = \gamma_{air} + \gamma_{SiO_2} = \frac{2\pi n_{2\,air}}{\lambda A_{eff\,air}} + \frac{2\pi n_{2SiO_2}}{\lambda A_{eff\,SiO_2}}, \tag{4.95}$$

where n_{2air}, n_{2SiO_2}, $A_{eff\,air}$, $A_{eff\,SiO_2}$ are nonlinear refractive indices of air and fused quartz, as well as the overlapping areas of the propagating radiation mode with the core and the cladding.

Figure 35 shows the dependences of the nonlinear coefficient for the fundamental mode of the PC fibre optical waveguide. It should be noted that the contribution of fused quartz to the nonlinear characteristics of the modes is insignificant because of the small overlap of the mode field with the photonic-crystal cladding. At the same time, the value of the nonlinear coefficient depends strongly on the contribution of the air filling the hollow cladding or another gas. In this connection, the nonlinear mode coefficient strongly depends on the diameter of the hollow cladding and the properties of the

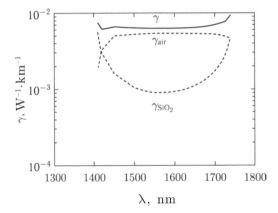

Fig. 35. Spectral dependences of the nonlinear coefficient for the fundamental mode of a PC fibre lightguide (solid line), having parameters $d/\Lambda = 0.5$, $2r = 3\Lambda$, and also for air and fused quartz (dashed lines) [28].

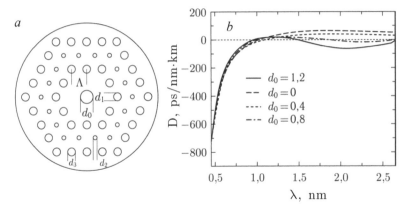

Fig. 36. The structure of a holey PC fibre lightguide with a hollow core (*a*) and the calculated spectral dependences of the group delay dispersion coefficient (*b*) for a fibre with parameters: $\Lambda = 1.5$ μm, $d_1 = 0.8$ μm; $d_2 = 0.4$ μm; $d_3 = 0.8$ μm, given for different values of the diameter of the core.

medium filling it. In this case, the value of the nonlinear coefficient turns out to be extremely small and can be slightly corrected by the choice of dimensions and the structure of the arrangement of the holes in the cladding.

As was shown above, the nonlinear optical properties of optical fibres depend significantly on the dispersion characteristics of the modes directed to them. Figure 36 shows the calculated dispersion profiles (DGV) for four PC-fibre lightguides with a hollow core. As can be seen, all optical fibres have two zero-dispersion wavelengths localized near the values of 0.96 and 1.45 μm. In connection with this, such optical fibres are able to exhibit nonlinear optical properties, and, in particular, can be used to generate a supercontinuum.

As a rule, hollow FC fibre fibres are actively used to increase the interaction length of laser pulses with a gaseous medium and to increase the efficiency of nonlinear optical processes. [1]. In this case, the photonic band gaps of the periodic envelope of hollow PC fibres provide a significant reduction in losses in such structures as compared to conventional hollow fibre fibres. Investigation of pulsed radiation transmittance of a titanium–sapphire laser with an energy of 0.5–5 mJ, pulse duration of 100–120 fs and a wavelength of 800 nm through PC-fibres with a core diameter of 14 µm has shown that due to the loss of radiation power in such structures their length should not be less than 11 cm.

In Fig. 37 the solid line shows the emission spectrum of a titanium-sapphire laser with an energy of 0.9 µJ and a duration of 100 fs, formed after the passage of the pulse through an air-filled holey PC fibre lightguide 8 cm in length and a hollow core diameter of 14 µm [1]. Since the wavelength of the laser lies in the region of normal dispersion of the fibre, the pulse spectrum undergoes a broadening due to the effect of phase self-modulation. At the same time, the nonlinear spectral transmission caused by the presence of photonic crystal zones in the optical fibre envelope influences the formation of the pulse spectrum (the transmission spectrum of a holey PC light guide is shown in Fig. 37 by a dashed curve). As can be seen from Fig. 37, within the allowed photonic crystal zone of the

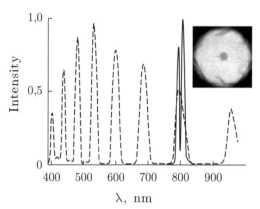

Fig. 37. Radiation spectrum of a titanium–sapphire laser (λ_p = 800 nm) at exit from an air-filled holey PC fibre lightguide 8 cm long with a diameter of the hollow core of 14 µm (top right – photograph of the cross section of the lightguide). Solid line – radiation spectrum, at exit of the lightguide, the dashed line – transmission spectrum of the lightguide [1].

optical fibre, the pulse spectrum is broadened and its width is 35 nm, which demonstrates the presence of sufficiently high nonlinear optical properties for the holey PC fibres. Note that in order to achieve an analogous broadening of the pulse spectrum using conventional hollow fibre fibres with an internal diameter of 100 μm, an increase in the radiation intensity would be approximately 100 times.

Proper selection of parameters for laser radiation, hollow a fibre-optic fibre and a gas filling the core makes it possible to achieve agreement between the width of the transmission peak of the optical fibre and the spectrum of the output pulse broadened as a result of its nonlinear optical interaction with the gaseous medium.

References

1. Zheltikov A.M., Usp. Fiz. Nauk, 2004. V. 174, No. 1. P. 73–105.
2. Agrawal G., Non-linear fibre optics. Moscow: Mir, 1996.
3. Voronin V.G., Nanii O.E., Fundamentals of nonlinear fibre optics. Moscow, Universitetskaya Kniga, 2011
4. Kandidov V.P., et al., Kvant. Elektronika, 2009. V. 39, No. 3. P. 205–228.
5. Ranka J.K., et al., Opt. Lett. 2000. V. 25. P. 25.
6. Dianov E.M., Kryukov P.G., Kvant. Elektronika. 2001. V. 31, No. 10. P. 877–882.
7. Lehtonen M., Application of microstructured fibres (Doctoral dissertation), Espoo, Finland: Helsinki University of Technology. 2005..
8. Hilligsqe K.M., Nonlinear Wave Propagation in Photonic Crystal Fibres and Bose-Einstein Condensates. Ed. University of Aarhus, Denmark, 2005.
9. Koshiba M., Saitoh K., Opt. Express. 2003. V. 11. P. 1746–1756.
10. Miret J.J., et al., Opt. Express. 2009. V. 17, No. 11. P. 9197–9203.
11. Gaponov D.A., Biryukov A.S., Kvant. Elektronika, 2006. V. 36, No. 4. P. 343–348.
12. Agrawal G.P., Application of Nonlinear Fibre Optics. Academic Press, Elsevier, 2008.
13. Dudley J.M., et al., Review of Modern Physics, 2006, Vol. 78, No. 4. P. 1135–1184.
14. Herrmann J., et al., Physical Review Lett., 2002. V. 88, No. 17. P. 173901.
15. Skryabin D. V., Gorbach A.V., Review of Modern Physics. 2011. V. 82, No. 2. P. 1287–1299.
16. Efimov A., et al., Physical Review Letter, 2005, V. 95, No. 21. Article ID: 213902.
17. Luan F., et al., Optics Express. 2006. V. 14, No. 21. P. 9844–9853.
18. Dudley J.M., et al., Reviews of Modern Physics. 2006. V. 78. P. 1135–1184.
19. Roy S., et al., Current Science. 2011. V. 100, No. 3. P. 321–342.
20. Chow K.K., et al., Electronics Letters. 2006. V. 42. P. 989–991.
21. Driben R., Zhavoronkov N., Optics and Photonics Journal. 2012. V. 2. P. 211–215.
22. Hansen K.P., et al., in: Proc. Optical Fibre Communication Conference & Exhibition, Anaheim, CA, 2002.
23. Liu X., et al., Opt. Lett. 2001. V. 26. P. 358–365.
24. Holzwarth R., et al., Physical Review Letters. 2000. V. 85. P. 2264–2271.
25. Genty G., et al., Opt. Express. 2004. V. 12, No. 15. P. 3471–3480.
26. Genty G., et. al., Opt. Express. 2002. V. 10. P. 1083–1098.

27.   Hilligshe K.M., et al., Opt. Express. 2004. V. 12. P. 1045–1048.
28.   Poli F., Photonic Crystal Fibers. Springer, 2007.
29.   Lee B., Optical FibreTechnology. 2003. V. 9. P. 57–71.
30.   Varrallyay Z., Szabo A., Szipo R., Optics Express. 2009. V. 17, No. 14. P. 11869–11873.

5

Fibre Lasers

Introduction

The creation of a radically new source of optical laser radiation, which made N.G. Basov, A.M. Prokhorov and Ch. Townes the Nobel Prize winners in physics in 1964, was an event that greatly affected the life of the human society. The uniqueness of the characteristics of laser radiation has served as a basis for the widespread use of lasers in various fields of scientific research and high technology. In any laser system, the most important part is the laser itself. At present, there are a huge number of articles and monographs devoted to the study of the whole variety of lasers and technological structures based on them, and this is confirmation that even today laser physics and technology do not stand still, but are continuously developed and constant improvement of all types of lasers occurs. It should be noted that the end of the 20th and early 21st centuries is characterized by the transition to the so-called diode pumping (pumped by semiconductor laser diodes) of solid-state lasers. For the work on the creation of semiconductor lasers on heterojunctions capable of work at room temperature and effectively used for pumping solid-state lasers, the Russian scientist Zh.I. Alferov was awarded the Nobel Prize in Physics in 2000.

The main advantages of diode pumping, in contrast to the lamp, are that energy consumption is significantly reduced, the service life is extended, the compactness of the laser system is increased and the requirements for cooling systems are reduced. Diode-pumped lasers, in turn, are the most popular fibre lasers. The creation of fibre lasers, which are practically ideal converters of the light energy of semiconductor laser pumping diodes into laser radiation with a record efficiency, was the result of many years of development of laser physics and fibre optics.

For the first time, the transmission of laser radiation over an optical fibre was demonstrated by E. Snitzer and J.W. Hicks in 1961 [1]. The main problem of their design was the high attenuation of radiation during the passage of the glass fibre. However, a few years later E. Snitzer created the first laser, as the working medium of which a fibre lightguide with a core doped with neodymium was used [2]. In 1966, K.C. Kao and G.A. Hockham created an optical fibre whose attenuation was about 20 dB/km, while other fibres existing at the time were characterized by attenuation of more than 1000 dB/ km [3]. In 2009, K.C. Kao was awarded the Nobel Prize in Physics for his outstanding contribution to the study of optical fibres for optical communication. Today, the progress in the production of optical fibres made the usual serial production of optic fibres with losses of 0.2 dB/km [4], which greatly expanded the possibilities of their use in communication systems and in the production of fibre lasers. The dependence shown in Fig. 1 illustrates the rapid development of fibre lasers, which began in the late 1980s [5, 9].

The main areas of research were related to experimentation in the use of various impurities in optical fibres to achieve specified parameters of the generated radiation. The continuous and pulsed regimes of the generation of fibre lasers and methods for their realization were studied in detail. The advantages of fibre laser systems have been identified, which include: a very high (up to

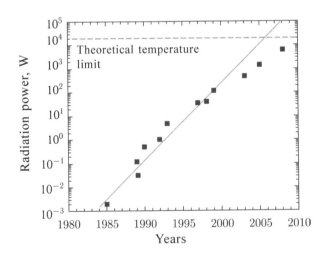

Fig. 1. Dynamics of the output power growth of continuous fibre lasers.

80%) efficiency of converting pumping radiation into coherent laser radiation achieved due to the long interaction length of pump radiation with the active medium and a small diameter of the light-conducting core (usually from 4 to 20 µm); to ensure the operation of fibre lasers water cooling is not required and air cooling is sufficient; practically independent of the radiation power, the high quality of the output beam, determined by the properties of the optical fibre, the simplicity and reliability of fibre lasers, the resonator of which can be completely executed on the basis of fibre elements, and the lightguide fibre can be compactly folded into rings and isolated from the external medium, which eliminates the need for alignment and constant maintenance of the laser system and, on the whole, increases the stability of its operation under various external conditions.

5.1. The principle of the fibre laser

A fibre laser consists of the following main structural elements: a pumping source (laser diodes), a fibre amplifier – active fibre doped with active matter, in which radiation is generated, and a resonator (the most common: Fabry–Perot type resonators, ring resonators, Bragg fibre gratings) (Fig. 2). The elements that make up the fibre laser are considered below.

5.1.1. Active optical fibres

The high transparency of quartz-base material for optical fibres is provided by saturated states of the atomic energy levels. The admixtures of rare-earth elements (REE): Nd, Yb, Er, Ho, Tm, introduced into the core material by doping, transform the optical fibre into an active medium. By selecting the wavelength and power of the pump radiation, it is possible to create an inverse state of the

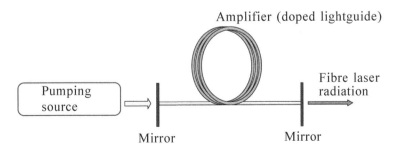

Fig. 2. Schematic diagram of a fibre laser.

populations of the impurity energy levels in such a medium. The applicability of active ions for the doping of optical fibres based on quartz glass is determined by the following factors: first, the active ions have radiative transitions in the near infrared region of the spectrum, where the quartz glass is the most transparent, and secondly, in the quartz glass the phonon energy is 400–1100 cm^{-1}, which does not allow them to influence the state of the energy levels of the impurity. The luminescence characteristics of REE Nd, Yb, Er, Ho, and Tm are given in Table 1, and Fig. 3 shows the schemes of energy levels of these ions and the used optical transitions are indicated [7].

Neodymium (Nd^{3+}). Neodymium ions in quartz glass have a number of strong absorption bands in the visible and near infrared ranges of the radiation spectrum. For selective excitation of ions to the $^4F_{3/2}$ level, semiconductor lasers with a radiation wavelength of 0.8 μm are most often used. The three main luminescence bands of the ion are located in the regions of 0.92; 1.06 and 1.34 μm. The most intense is the luminescence band in the 1.06 μm region corresponding to the $^4F_{3/2} \rightarrow {}^4I_{11/2}$ transition. The lifetime of this excited state is ~0.5 ms. At this wavelength, the laser operates on a four-level

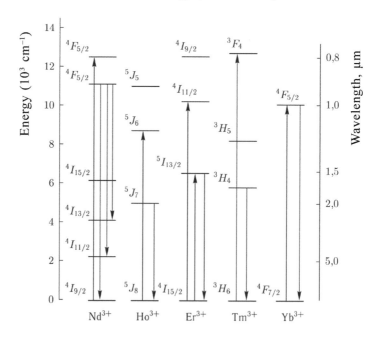

Fig. 3. Scheme of energy levels of rare earth ions used to create active optical fibres [3].

Table 1. Rare-earth elements used in active optical fibres and spectral regions of their luminescence [7]

Active ion	Luminescence region, μm
Nd^{3+}	0.92–0.94 1.05–1.1 1.34
Ho^{3+}	1.9–2.1
Er^{3+}	1.53–1.6
Tm^{3+}	1.7–1.9
Yb^{3+}	0.98–1.16

generation scheme. This explains the fact that neodymium was the first impurity to create active optical fibres based on quartz glass.

Realization of fibre lasers using the optical transition $^4F_{3/2} \rightarrow ^4I_{9/2}$ ($\lambda \sim 0.92$ μm) is hampered by competition from the luminescence in the 1.06 μm region, and generation in the $^4F_{3/2} \rightarrow ^4I_{13/2}$ ($\lambda \sim 1.34$ μm) transition region is not realized due to absorption from the excited state of the ion.

Holmium (Ho^{3+}). Has a $^5J_7 \rightarrow ^5J_8$ laser transition with the corresponding emission wavelength ~2 μm. The lifetime at the excited level is about 0.5 ms. In this case, the holmium fibre laser must operate in a three-level scheme, since there is absorption from the ground state at the same wavelength. Therefore, for such lasers, to achieve inversion of the populations of the energy levels, a necessary condition is to pump into the intense absorption band of the ion. The use of the absorption band in the 0.9 μm region corresponding to the $^5S_8 \rightarrow ^5J_5$ transition for pumping by a semiconductor laser has not been used due to the weak intensity of this spectral band. At the same time, the presence of an intense absorption band in the 1.15 μm region corresponding to the $^5S_8 \rightarrow ^5J_6$ transition allows it to be used to create a holmium fibre laser.

Erbium (Er^{3+}). Erbium ions in quartz glass possess a $^4I_{13/2} \rightarrow ^4I_{15/2}$ laser transition corresponding to radiation in the region of 1.53–1.6 μm. The lifetime at the metastable level is 10^{-12} msec. This spectral range coincides with the region of minimal losses in quartz fibres. In combination with the ability to use semiconductor pumping sources at 0.98 and 1.45–1.48 μm this led to the widespread use of erbium-doped fibre optics to create fibre lasers for communication systems. It should also be noted that the quantum pumping efficiency for these fibres is close to 100%, and the long lifetime at the metastable level, about 10 ms, allows achieving high gain factors.

Thulium (Tm^{3+}). Laser generation in thulium-doped quartz glass is associated with the $^3H_4 \to {}^3H_6$ transition when pumped into the energy band due to $^3H_4 \to {}^3H_4$ ($\lambda \sim 0.79$ μm) and $^3H_4 \to {}^3H_5$ ($\lambda \sim 1.06$–1.5 μm) transitions. The spectral range of possible lasing is 1.85–2.1 μm, and the lifetime at the metastable level is ~0.2 ms.

Ytterbium (Yb^{3+}). Its energy structure consists of two sublevels: ground $^2F_{7/2}$ and excited $^2F_{5/2}$. As a result, only one absorption band is observed in the absorption spectrum of the ytterbium-doped quartz optical fibre, which has a complex shape due to the Stark splitting of the sublevels. This absorption band has two maxima with centres near the wavelengths of 0.915 and 0.976 μm. Therefore, pumping is used in semiconductor lasers emitting in these spectral ranges. The luminescence spectrum due to the $^2F_{5/2} \to {}^2F_{7/2}$ transition, has maxima in the range of 0.978–0.982 μm and 1.03–1.04 μm, extending to the region of 1.15–1.2 μm. The lifetime for an optical fibre additionally doped with aluminum ions is ~0.8 ms. This makes it possible to implement on its basis a wide range of radiation sources in the range of 0.98–1.2 μm.

The composition of the material of active fibre lightguides includes ions of elements possessing optical transitions, and additional impurities correcting the refractive index and the spectral characteristics of the material. In this case, the active ion can be introduced both into the core of a quartz fibre lightguide and into its reflecting cladding if a significant part of the optical power of the channeled radiation propagates in it. A number of technological processes have been developed for the production of active optical fibres: MCVD (modified chemical vapour deposition), OVD (external gas phase deposition), VAD (axial deposition from the gas phase), deposition using plasma PCVD (plasma chemical vapour deposition) and SPCVD (surface plasma chemical vapour deposition). To introduce the active impurity in these processes, the impregnation method is most widely used, when the non-melted porous core material is impregnated with a solution of the active additive salt and doping from volatile compounds is carried out.

The requirement for active optical fibres is the presence of sufficiently low non-resonant optical losses. The acceptable loss is 5 to 20 dB/km. Figure 4 shows an example of the optical loss spectrum for an optical fibre with a core doped with Yb^{3+} ions with a concentration of $8 \cdot 10^{19}$ cm^{-3} [7].

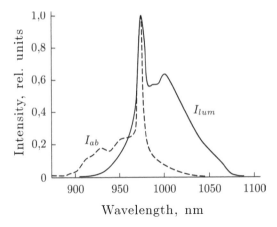

Fig. 4. Absorption (I_{ab}) and luminescence (I_{lum}) spectra of an optical fibre doped with Yb³⁺ ions.

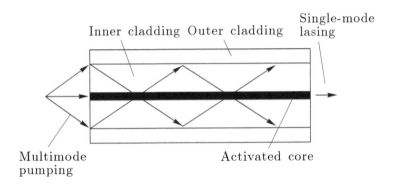

Fig. 5. Scheme of pumping a fibre laser.

The creation of fibre lasers required the development of special optic fibres with a double cladding (Fig. 5) [7]. The active medium of a fibre laser is a single-mode quartz core doped with an active rare-earth impurity, as well as impurities that form the profile of the refractive index. This core is surrounded by an inner cladding of quartz glass and an outer cladding with a refractive index lower than the refractive index of quartz glass. The inner cladding together with the outer cladding form a multimode fibre along which the pump radiation propagates. It has a typical size of 0.1–1 mm. When propagating through such a multimode fibre, the pump radiation is absorbed by rare-earth impurity ions in the activated core, causing luminescence, which in the presence of positive feedback can develop into laser generation.

Both individual laser diodes and laser diode systems (matrices and rulers) with fibre output are used as sources of pumping of fibre lasers. A number of firms produce laser pumping modules with an output power of up to several hundred watts. The output optical fibre of such modules has a core with a diameter of more than 200 μm and a numerical aperture of about 0.2. It is quite obvious that the effective input of pumping radiation into the core of an active light guide with a diameter of 5–20 μm and a numerical aperture ~0.1 is a very complex problem and requires the development of a special optical fibre design. Therefore, to ensure effective coupling of the modes of the inner cladding to the activated core, it is necessary to use optical fibres with a non-circular shape of the inner cladding, since otherwise a large portion of the pump power propagates in modes that do not cross the core region.

Figure 6 *a* shows the scheme of the so-called optical fibre with a double cladding [5]. Its core, containing an admixture of a rare-earth element, is surrounded by a non-round first cladding of pure quartz glass with a diameter of several hundred micrometers. This first cladding is in turn surrounded by a second cladding, usually of a polymeric material with a refractive index lower than that of quartz glass. Thus, the first cladding is a multimode fibre, which is efficiently excited by pump radiation due to the large transverse dimension and high numerical aperture. At certain values of the diameters of the core and the first cladding as well as the length of the optical fibre, the pump radiation effectively excites the ions of the rare-earth elements located in the core.

The most common type of optical fibres with a double cladding uses silicone rubber as the material of the outer cladding that provides a numerical aperture of the multimode fibre $NA = 0.38$,

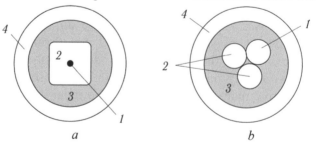

Fig. 6. Diagrams of an active fibre for matching with powerful diode pumping systems: a – *1* – core containing a rare-earth element; *2* – first cladding; *3* – second cladding; *4* – protective cladding; b – *1* – active lightguide; *2* – passive pumping lightguides; *3* – general second cladding; *4* – protective cladding.

or Teflon of the AF brand, which allows increasing the numerical aperture to 0.6.

The disadvantages of using a silicone coating include a high level of optical losses of pump radiation which are 50 dB/km or more. The use of Teflon makes it possible to obtain fibres with losses in the cladding of ~10 dB/km. In addition, Teflon has a high transmission for UV radiation, which allows recording refractive index gratings in the core without removing the polymer. However, this coating has a small thickness of ~10–20 μm, which increases the probability of mechanical damage to the optical fibre.

Table 2. Influence of the geometry of the inner cladding on the efficiency of absorption of pump radiation in an active optical fibre doped with Yb^3 ions [7]

Cladding geometry	Absorption at $\lambda = 978$ nm, dB/m	
	Straight fibre	Figure 8-shaped
Circular	0.3 ± 0.05	0.6 ± 0.05
D-shaped	2.2 ± 0.05	2.2 ± 0.05
Rectangular	3.5 ± 0.05	3.5 ± 0.05
Square	3.3 ± 0.05	3.3 ± 0.05

In work [7], studies are made of the effect of various profiles of the cross section of the inner cladding of an active fibre on the efficiency of absorption of pump radiation. Table 2 shows the effect of the geometry of the inner cladding on the efficiency of absorption of pump radiation in an active optical fibre doped with Yb^{3+}. These fibres had different shapes and parameters of the inner cladding: round (diameter 125 μm), D-shaped with one ground surface (125 × 100 μm), rectangular (150 × 75 μm) and square (125 × 125 μm). Absorption was measured for two configurations of light guides: linear and bent in the form of a 'figure eight' with a bending radius of 1 cm. The results of the studies showed that bending has virtually no effect on the amount of absorption, and any non-circular shape of the inner cladding makes it possible to obtain a pump absorption efficiency close to 100%.

Such designs of an active fibre lightguide give good results when creating medium-power fibre lasers (~100 W). For powerful fibre lasers, another design of an optical fibre is proposed in which pump radiation propagates through discrete optical fibres made of quartz glass in optical contact with the active fibre. In this case, all optical fibres are surrounded by a single cladding of polymer material (Fig. 6 *b*) [5]. In such a structure, distributed pumping of the active fibre through its lateral surface is realized due to evanescent waves penetrating from the pump waveguide into the active fibre. In this case, the number of passive fibres transmitting the pump radiation can be much greater than three. Pumping radiation can be introduced through both ends of a passive fibre, so the number of possible points for the injection of pumping radiation is equal to twice the number of passive fibres. In the literature, such a structure has its own abbreviation (GTW).

Another option to increase the interaction of radiation from laser diodes with the active medium is the use of an active fibre with a microstructured cladding [11]. The characteristic value of the numerical aperture of such optical fibres is 0.5, which, unlike traditional automatic devices of the same diameter, makes it possible to increase the pump power by several times using identical radiation sources.

Other methods for introducing pumping radiation into the active fibre are known, using volumetric elements such as lenses and mirrors. But in this case the laser design is not completely fibre.

5.1.2. Resonators of fibre lasers

5.1.2.1. Fabry–Perot type resonators

In the first fibre lasers, dielectric mirrors were used to create the Fabry–Perot resonator (Fig. 7 *a*). The use of dichroic mirrors made it possible to make them practically transparent at the wavelength of the pump wave and to maintain a high reflection coefficient at the lasing wavelength. Initially, the active fibre was placed between the mirrors of the resonator, but this construction was difficult to adjust. A partial solution to the problem consisted in applying dielectric mirrors directly to the ends of the optical fibres. However, this increased the risk of their damage by powerful focused pump radiation and

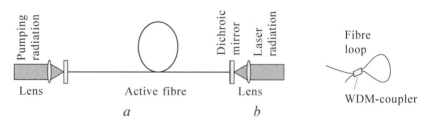

Fig. 7. Diagram of a fibre laser with a Fabry–Perot interferometer (*a*) and a mirror based on a loop of a fibre lightguide and a WDM coupler (*b*).

tightened the requirements for processing the ends of optical fibres. Therefore, the problems of aligning and protecting mirrors began to be solved using WDM (wavelength-division multiplexing)-couplers (Fig. 7 *b*).

Mirrors of a fibre laser resonator provide a multipass passage of laser radiation through an amplifying medium (active lightguide), and if the generation threshold is reached (the amplification factor exceeds the total losses with a double pass between the mirrors), the radiation power increases sharply. Nevertheless, the radiation power can not grow to infinity and its value stabilizes at a certain stationary level, determined by the saturation effect: when the amplification of the radiation becomes equal to its losses in the resonator. It should be noted that if the resonator mirrors are placed on the end of the active fibre, then, unlike other laser sources of radiation, there are no diffraction losses of power in the fibre laser.

The mirror-shaped fibre laser cavity supports laser-generated modes corresponding to standing waves of a given resonator. If the length L of the resonator contains an integer number of half-waves:

$$2nL = N\lambda, \tag{5.1}$$

where λ is the radiation wavelength, n is the refractive index of the material of the active fibre, then such waves, passing through the resonator, do not change their phase and, due to interference, amplify each other. All other closely located waves gradually quench each other.

Thus, the spectrum of natural frequencies of the optical cavity is determined by the relation

$$v_N = \frac{c}{2nL} N, \tag{5.2}$$

where c is the speed of light in a vacuum. The intervals between neighboring resonator frequencies are the same and equal

$$\Delta v = \frac{c}{2nL}. \tag{5.3}$$

Since the lines in the luminescence spectrum of the active optical fibre are broadened and have a certain width Δv_1, then there can be situations when several eigenfrequencies of the resonator fit into the width of the spectral line (Fig. 8). In this case, the radiation from the fibre laser will be multimode. If the condition $\Delta v_1 < \Delta v$ is satisfied,

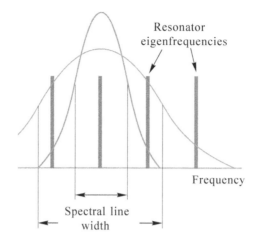

Fig. 8. Illustration of single-mode and multimode operating modes of a fibre laser. (Green and violet colours show different realizations of luminescence spectra).

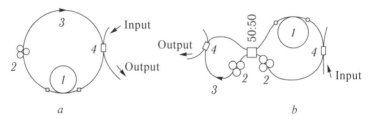

Fig. 9. Schemes of ring fibre resonators: *a* – ordinary ring resonator built into the fibre laser: Input – radiation of the pump; Output – output radiation; *1* – active fibre; *2* – polarizer; *3* – optical insulator; *6* – WDM-coupler; *b* – fibre laser with a ring resonator in the form of the figure eight: Input – radiation pumping; Output-output radiation; *1* – active fibre; *2* – polarizer; *3* – optical insulator; *4* – WDM-coupler; 50:50 – 50/50 divider.

only one frequency will be observed in the emission of a fibre laser, that is, it will operate in the single-mode or single-frequency mode.

5.1.2.2. Ring fibre resonators

In fibre lasers, most often pulsed, special constructions of fibre ring resonators are used (Fig. 9).

The simplest design of the ring resonator is the connection of both ends of the WDM-coupler to the active fibre (Fig. 9 *a*). A feature of fibre ring resonators is the transmission of light in only one direction, regardless of frequency, with the exception of certain resonance frequencies. Usually, additional elements are used in ring resonators:

insulators and polarizers that ensure the safety of polarization of radiation and unidirectionality of its propagation.

For pulsed mode-locked fibre lasers, the so-called *figure-of-eight lasers* are used, called so for the shape of the fibre connection. Both resonator loops, made in the form of a figure eight, serve as Sagnac loops [12]. The active fibre is placed asymmetrically with respect to the resonator loops, which creates a nonlinear phase difference between counterpropagating waves and ensures that the modes are synchronized when a certain threshold pumping power is exceeded.

5.1.2.3. Resonator based on fibre Bragg gratings

Recently, in many fibre lasers, fibre Bragg gratings (FBG) are used as mirrors of resonators, formed directly in the active fibre or in an undoped optical fibre welded with an active fibre. Figure 10 shows a diagram of a fibre laser with Bragg mirrors. The resonator inside the optical fibre is created by pairs of intra-fibre Bragg gratings of the optical waveguide, in which a structure with a modulated refractive index is created.

The fibre Bragg grating is a segment of an optical fibre in the core of which, with the help of ultraviolet radiation, a periodic change in the refractive index with an amplitude of $\sim 10^{-5}-10^{-3}$ with dashes oriented along the normal to the fibre axis and a period of the order of the wavelength of the propagating radiation. In its essence, FBG is a one-dimensional photonic crystal (see Chapter 3), effectively reflecting radiation with a wavelength close to the Bragg wavelength of the crystal. The main characteristics of FBG are: the modulation period of the refractive index Λ, the amplitude of

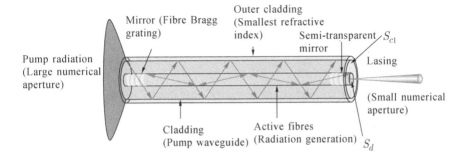

Fig. 10. Schematic of a fibre laser with mirrors based on a FBG. S_d and S_{cl} are the cross-sectional areas of the doped region and the first cladding.

the induced change in the refractive index δn, the number of lattice dashes N_p, and its length L. Two modes propagating in the optical fibre with propagation constants β_1 and β_2 effectively interact at the FBG if the phase matching condition is satisfied for them [13]

$$\beta_2 - \beta_1 = \frac{2\pi}{\Lambda} m, \tag{5.4}$$

where m is an integer.

Since the FBG should perform the role of a mirror and effectively reflect the incident wave in the opposite direction ($\beta_2 = -\beta_1$), then for this case the phase synchronism condition (5.4) will have the form

$$2\beta_1 = \frac{2qm}{\Lambda}. \tag{5.5}$$

From (5.5), we can obtain an expression for the resonance (Bragg) wavelength of the radiation of FBG (λ_B), for which an effective reflection will be observed:

$$2n_1^* \Lambda = \lambda_B, \tag{5.6}$$

where n_1^* is the effective refractive index of the propagating mode of the light guide (for the fundamental mode of the single-mode fibre, $n_1^* = n$, where n is the refractive index of the core material of the fibre).

The reflection coefficient of a FBG of length L is described by expression

$$R = \text{th}^2 (\kappa L), \tag{5.7}$$

where κ is the coupling coefficient of the modes:

$$\kappa = \frac{\pi \eta \delta n}{\lambda_B}, \tag{5.8}$$

and η is the coefficient of overlap of the mode fields in the region of FBG.

The spectral width of the FBG at half-height of the resonance is determined by the following relation:

$$\Delta\lambda = \lambda_B \alpha \left[\left(\frac{\delta n}{n}\right)^2 + \left(\frac{1}{N}\right)^2 \right]^{1/2}, \tag{5.9}$$

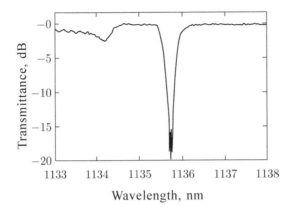

Fig. 11. The transmission spectrum of FBG [7].

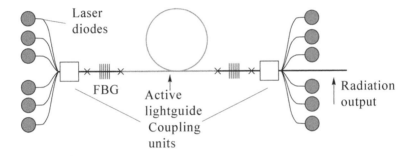

Fig. 12. Scheme of a fully fibre laser pumped from an array of multimode laser diodes.

where α is the lattice depth parameter, taking the values: ~1 for deep gratings and ~0.5 for gratings of small depth.

Figure 11 shows the experimentally measured transmission spectra of FBG with the following parameters: $L = 5$ mm, $\delta n = 8 \cdot 10^{-4}$, $\Lambda = 0.4$ μm [7]. As can be seen, FBG provides a reflection coefficient $R \sim 0.99$ at a wavelength of 1.136 μm with a reflection spectrum width of ~0.4 nm. This result convincingly demonstrates the possibility of using the FBG as effective mirrors for fibre laser resonators.

Thus, by varying the parameters of the FBG, it is possible to create fully fibre mirrors with different reflectance widths (0.05–5 nm) and different reflection coefficients (1–99.9%), thereby forming a completely fibre resonator at the desired wavelength and the desired quality factor. Therefore, with the development of the technology of production of active optical fibres with a double cladding powerful one-mode fully fibre lasers pumped into the first cladding with the help of multimode laser diodes with a fibre output have been created (Fig. 12). The selectivity to the frequency of Bragg gratings makes it possible to obtain a fibre laser operating on one longitudinal mode with a narrow frequency band of generation.

We note that in practice the Bragg grating created inside the fibre has slightly different parameters, since the very creation of this fibre changes the effective refractive index of the mode at the location of the optical fibre and thus changes the resonant wavelength.

For intra-fibre gratings, the danger is a high ambient temperature. Although in general the grating failure temperature depends essentially on the method of its creation and the fibre material, most often the critical temperatures lie in the range 300–600°C.

5.1.3. Special features of active lightguides as a medium of amplification of radiation

Active fibre lightguides as a radiation amplification medium have a number of features in comparison with traditional laser crystals and glasses. Thus, the waveguide nature of the propagation of amplified radiation and pump radiation, as a rule, excludes their losses due to radiation through the side surfaces of the medium. An exception is the case of the bending of the active fibre. This is an undoubted advantage of fibre lasers in comparison with conventional solid-state lasers. Since the characteristic length of the active medium of fibre lasers is one or tens of meters, in analyzing their efficiency it is essential to take into account the nonresonant optical losses of both the pump radiation and the emission of the signal wave.

Other features of fibre lasers are due to the fact that in a single-mode optical fibre, the radiation propagates not only in the core, but also in the reflecting cladding. In this case, the fraction of the radiation power propagating in the core and in the envelope can be comparable. In addition, the active impurity can be introduced into the core and into the cladding both separately and simultaneously. It is also possible to dope different regions of the fibre with different impurities. Due to the low concentration of doping additives, the quantum-mechanical interaction of impurity ions is absent, and only their optical coupling exists. As a result, the classical expressions describing the process of amplification of radiation in laser crystals, for the description of amplification in active optical fibres, should be modified [7, 15, 16]. Let us briefly consider the relationships for the evolution of signal power and pumping in a three-level amplification scheme, over which the majority of existing fibre lasers operate [7]:

$$\frac{dP_s}{dz} = \chi_s \left[N_1 \sigma_p(\lambda_s) - N_0 \sigma_a(\lambda_s) \right] P_s - \sigma_p P_s, \tag{5.10}$$

$$\frac{dP_p}{dz} = -\chi_p N_0 \sigma_a(\lambda_p) P_p - \alpha_p P_p, \tag{5.11}$$

where P_s and P_p, λ_s and λ_p are the powers and wavelengths of the signal and pumping, respectively; σ_a and σ_p are the absorption and luminescence cross sections; N_0 and N_1 are the populations of the ground and metastable levels; α_s and α_p are nonresonant loss coefficients for the signal and pump; χ_p are relative overlap integrals of the signal and pump radiation fields with the active region.

In the case of pumping into the first sheath of the fibre χ_p can be roughly estimated as

$$\chi_p \cong \frac{S_d}{S_{c1}}, \tag{5.12}$$

where S_d and S_{c1} is the area of the cross section of the doped region and the first envelope over which the pump radiation propagates, respectively (see Fig. 10). We note that relation (5.12) is approximate, since in the general case the overlap integral for the pump radiation depends on the distribution of the radiation intensity of the pump source, the method of exciting the fibre, the shape of the first cladding, etc. Nevertheless, it becomes obvious that the use of pumping in the first cladding is equivalent to a substantial decrease in the absorption cross section at the pump wavelength.

5.2. Continuous fibre lasers

The beginning of the 21st century is characterized by increasing interest in the high-energy (pumped through the cladding) continuous fibre lasers that are replacing conventional solid-state lasers.

Such lasers begin to find wide application in various fields of science and technology, because, in contrast to solid-state lasers, they are more compact, reliable and temperature-stable. Indeed, the thermal flux produced by optical pumping of the active fibre is distributed along a long fibre length. This allows to effectively dissipate heat to the environment and significantly reduce the risk associated with thermal damage, which is especially important for continuous generation of radiation.

The use of multicladding structures and single-mode active fibres in fibre lasers allows not only to achieve the high efficiency of conversion of pump radiation into laser radiation, but also to ensure high quality of generated radiation. A number of recent publications report the creation of lasers with a single-mode power of more than 100 W and multimode lasers with a power of ~1000 W. Significant results were achieved in the spectral reconstruction of the radiation generated by fibre lasers in the regions of 1 μm, 1.5 μm, and 2 μm (Table 3) [16].

Table 3. Comparative characteristics of tunable fibre lasers

Dopant	Tuning range, nm	Maximum radiation power, W	Effic-iency, %	Thres-hold, W	Active lightguide length, m	Pump radiation, nm	Material of matrix of active lightguide
Yb	1070–1106	6.6	24	1,6	30	915	Alumo-silicate
Nd	1057–1118	0.83	15	0,2	5	808	Alumo-silicate
Tm	1860–2090	7.0	26	5	3.8	787	Alumo-silicate
Er-Yt	1533–1600	6.7	28	0,5	3.3	913	Alumo-silicate

In this section, we consider the characteristics of continuous fibre lasers based on active fibres doped with various rare-earth elements.

5.2.1. Fibre lasers based on active fibres doped with neodymium (Nd³⁺)

Fibre lasers based on an active fibre doped with Nd^{3+} ions emitting in the 1.06 μm region were the first lasers using pumped in the cladding. This was due to the fact that the first powerful pump systems based on semiconductor lasers were of low brightness. As a result, active light guides with an inner cladding size of several hundred micrometers were needed to enter the pump radiation. Consequently, the population inversion was also small, and lasing was possible only in systems operating in a four-level scheme where there is no reabsorption of the signal wave.

Figure 13 presents typical emission characteristics of a neodymium laser [17]. In this case, the neodymium-doped core had a diameter of 5 μm, and the square first cladding had a size of 290 × 290 μm.

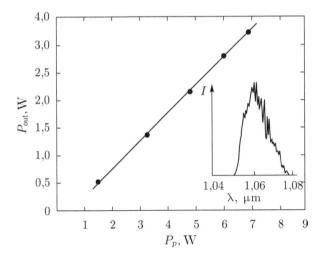

Fig. 13. Dependence of the output power of a neodymium fibre laser on the power of pump radiation. The inset – spectrum of the lasing of a fibre laser [17].

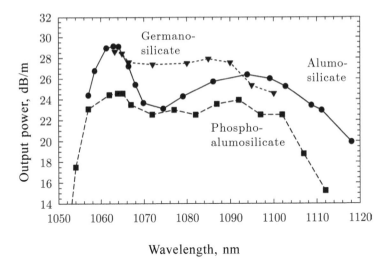

Wavelength, nm

Fig. 14. Spectral characteristics of active fibres pumped into the cladding with an alumo-silicate, germanium–aluminosilicate and phosphorus–alumosilicate core doped with Nd^{3+} ions [18].

A diode array with a radiation wavelength of 0.81 μm was used for pumping.

The doping of the core of the active fibre with additions of aluminum and germanium leads to the fact that the active ions fall into different environments, as a result of which they have different luminescence spectra (Fig. 14) [16]. This phenomenon was used to create a fibre laser emitting simultaneously at wavelengths of 1.06 and 1.09 μm [18].

Of considerable interest is the creation of a neodymium fibre laser with a lasing wavelength of 0.92 μm [15]. Such lasers can be used for pumping ytterbium fibre amplifiers, as well as in frequency doubling circuits for obtaining radiation with a wavelength in the blue region of the spectrum.

5.2.2. Lasers based on active lightguides of doped with ytterbium (Yb3$^+$)

Ytterbium fibre lasers are currently one of the most common types of fibre lasers pumped into a cladding [7]. The energy system of the levels of ytterbium ions in quartz glass is extremely simple (Fig. 3). The presence of only the ground and excited levels makes it possible, in contrast to erbium and neodymium, to substantially increase the concentration of the active impurity in the core. This in turn makes it possible to reduce the length of the active fibre and, accordingly, to reduce the effect of additional optical losses.

The absorption spectra for such active fibres are characterized by the presence of a complex absorption band with maxima at 0.915 and 0.976 μm. The luminescence spectrum consists of a narrow line with a maximum at 0.98 μm and a fairly wide band extending up to 1.2 μm with a maximum at 1.035 μm (Fig. 4).

Figure 15 *a* shows a typical dependence of the output power of an ytterbium fibre laser (emission wavelength 1.1 μm) on the pump power. The efficiency of using pumping in such lasers can reach 80% (while the quartz glass of the core is usually additionally doped

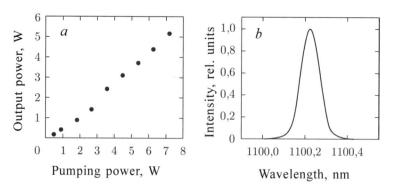

Fig. 15. Dependence of the output power of an ytterbium fibre laser on the pumping power (*a*) of the laser emission spectrum (*b*) [19].

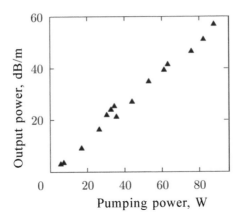

Fig. 16. Dependence of the output power of an ytterbium fibre laser on the pump power through passive optical fibres [7].

with aluminum and germanium), and the width of the spectrum of the generated radiation is 0.1 nm (Fig. 15 *b*) [7.19]. By varying the composition of the quartz glass impurities, it is possible to obtain the radiation of an ytterbium fibre laser at a wavelength of 0.98 μm [7].

Recently, the fibre lasers with increased output power have been constructed on an increasing using lasers in which the active optical fibre is combined with passive pump lightguides operating according to the scheme in Fig. 6 *b*. As shown in [7], the use of four pump sources with a total power of up to 100 W made it possible to achieve the power of a 65 W in a continuous fibre laser at a wavelength of 1.072 μm. The dependence of the output power of this laser on the pump power is shown in Fig. 16.

5.2.3. Fibre-based lasers based on active optical fibres doped with erbium ions (Er^{3+})

Er^{3+} ions in quartz glass have a luminescence band with a maximum at a wavelength of 1.53 μm, which makes it possible to create fibre lasers for the spectral range of 1.53–1.6 μm (Fig. 17) [7].

The lasers based on optical fibres doped with erbium ions work in a three-level scheme (Fig. 3). Since the absorption band maximum practically coincides with the luminescence maximum, a high degree of population inversion is required to provide lasing at a wavelength of 1.54 nm. Therefore, the first erbium fibre lasers pumped into the cladding used fibres with a small diameter of the inner cladding, which made it possible to achieve high levels of population inversion

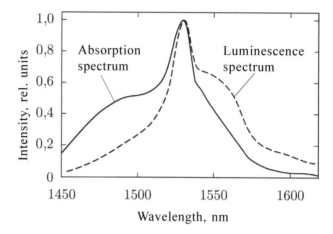

Fig. 17. Absorption and luminescence spectra of erbium ions in an alumosilicate quartz fibre [7].

[7]. In particular, for laser fibres with a numerical aperture of 0.18 and an inner cladding diameter of 22 µm with a pumping power of 900 mW ($\lambda = 980$ nm), laser generation with a power of only 300 mW was obtained. That is, the efficiency of the pump power conversion was 30%, and the output power is limited by the possibilities of inputting the pumping power into the inner cladding.

At the same time, in the spectral range 1.56–1.6 µm in an active fibre, luminescence predominates over absorption. This makes it possible to reduce the requirements for population inversion and use active lightguides with a large diameter of the inner cladding, and, therefore, to use more powerful pump radiation sources.

Lasing in the range 1.53–1.6 µm is produced in most cases using optical fibres doped simultaneously by ytterbium (Yb^{3+}) and erbium (Er^{3+}) ions. In such optical drives, the effective absorption of pumping at a wavelength of 0.976 µm is provided by Yb^{3+} ions, which transfer energy to Er^{3+} ions. Such energy transfer is possible due to the proximity of the energy levels of $^2F_{5/2}$ ytterbium ions and $^4I_{11/2}$ levels of erbium ions (Fig. 3) [20]. One of the problems of implementing such a fibre is the choice of the correct chemical composition of its core. In order to reduce the probability of reverse energy transfer, it is necessary to decrease the lifetime of the Er^{3+} ions in the excited state, which is achieved using the phosphosilicate quartz glass for the core material. In addition, it is important to choose the ratio of the concentrations of active ions. It was established [7] that the maximum efficiency of lasing is 50% when the ytterbium ion

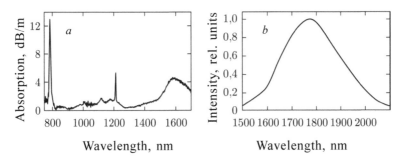

Fig. 18. Absorption (*a*) and luminescence (b) spectra of an active fibre doped with thulium ions.

concentration exceeds the concentration of erbium ions by a factor of 30. Currently, IPG company produces commercially a series of continuous fibre lasers with an output power of more than 100 W with the wavelengths of radiation in the range 1.53–1.6 μm.

5.2.4. Fibre lasers based on active optic fibres, doped with thulium ions (Tm^{3+})

For thulium-doped Tm^{3+} active optical fibres, the effective use of pumping radiation is complicated by the fact that lasing in the region 1.8–2 μm occurs according to a three-level scheme (Fig. 3). However, the situation with the pumping of the active fibre is significantly facilitated by the presence of powerful absorption bands with maxima at wavelengths of 0.787, 1.21, and 1.16 μm (Fig. 18 *a*) due to the $^3H_4 \rightarrow {}^3F_4$ transition.

For example, as can be seen from Figs. 18 *a* and *b*, if the maximum of the band absorption of active fibres doped with thulium ions (Tm^{3+}) is in the region of 1.6 μm, the maximum of the luminescence band is in the region of 1.8 μm. The width of the luminescence line is about 300 nm [21]. All this makes it possible to obtain a sufficient degree of population inversion, in which the reabsorption of the laser-generated wave in the spectral region of more than 1.8 μm is insignificant.

In recent years, many works devoted to the development of lasers based on active fibres doped with thulium ions have been published. For example, in [22] it was reported about the creation of a continuous laser with a power of 14 W with a wavelength of generation of 2 μm at a pumping power of ~37 W at a wavelength of 0.787 μm. In Ref. [23], a tunable laser was constructed in the wavelength range 1.86–2.09 μm.

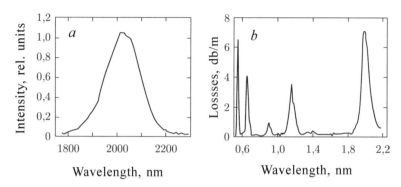

Fig. 19. Luminescence (*a*) and absorption (*b*) spectra of an active fibre doped with holmium ions [17].

5.2.5. Fibre-based lasers based on active optical fibres, doped with holmium ions (Ho³⁺)

In the energy spectrum of Ho^{3+} there is a transition $^5I_7 \rightarrow {}^5I_8$ (Fig. 3), which can be used to obtain lasing in the region of 2 μm. Figure 19 *a* shows the luminescence spectrum of an optical fibre doped with Ho^{3+} ions when excited by a crypton laser with a wavelength of 0.676 μm corresponding to the $^5I_8 \rightarrow {}^5I_4$ transition with a luminescence time constant of 0.5 ms.

Since the holmium fibre lasers operate in a three-level scheme (Fig. 3), in order to obtain an inverted population, it is necessary to ensure effective absorption of pump radiation. The strongest absorption bands are in the visible region of the spectrum. Therefore, the first holmium fibre laser [24] was pumped by irradiation of an argon laser ($\lambda = 0.458$ μm). The laser had a low efficiency: the radiation power was 0.67 mW with an absorbed power of 85 mW.

Figure 19 *b* shows the absorption spectrum of the active fibre, doped with holmium ions. As can be seen, the presence of strong absorption lines in the infrared region makes it possible to use powerful ytterbium fibre lasers for their pumping. In Ref. [17] the authors demonstrated the development of a holmium fibre laser pumped by an ytterbium fibre laser over a wavelength of $\lambda = 1.15$ μm corresponding to the $^5I_8 \rightarrow {}^5I_6$ transition generating radiation at a wavelength of 2.001 μm with a power of 3 W.

In the last few years, the holmium fibre lasers have been constructed using additional doping of the core with thulium ions [25] and ytterbium [26]. Pumping was carried out in the absorption bands of these elements with subsequent transfer of excitation to holmium ions. For the holmium lasers with active lightguides additionally doped with thulium ions (Tm^{3+}) continuous lasing was produced at a wavelength of about 2.1 μm with a power of 5 W at a pumping power of 20 W. For holmium lasers with active light guides additionally doped with ytterbium ions (Yb^{3+}), The output power was 0.85 W with a pumping power of 11 W.

5.3. Fibre lasers based on stimulated Raman scattering of radiation (fibre SRS lasers)

The above-described fibre lasers based on active fibres doped with rare-earth ions can effectively generate radiation only in certain regions of the spectrum without filling the entire near-IR range. To overcome this problem, it has been proposed to use the phenomenon of stimulated Raman scattering in optical fibres, which makes it possible to create effective wavelength-of-radiation converters of lasers and to obtain lasing at practically any pre-determined wavelength of the near infrared range [7].

5.3.1. The phenomenon of stimulated Raman scattering of radiation in optical fibres

When a substance is illuminated by laser radiation, scattered radiation waves are observed at the output the frequencies of which have a Stokes and antiStokes shift relative to the laser radiation frequency by some amount characteristic of the given substance. This effect, due to the interaction of radiation with the vibrational energy levels of the matter molecules, is called the Raman scattering of light. When the power of the radiation introduced into the substance is increased, when it exceeds a certain threshold value, the effect of an intense conversion of the laser radiation power into the radiation of the region of the lower-frequency spectrum (Stokes scattering) (Fig. 20), which reaches saturation when a certain characteristic value of power P_{sat} is reached. Such a process, which is a consequence of the nonlinear-optical interaction of radiation with the medium, is called stimulated Raman scattering (SRS) of radiation [28].

Since anti-Stokes scattering is determined by molecules in an excited state, and at not very high temperatures, the population of the first vibrational level is small (at room temperature the vibrational frequency is ~1000 cm^{-1} and at the first vibrational level is only 0.7%

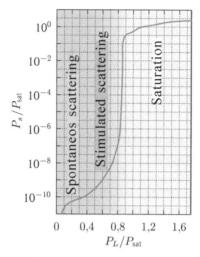

Fig. 20. The nature of the dependence of the efficiency of the SRS conversion of laser radiation into the Stokes wave.

of molecules), then the anti-Stokes intensity of the SRS scattering of tradiation is small.

Fibre-optic lightguides, using the phenomenon of stimulated Raman scattering, make it possible to efficiently convert laser pumping radiation into the radiation of lower frequencies. The SRS of laser radiation in an optical fibre was first observed in 1971 [29].

In the case of continuous pumping and in the absence of losses in the optical fibre, the increase in the power of the Stokes scattering wave (signal wave), when the pump wave propagates in the direction of the z axis, is approximately described by the relation [7]

$$\frac{dP_s}{dz} = GP_pP_s, \tag{5.13}$$

where P_s is the Stokes wave power; P_p is the power of the pump wave; G is the gain coefficient of the Stokes wave.

The peculiarity of optical fibres as a stimulated Raman scattering medium is a relatively small gain. For example, for fused quartz, which is the main material of optical fibres, it is $\sim 10^{-13}$ m/W. In addition, the number of additives that change the spectrum of stimulated Raman amplification is limited by the technological capabilities and the requirement of maintaining low losses for propagating radiation. Also, optical fibres based on quartz glass, due to their long length, make it possible to reduce the thresholds for obtaining stimulated Raman scattering by increasing the interaction

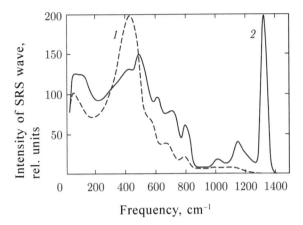

Fig. 21. Spectrum of stimulated Raman scattering in an optical fibre with a core of germanosilicate (*1*) and phosphorosilicate glass (2) [7].

length of the pump radiation with the medium. Quartz glass itself, as a non-crystalline material, has a wide spectrum of Raman scattering with a maximum near the frequency 440 cm^{-1}. The addition of the GeO$_2$ impurity slightly changes the shape of the Raman spectrum. The corresponding SRS spectrum for a light guide with a germanosilicate and phosphosilicate glass core is shown in Fig. 21.

Initially, the effect of stimulated Raman amplification was used to create amplifiers for radiation in telecommunication lines. Figure 22 shows the scheme of an SRS (Raman) amplifier, which makes it possible to convert the pump radiation energy into energy propagating in the fibre waveguide signal [28].

For real propagation conditions for continuous radiation in an optical fibre, the interaction between the pump wave and the signal (Stokes) wave is described by the following system of equations [28]:

$$\frac{dP_s}{dz} = \frac{g_R}{A_{eff}} P_p P_s - \alpha_s P_s, \qquad (5.14)$$

$$\frac{dP_p}{dz} = -\frac{\omega_p}{\omega_s} \frac{g_R}{A_{eff}} P_p P_s - \alpha_p P_p, \qquad (5.15)$$

where α_s and α_p are the absorption coefficients of the signal wave and the pump wave in an optical fibre; ω_s and ω_p are the cyclic frequencies of the signal wave and pump waves; g_R is the SRS gain

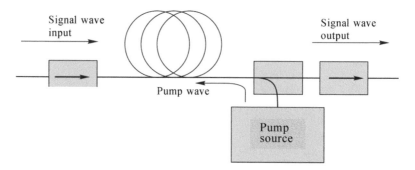

Fig. 22. Scheme of an SRS amplifier that converts the energy of pump radiation into a signal wave.

coefficient; A_{eff} is the effective area occupied by the mode in the core of the lightguide. Equations (5.14), (5.15) are valid for waves having the same polarization state.

To simplify the solution, the first term on the right-hand side of equation (5.15) is usually neglected. This allows one to obtain an expression for describing the evolution of the power dynamics of a signal wave propagating in a fibre, [28]:

$$P_s(z) = P_s(0)\exp\left(\left(\frac{g_R}{A_{\text{eff}}}\right)P_p(0)z_{\text{eff}} - \alpha_s z\right), \qquad (5.16)$$

$P_p(0)$ and $P_s(0)$ are the powers of the pump wave and the signal wave at the input of the optical fibre, respectively, and

$$z_{\text{eff}} = \left(1 - \exp(\alpha_p z)\right)/\alpha_p. \qquad (5.17)$$

Since the parameters g_R and A_{eff} can not be measured accurately in practice, an easily measured integral characteristic, called the gain of the Stokes wave for a given type of fibre at a given wavelength, is often used for calculations [7, 28]:

$$G = \frac{g_R}{A_{\text{eff}}}. \qquad (5.18)$$

If an active amplifying medium is placed between the mirrors of the resonator, then an SRS laser can be obtained. The use of fibre couplers and Bragg gratings made it possible to substantially simplify the construction of an SRS laser and make it acceptable for practical use. To simplify even further the design of an SRS laser, the use of an optical fibre with a core doped with phosphorus oxide as an active medium has been made possible. As can be seen from Fig. 21, such a phosphosilicate fibre has a Stokes shift at a frequency of 1330 cm^{-1}, which is three times larger than that of a germanosilicate fibre.

Significant progress in the development of stimulated Raman lasers was initiated by their use as multistage pump sources for Raman amplifiers [7, 28].

Figure 23 shows a scheme of a multistage SRS laser, which consists of a fibre lightguide and a set of Bragg gratings of the refractive index corresponding to the Stokes shifts of the frequency of propagating radiation in the fibre material. The gratings tuned to intermediate wavelengths have reflection coefficients close to 100%. Many studies have been devoted to the theoretical analysis of stimulated Raman lasers [7]. For the case of using a single-mode optical fibre, under the assumption of the absence of stimulated Mandel'shtam–Brillouin scattering and the use of broadband pumping, the operation of a multistage SRS laser can be described by a system of equations [7]:

$$\frac{dP_p^{+,-}}{dz} = \mp\alpha_p P_p^{+,-} \mp \frac{\omega_p}{\omega_{s1}} \frac{g_R^1}{A_{\text{eff}}^1}\left(P_{s1}^+ + P_{s1}^-\right)P_p^{+,-}, \qquad (5.19)$$

$$\frac{dP_{sk}^{+,-}}{dz} = \mp\alpha_{sk} P_{sk}^{+,-} \mp \frac{\omega_{sk}}{\omega_{sk+1}} \frac{g_R^{k+1}}{A_{\text{eff}}^{k+1}}\left(P_{sk+1}^+ + P_{sk+1}^-\right)P_{sk}^{+,-} \pm$$

$$\pm \frac{g_R^k}{A_{\text{eff}}^k}\left(P_{sk-1}^+ + P_{sk-1}^-\right)P_{sk}^{+,-}, \qquad (5.20)$$

$$\frac{dP_{sn}^{+,-}}{dz} = \mp\alpha_{sn} P_{sn}^{+,-} \pm \frac{g_R^n}{A_{\text{eff}}^n}\left(P_{sn-1}^+ + P_{sn-1}^-\right)P_{sn}^{+,-}, \qquad (5.21)$$

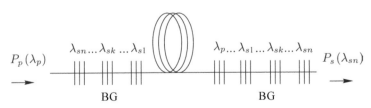

Fig. 23. Scheme of a multistage SRS laser: $P_p(\lambda_p)$ – pump power with a wavelength λ_p at the entrance to the optical fibre; $P_s(\lambda_{sn})$ – power of the Stokes wave (λ_{sn}) at the output of the n-th cascade of an SRS laser; BR – fibre-optic Bragg gratings with corresponding resonant frequencies.

where $k = 1, 2, 3,...,n$; n is the number of laser cascades; P_p is the power of the pump wave; P_{sk} is the power of the signal wave at the intermediate frequency; P_{sn} is the power of the signal wave at the output stage; α_p is the coefficient of pump wave loss; α_{sk} is the coefficient of signal wave loss at the intermediate frequency; ω_p, ω_{sk} are the cyclic frequencies of the pump wave and the signal wave at the intermediate frequency, $(+, -)$ are the indices denoting the direction of the wave movement from left to right and right to left, respectively.

The above system of equations is used to numerically calculate and optimize the parameters of multistage SRS lasers, including the choice of the fibre length and the reflection coefficients of Bragg mirrors.

5.3.2. Fibre-optic SRS lasers

The use of the stimulated Raman scattering phenomenon in optical fibres makes it possible to create effective converters of the wavelength of radiation for laser sources and, in contrast to lasers based on active fibres doped with rare-earth impurities, to obtain lasing at practically any wavelength of the near infrared spectral range. In this case, the configuration of the SRS laser and the characteristics of the lightguide used depend on the wavelengths of the pump radiation and the signal wave. As a pumping source, due to high generation efficiency and a wide spectral range of radiation, ytterbium fibre lasers are most often used.

5.3.2.1. Single-stage SRS lasers

Single-stage SRS lasers are the simplest type of fibre lasers, since only one cascade of pump radiation conversion is used. The light guide with a germanosilicate glass core with a frequency spectrum in Fig. 21 converts the radiation of an ytterbium fibre laser into radiation with a wavelength in the region of 1.1–1.22 μm.

In [30], the pump radiation transformation of an ytterbium laser with a wavelength of 1.09 μm into a signal wave radiation $\lambda = 1.15$ μm in an optic fibre with a length of 500 m with a loss of 0.8 dB/km at a wavelength of 1.06 μm, in which Bragg gratings with a reflection coefficient of 20% were produced, were studied. Figure 24 shows the experimental dependence of the output power of the signal wave on the power of the pump wave and the calculated dependence of

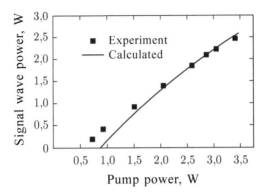

Fig. 24. Signal wave power of a single-stage SRS laser based on a germanosilicate fibre and a ytterbium fibre laser on the pump power [30].

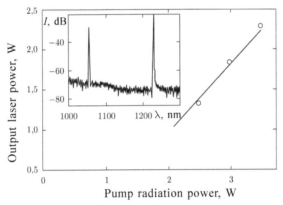

Fig. 25. Dependence of the output power of a single-stage SRS laser based on a phosphosilicate fibre on the pumping power of an ytterbium laser. (inset – spectrum of the radiation of an SRS laser) [31].

the conversion efficiency obtained using relations in section 5.3.1. The conversion efficiency of the ytterbium fibre laser was ~70%.

The optic fibres with a core based on phosphosilicate glass are characterized by the presence of an additional peak of Raman scattering with a maximum at 1330 cm^{-1} (Fig. 21)

The single-stage SRS lasers based on such fibres, using the pump radiation of an ytterbium fibre laser, make it possible to obtain a signal wave radiation in the region of 1.22–1.35 μm [31].

The use of a neodymium fibre laser as a pumping source makes it possible to create SRS lasers with a radiation wavelength in the region of 1.24 μm with a conversion efficiency of about 70% [32]. Figure 25 shows the dependence of the output power and the emission spectrum (inset) of an SRS laser at a wavelength of 1.234

µm on the power of a pump laser emitting at a wavelength of 1.06 µm.

5.3.2.2. Multistage SRS lasers

Multistage SRS lasers are used to produce radiation with wavelengths greater than 1.35 µm. In particular, the use of a four-stage Raman laser based on a germanosilicate fibre made it possible to generate radiation in the range 1.35–1.45 µm when pumped with an ytterbium fibre laser. However, an increase in the number of cascades significantly complicates the design of the laser, as the number of pairs of Bragg gratings increases and their spectral characteristics must be maintained with high accuracy.

In addition, the Bragg gratings introduce excess losses (about 0.05 dB per grating), which reduces the efficiency of radiation conversion. In this connection, lightguides with a phosphosilicate core were used in the SRS lasers, in which the amplification is observed not only at a Stokes shift frequency of 1330 cm^{-1} but also at a frequency with a spectral shift of 440 cm^{-1} corresponding to the maximum of stimulated Raman scattering in pure fused quartz, which is the main material of the fibre.

This allowed simultaneous use of two different spectral shifts in the same optical fibre. A stimulated Raman laser with a radiation wavelength of 1.407 µm was used in Ref. [33], using a single 1330 cm^{-1} frequency shift and two 440 cm^{-1} frequency shifts. The maximum power of the converted radiation was 1 W with a conversion efficiency of 25%. A similar laser configuration was used to create an SRS laser at a wavelength of 1.43 µm [34].

Figure 26 shows the emission spectrum of such a laser with a maximum radiation output power at a wavelength of 1.43 µm about 1.3 W.

5.3.2.3. Composite SRS lasers

The long-wavelength boundary of two-stage SRS lasers is 1.6 µm, which is due to the properties of phosphosilicate fibres and the long-wave edge of the lasing of ytterbium lasers (1.12 µm), where the efficiency of lasers exceeds 60%. At the same time, the spectral range of more than 1.6 µm is of considerable interest for a number of practical applications.

Unfortunately, phosphosilicate fibre lightguides have considerably greater losses in comparison with germanosilicate lightguides: 0.8–1

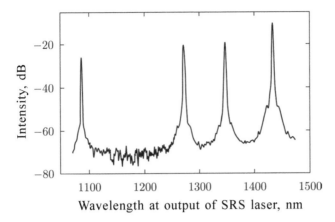

Fig. 26. Spectrum of the radiation of an SRS laser based on a phosphosilicate fibre with the use of Stokes shifts of 440 and 1330 cm^{-1} [33].

dB/km and 0.2 dB/km at a wavelength of 1.5 µm, respectively. Moreover, with an increase in the wavelength, the optical losses in phosphosilicate fibres grow faster than in germanosilicate fibres. Therefore, germanosilicate fibre fibres are preferred for the creation of SRS lasers.

However, the use of optical fibres based on germanosilicate glass as an active medium requires the use of a seven-stage conversion to produce radiation with a wavelength of more than 1.6 µm. Such a circuit turns out to be difficult to manufacture and inefficient due to the high pumping power [7]. Therefore, in [34] it was proposed to use a composite SRS laser, the scheme of which is shown in Fig. 27.

As can be seen, the laser consists of two parts: a two-stage SRS laser based on a phosphorus silicate fibre and a single-stage Raman laser based on a germanosilicate fibre. The first part of the composite SRS laser makes it possible to obtain output radiation

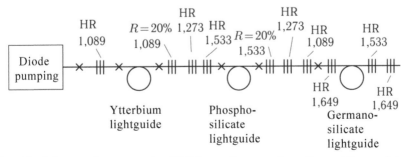

Fig. 27. Scheme of a composite SRS laser. Crosses show the welding points of lightguides; near each Bragg grating there is its resonant wavelength (µm) is indicated; HR − R = 100% (Bragg gratings with high reflection coefficient).

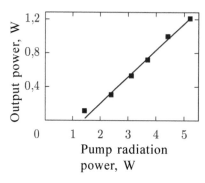

Fig. 28. Dependence of the output power of a composite SRS laser at a wavelength of 1.649 μm on the power of an ytterbium pumping laser λ = 1.089 μm [34].

at a wavelength of 1.533 μm, and the second at a wavelength of 1.649 μm. The maximum radiation power at a wavelength of 1.633 μm was 2.07 W at the power of a semiconductor pump laser is 8 W. The germanosilicate optical fibre of the second part of the stage of the composite SRS laser was in the form of a standard telecommunication lightguide with a wavelength of zero chromatic dispersion shifted to 1.55 μm.

Figure 28 shows the dependence of the output radiation power a composite SRS laser at a wavelength of 1.649 μm (width of the generation line 0.35 nm) from the pump radiation power of an ytterbium fibre laser (λ = 1.089 μm). The power conversion efficiency of the pump laser was 32%.

The use of a composite configuration of stimulated Raman lasers makes it possible to create radiation sources that cover a spectral range of 1.6–1.75 μm. To obtain laser radiation in the wavelength region near 2 μm or more, it is necessary to use optical fibres based on GeO_2, since the glass containing germanium dioxide has minimum of losses in this region.

A composite SRS laser was constructed in [7], consisting of from a four-stage SRS laser based on a germanate fibre, generating radiation at a wavelength of 2.06 μm. A two-stage SRS laser based on a phosphosilicate fibre lightguide with a wavelength of 1.472 μm was used as the pump radiation source, which in turn was pumped by pumping radiation of an ytterbium fibre laser with a wavelength of 1.057 μm.

5.3.2.4. Fibre-optic SRS lasers with random distributed feedback

As was shown in [35], if the fibre length between the Bragg gratings is increased to 270 km, the structure of the longitudinal modes with intermode distance $c/2nL \sim 400$ Hz is observed in the spectrum of the radiation generated by the stimulated Raman laser. With increasing fibre length between Bragg gratings up to 300 km, lasing continues to be observed, but already in the 'modeless' mode. In this case, the lasing is facilitated by Rayleigh scattering, which arises on submicron irregularities of the refractive index of the optical fibre.

Although the scattering of radiation in the optical fibre occurs in all directions, part of the backscattered radiation comes back into the lightguide and begins to propagate in the opposite direction. Despite the fact that the efficiency of Rayleigh reflection is very small, $\sim 0.1\%$, at resonator lengths ~ 300 km, due to the attenuation of the light wave upon approach to the resonator, it becomes larger than the reflection from the second mirror. In this case, the effect of distributed Rayleigh reflection allows lasing in a standard telecommunication fibre even in the absence of resonator mirrors, if by means of stimulated stimulation it creates distributed amplification in it. Such a laser, without a resonator, operating only due to random distributed feedback (RDFB) in Rayleigh scattering, was first demonstrated in 2010 and was called the fibre RDFB laser [36].

A schematic diagram of the RDFB fibre laser illustrating the principle of its operation is shown in Fig. 29 [37].

The active medium of the laser is represented by a standard telecommunication single-mode fibre with a length of $L \sim 100$ km.. The attenuation coefficient in an optical fibre is ~ 0.2 dB/km at a wavelength of 1.55 μm. The photons propagating in a long fibre waveguide are almost completely lost due to elastic Rayleigh

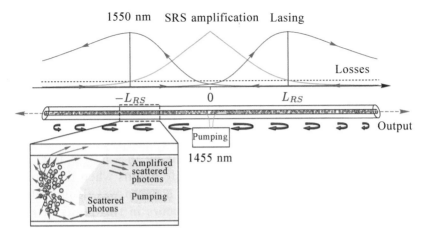

Fig. 29. Diagram of the device and operation of a RDFB fibre laser [37].

scattering by microscopic inhomogeneities of the refractive index (see the inset in Fig. 29). Only a small part of them, ~0.001, is captured by the light guide and comes back. If pumping radiation λ_p = 1455 nm is introduced in the middle of the fibre, then, because of the SRS effect, a distributed amplification is created for both direct and scattered photons with a wavelength maximum of λ = 1560 nm. When the condition that the integral gain of photons exceeds the losses in one complete pass, lasing and emission wavelength of ~1560 nm arise.

With allowance for damping, the main contribution to Rayleigh reflection is recruited back at a length of ~100 km, with the integral reflection coefficient being very small, ~0.1%. If the scattered radiation in a fibre lightguide is amplified by the effect of stimulated Raman scattering, the situation changes fundamentally. The introduction of pumping radiation into the optical fibre at a wavelength of ~1.45 μm due to the stimulated Raman scattering effect leads to the appearance of radiation at a wavelength of 1.55 μm, which, amplifying, propagates through the optical fibre and is scattered backward. Since the pump radiation also attenuates and is stronger, since the Rayleigh scattering efficiency is proportional to 1/4, in order to obtain a more uniform SRS gain (proportional to pump power P_0), the pump radiation is introduced in the middle of the fibre in two opposite directions using directed couplers.

As a result, on the sections from the middle of the optical fibre up to the points indicated in Fig. 29 how $|z| = L_{RS}$, the SRS gain $g_R P_0$ for the Stokes wave exceeds the loss, that is, the $2L_{RS}$ interval corresponds to the amplifying region for both the generated Stokes wave and backscattered radiation. When the distance $|z| = L_{RS}$ is exceeded, the generated wave simply damps down to the exit from the fibre lightguide. Since for a telecommunication fibre waveguide $g_R \sim 0.4$ km^{-1} W^{-1}, at a pump power $P_0 \sim 1$ W an integral gain is achieved $G \sim \exp(2g_R P_0 L_{RS}) \gg 10^6$, which is fully sufficient for overcoming the lasing threshold.

Figure 30 shows the experimentally obtained dependence of the radiation power of a symmetric SPS-laser on the power of the pump radiation. As can be seen, the lasing in a RDFB laser based on a telecommunication fibre waveguide begins when the threshold pump power level reaches ~1.5 W. At a pump wavelength of 1467 nm, random lasing in a RDFB laser occurs in two peaks of Raman amplification in a lightguide corresponding to stimulated Raman scattering for quartz glass in the vicinity of 1557 and 1567 nm

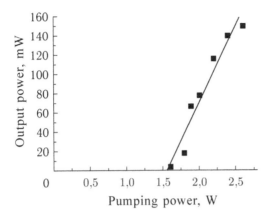

Fig. 30. Dependence of the power of the RDFB laser radiation on the pumping power [37].

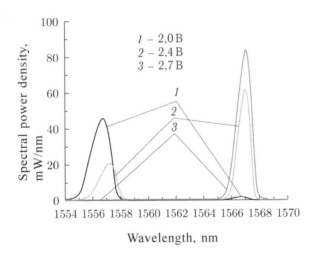

Fig. 31. The power density spectrum of the RDFB laser radiation for different pump power values (*1* – 2 W, *2* – 2.4 W and *3* – 2.7 W) [37].

wavelengths (Fig. 31). A further increase in the pump power leads to the disappearance of a maximum with a wavelength of 1557 nm.

The results of an experimental study of the emission spectrum of the RDFB laser showed that the laser emits a continuum of random spectral components near the maximum generation frequency [37]. This indicates the modeless generation of RDFB laser radiation.

The lasing threshold for the fibre RDFB laser with a fibre length of ~100 km is much higher than for an SRS laser with a fibre Bragg grating (FBG) resonator (Fig. 32). However, in contrast to the SRS

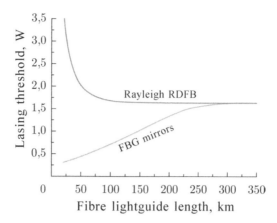

Fig. 32. Calculated dependences of generation thresholds for a RDFB laser and a laser based on a FBG mirror on the length of a fibre waveguide [37].

laser, the threshold power of pumping with a fibre length does not increase, but decreases with the subsequent output to saturation. For large optical fibre lengths (>100 km), the differential efficiency of the lasing of a fibre RDFB laser (when radiation from both ends of the fibre is taken into account) is ~30%, which is comparable with the efficiency of stimulated Raman lasers.

The magnitude of the generation threshold for a fibre SRS-laser can be described by the expression [37]

$$P_{th} = 2\frac{\alpha}{g_R}\left(1 + \ln\left(\frac{g_R P_{th}}{\alpha}\right)\right) + \frac{\alpha_P}{g_R}\ln\left(\frac{1}{Q}\sqrt{\frac{\alpha_P}{\pi\alpha}}\right), \qquad (5.22)$$

where α is the absorption coefficient for the generated radiation; α_P is the absorption coefficient for pumping; g_R is the coefficient of SRS amplification in an optical fibre.

Since $\alpha_P \sim 0.3$ dB/km, then, according to (5.22), the lasing threshold is ~1.8 W, which is close to the experimentally obtained values.

As can be seen from Fig. 32, the lasing thresholds for the fibre RDFB laser and a laser with a FBG resonator are equalized when the length of the optical fibre reaches ~300 km and then does not grow. This indicates that, with an increase in the length of the optical fibre to 300 km, owing to the attenuation of pump radiation and stimulated Raman scattering FBG ceases to play any noticeable role in the reflection process. In this connection, the lasing in both types

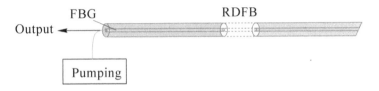

Fig. 33. Diagram of a semisymmetric fibre RDFB laser with a resonator formed by a randomized distributed feedback.

of lasers is provided only by the feedback due to Rayleigh scattering radiation in an optical fibre.

Of great interest for practical use are the semisymmetric schemes of fibre RDFB lasers (Fig. 33). In this scheme, a fibre Bragg grating is installed from one end of the optical fibre, performing the function of a 100% reflector.

In such a laser, the generation threshold is two times lower than in a laser with a symmetrical circuit (Fig. 32), since in this case the pump radiation propagates only in one direction (Fig. 34).

The generation spectrum of a RDFB laser in a semisymmetric scheme is determined by the resonance wavelength of the optical fibre and is practically independent of the length of the fibre in the range 41–165 km [37]. This makes it possible to obtain a multiwave lasing if one uses optical fibres with consecutively recorded FBGs with different resonant wavelengths [37]. In this case, each of the FBGs generates an independent resonator with a distributed Rayleigh mirror, and the laser generates an independent sequence of wavelengths.

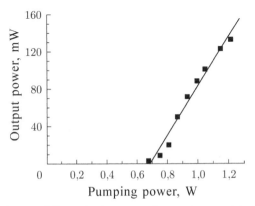

Fig. 34. Dependence of the output power of a semisymmetric RDFB laser on the pumping power [37].

Similarly, if an acousto-optical modulator performing the functions of a spectral filter is placed in the middle of a symmetrical scheme of a fibre SPS-laser, it is possible to ensure a smooth reconstruction of the laser emission spectrum (Fig. 35 *a*) [37].

For a laser with an optical fibre length of ~40 km the lasing power is ~2W with the tuning of the emission wavelength in the range 1535–1570 nm with a variation in the radiation power of less than 3% (Fig. 35 *b*).

5.4. Pulsed fibre lasers

The generation of laser radiation in the form of short pulses is an extremely important problem. The significance of this problem is due to two reasons. First, the pulse duration determines the time interval for the interaction of laser radiation with the medium, which is extremely important in the study of fast processes. Secondly, the energy of laser radiation, being concentrated in a short laser pulse, causes a large pulse power and high intensity of the electromagnetic

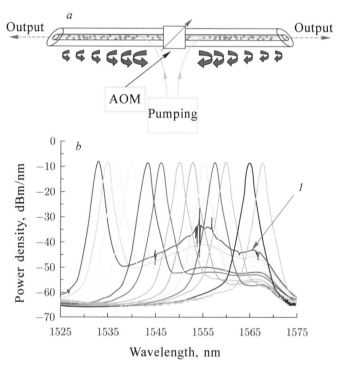

Fig. 35. Scheme of a symmetric fibre RDFB laser with an acoustooptic modulator (AOM) (*a*) and the spectral density of tunable radiation power (*b*), *1* – dependence of the laser tuning characteristic [37].

field of its light wave. At present, lasers, including fibre ones, are capable of generating ultrashort pulses with a duration of only a few femtoseconds. During this time, the pulse can pass a very short distance of several micrometers in length. But even small changes in its energy during this time can lead to significant changes in the power and strength of the pulse field.

At the present stage of development of laser physics for the generation of ultrashort light pulses it is necessary to use continuous laser radiation, since this allows to have highly stable pulse sequences [38].

Fibre lasers are increasingly being used among the generators of ultrashort pulses (USP). As for traditional laser generators, ultrashort pulses in fibre lasers use widely the resonator Q-modulation methods, active and passive mode synchronization, and a combination of these methods [39].

5.4.1. Methods for obtaining pulsed radiation from fibre lasers

5.4.1.1. Modulation of the quality factor of fibre lasers

The simplest method of modulating the quality factor (Q-factor) of a laser is periodic overlapping of the beam in the resonator. In this mode, a gate is placed in the resonator of the laser, overlapping the light radiation flux between the mirrors. In the pumping process, energy accumulates at the upper excited laser level, which, due to the absence of a feedback channel, can not be induced to be emitted. As a result, the population inversion in the laser becomes very large. If the shutter is rapidly opened at a certain instant, this will lead to a sharp change in the Q-factor of the resonator from low to high values. A consequence of this will be the emission of energy in the form of a giant laser pulse. Such a method is called Q-modulation or Q-switching. This method can be used if the lifetime of the excited energy level of the laser is sufficiently large: 10^{-5}–10^{-3} s.

Figure 36 shows a time diagram of the development of a giant pulse in a laser with Q-switching [40]. The following conditions must be fulfilled for successful implementation of Q-switching: the lifetime of the excited state should be large enough so that during the Q-switching period the population inversion can reach large values and the Q-switching of the resonator must occur very rapidly so that the pump energy is not lost due to the spontaneous relaxation of the excited states. When the quality factor of the resonator is switched on, the laser gain greatly exceeds the losses, and the number of photons increases sharply from the initial value to that established by spontaneous emission. As a result of an increase in the number of photons, population inversion $N(t)$ starts to decrease from its maximum value N_{max} and by the time $t = t_p$ it will reach the threshold population inversion N_p. This means that at the time $t = t_p$ the laser pulse will have the maximum power. Further, with increasing time, the losses will increase and the pulse power will decrease to zero.

Two types of Q-switching of lasers are known: active (Fig. 36a) and passive (Fig. 36b). The Q-switched modulators that use special control devices (opto-mechanical, acousto-optical or electro-optical shutters) are called active. These control devices function as optical shutters. Under the influence of the control signal, the transmission of the gate changes, which makes it possible to quickly switch on or off the Q-factor of the resonator. The process of generating a giant pulse in this case consists of two stages: a relatively slow rise of ~100 ns and a short stage of rapid development ~10 ns. Almost all of the pulse energy radiates in the second stage. In pulsed fibre lasers, the most widely used gates are based on the use of acousto- and electro-optic modulators.

In the case where the Q-switching is performed automatically, without external control devices, the modulators are called passive. The principle of operation of the passive modulators is based on the use in the resonator of nonlinear elements with the effect of

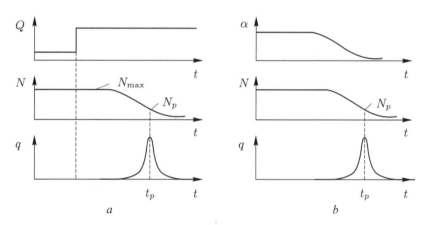

Fig. 36. Diagram illustrating the process of generation of a giant pulse in the radiation of a continuous laser for active (*a*) and passive (*b*) Q-switching of the resonator. Q is the Q-factor of the resonator; α is the absorption coefficient in a saturable absorber; N is the inverted population in the laser; N_p is the threshold inverse population; q is the the number of photons in the laser pulse.

saturation of absorption at the wavelength of radiation generated in the active element.

Figure 36 *b* shows the process of development of a giant pulse in the case of passive *Q*-switching of the resonator. The curve $q(t)$ demonstrates the time variation in the power flux of the generated radiation. The initial state corresponds to a practically non-clarified filter. In this state, the absorption and magnitude of the inverse population have the maximum values. As the pump of the active element increases, the intensity of generated radiation will also grow. As soon as it reaches the threshold value, the transmission of the filter will start to increase sharply, which triggers the generation of a giant pulse. As in the previous case, the process of forming a giant pulse consists of two stages: slow (due to high radiation losses) ~1 µs of the pulse development process and fast ~10 ns of the decay time of the basic pulse energy. After the giant pulse is displayed, the absorption coefficient of the filter again increases and during its relaxation the passive *Q*-factor returns to its initial state.

The materials for passive *Q*-switches include the solutions of organic dyes, semiconductors with quantum dots, materials containing nanotubes, graphenes, etc., as well as nonlinear–optical cells based on the Mandel'shtam–Brillouin stimulated scattering effect and semiconductor mirrors with a reflection coefficient which depends on the intensity of the radiation incident on them.

5.4.1.2. Generation of pulsed radiation due to mode-locking

If the laser pulse has a duration τ, then its spectrum width is $\Delta v \sim 1/\tau$. Thus, the shorter the pulse duration, the wider its spectrum. Therefore, a necessary condition for the generation of ultrashort laser pulses is the use of an active medium with a wide gain band. And also, due to the need to provide a wide spectrum of radiation generation, the operating mode of the laser must be multimode. Proceeding from this, in order to generate short laser pulses, it is necessary to ensure the coherent addition of the radiation of the modes supported in the resonator. To do this, we use the method of synchronization of laser modes, in which the amplitudes and phase differences of arbitrary resonator modes remain constant.

The method of mode-locking makes it possible to obtain ultrashort pulses with a duration of ~10 fs and a very high peak power of $\sim 10^{10}-10^{11}$ W. In this case, the laser mode can be locked in active, passive and mixed (active–passive) modes [39].

Active mode-locking in the fibre laser. Active mode-locking is achieved by introducing an optical element into the resonator the loss or refractive index of which is modulated by an external field.

As shown above (section 5.1.2), when a resonator is excited, the integer number (n) of half-waves of electromagnetic radiation should form on its length: $L = n\lambda/2$, where λ is the wavelength of the radiation in the medium of the resonator. In this case, the frequencies of the waves supported by the resonator form an equidistant spectrum with an intermode distance $\Delta v = V/2L \equiv \Omega/2\pi$, which depends only on the length of the resonator (V is the speed of light in the resonator medium).

Suppose that N longitudinal modes are excited in the laser cavity; these modes are equidistant in frequency, that is, the cyclical frequencies of neighbouring modes differ by an amount of Ω, and are synchronized in phase, that is, the phase difference of the two neighboring modes is constant and equal to Φ. Then if we enter the numbering of the modes with the help of the integer index j which takes the values:

$$-(N-1)/2, \; - (N-1)/2 + 1,...,0,...(N-1)/2 - 1, \; (N-1)/2,$$

the frequerncy and phase of an arbitrarily selected mode can be written in the form

$$\omega_j = \omega_0 + j\Omega, \tag{5.23}$$

$$\varphi_j = \varphi_0 + j\Phi, \tag{5.24}$$

where ω_0 and φ_0 are respectively the cyclic frequency and the phase of the central mode with the index $j = 0$.

Then, according to the principle of superposition, the resultant intensity of electromagnetic radiation at some point of space in the resonator can be represented in the form

$$E(t) = \sum_{j=-(N-1)/2}^{j=(N-1)/2} E_j \exp\left[i(\omega_0 + j\Omega)t + (\varphi_0 + j\Phi)\right]. \tag{5.25}$$

If we assume without the loss of generality that the amplitudes for all modes are the same: $E_j = E_0$, then expression (5.25) becomes simpler:

$$E(t) = E_0 \sum_{j=-(N-1)/2}^{j=(N-1)/2} \exp[i(\Omega t + \Phi)j] \exp[i(\omega_0 t + \varphi_0)]. \tag{5.26}$$

Applying to (5.26) the sum rule of a geometric progression, and under the assumption that $\Phi = 0$, we obtain the expression for the field strength of the electromagnetic wave in the resonator:

$$E(t) = E_0 \frac{\sin(N\Omega t / 2)}{\sin(\Omega t / 2)} \exp\left[i\left(\omega_0 t + \varphi_0\right)\right]. \qquad (5.27)$$

As follows from (5.27), the field in the resonator formed by the addition of N synchronized (mode-locked) longitudinal modes turns out to be the amplitude-modulated interference term. According to (5.27), the modulation period of the radiation field T is determined by the frequency difference between the neighbouring resonator modes:

$$T = \frac{2\pi}{\Omega} = \frac{2L}{V}. \qquad (5.28)$$

Figure 37 shows the result of the addition of several axial modes. As can be seen from the figure and the expression (5.27), with the number of locked modes N in the resonator increasing the field takes the form of pulses with an amplitude proportional to the number of longitudinal modes ($N \cdot E_0$) and a duration of the order of T/N. These pulses follow each other after a time interval T. The greater the number of longitudinal modes supported in the resonator, the greater the amplitude of the light pulses and the shorter their duration.

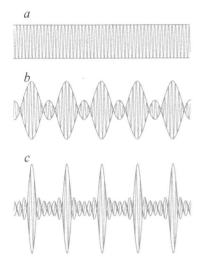

a

b

c

Fig. 37. An example of the formation of ultrashort pulses as the result of in-phase addition of the longitudinal modes of the Fabry–Perot resonator: *a* – one mode; *b* – three modes and *c* – seven modes [41].

Thus, if there was no mode-locking of the resonator, the laser radiation power would be composed of the power of the individual modes and would be proportional to their number:

$$P_L \sim N E_0^2. \tag{5.29}$$

In the case of mode-locking, the peak power of the laser radiation is N times greater:

$$P_L = (NE_0)^2 = N(NE_0)^2. \tag{5.30}$$

So, the interference of N locked longitudinal modes with a sufficiently large number of them leads to the fact that the radiation of a continuous laser acquires the character of a sequence of short and powerful light pulses that follow one after another in a time interval T (expression (5.28)), and the duration of each individual pulse is $\sim T/N$.

As can be seen, the necessary condition for the realization of the longitudinal mode-locking method for obtaining ultrashort light pulses is the condition of simultaneous excitation of a large number of such modes. In this case, in principle, a multifrequency laser generation mode is required. If the width of the gain line in the laser is equal to $\Delta\omega$ then the number of modes generated by a laser is determined by the relation

$$N = \frac{\Delta\omega}{\Omega} = \frac{\Delta\omega \, L}{\pi V}. \tag{5.31}$$

With the expression (5.31) taken into account, the expression for the duration of ultrashort pulses can be represented in the form:

$$\tau \cong \frac{T}{N} \cong \frac{2\pi}{\Delta\omega}. \tag{5.32}$$

So, if the width of the gain loop of a multimode laser is $\Delta\omega \sim 10^{12}$ s^{-1}, then under the condition of locking of its longitudinal modes it is possible to generate light pulses of duration ~ 1 ps.

In order to relate the phases of the longitudinal modes, it is necessary to provide periodic modulation of the resonator parameters with a frequency equal to or a multiple of the frequency difference of neighbouring longitudinal modes. Since this modulation is performed under the influence of external fields which lead to the modulation of the phase or amplitude of the light field in the resonator, it is customary to speak of the active mode-locking. Such modulation can be performed using electro- or acousto-optic modulators placed inside the laser cavity. Suppose that the amplitude of the central mode of the resonator is modulated periodically with frequency Ω. In this case, the field strength of the considered mode can be represented in the form [42]

$$E_0(t) = E_0(1 + \gamma \cos \Omega t) \exp\left[i(\omega_0 t + \varphi_0)\right], \tag{5.33}$$

where γ is the depth of modulation ($0 < \lambda < 1$).

Equation (5.33) can also be written in the form

$$E_0(t) = E_0 \exp[i(\omega_0 t + \varphi_0)] +$$
$$+ \frac{\gamma E_0 \exp(i\varphi_0)}{2}\left\{\exp\left[i(\omega_0 + \Omega)t\right] + \exp\left[i(\omega_0 - \Omega)t)\right]\right\}. \tag{5.34}$$

As follows from (5.34), the periodic modulation of the amplitude of the central mode of the resonator with frequency Ω leads to the appearance, along with radiation at a carrier frequency ω_0, of two additional components at frequencies $\omega_0 + \Omega$ and $\omega_0 - \Omega$. If the modulation frequency is chosen equal to the difference between the frequencies of neighbouring modes excited in the laser, the lateral frequencies $\omega_0 + \Omega$ and $\omega_0 - \Omega$ will coincide with the frequencies of the modes that are adjacent to the central mode. In this case, the second and third terms in (5.34) will play the role of a driving force for these two modes. As a result, these neighbouring modes will be generated in phase with the central mode of the laser. In turn, these modes will influence the modes adjacent to them and so on. In connection with this, as a result of parametric excitation, all the laser modes will become synchronized in a sufficiently short time interval.

In the case when the modulation frequency is chosen equal to the n-fold frequency difference of neighbouring modes, the modes with frequencies differing by an amount exceeding the intermode interval by a factor of n will be synchronized.

The main drawback of this method is that only a relatively small number of modes falling into the gain band of the active medium can be synchronized by modulation at a fixed frequency. This is due to the fact that because of the dispersion in the material of the active medium, the difference between the intermode frequencies changes

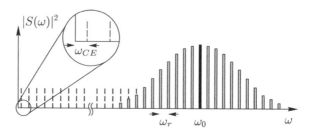

Fig. 38. Frequency representation of a pulse train generated by a continuous mode-locked laser: the repetition rate of the pulses ω_r; carrier frequency ω_0.

somewhat as the frequency of the mode from the centre of the gain band of the active medium is removed.

The change in the difference between the intermode frequencies accumulates from mode to mode with distance from the central mode. As a result, the process of parametric build-up comes out of resonance and terminates long before all the modes falling into the gain band are synchronized. In this connection, active-mode-locked lasers are not able to generate pulses shorter than several tens of picoseconds.

The radiation of a continuously operating laser with mode-locking is characterized by a set of equidistant frequencies (Fig. 38). A generator emitting a set of such frequencies is called a comb-generator. A femtosecond laser operating in the comb-generator mode can be used to solve a number of important applications of metrology and optical communication. However, in the resonators of lasers the pulses propagating in them are subject to the action of dispersion. As a result, a nonlinear dependence of the wave vector of the light pulse on the frequency $k(\omega)$ appears, which can be described using the Taylor series [41]:

$$k(\omega) = \frac{\omega_0}{V_{ph}} + \frac{\omega - \omega_0}{V_{gr}} + \frac{1}{2}\beta_2(\omega - \omega_0)^2 + ..., \qquad (5.35)$$

where V_{ph} is the phase velocity of the pulse; V_{gr} is the group velocity of the pulse; β_2 is the dispersion of group velocity; ω_0 is the frequency of the central mode of the resonator.

The dispersion of radiation in the cavity, in the case of the laser mode-locking process, can lead to its disbalance. To show this, we restrict ourselves to taking into account only the first two terms in equation (5.35):

$$k(\omega) = \frac{\omega_0}{V_{ph}} + \frac{\omega - \omega_0}{V_{gr}}. \tag{5.36}$$

The form of equation (5.36) shows that the light pulse propagates in the medium of the laser resonator with a group velocity different from its phase velocity corresponding to the phase velocity of the central modeller. This leads to an imbalance in the active laser mode-locking.

The dependence of the wave vector of the longitudinal mode with number j supported by a Fabry–Perot resonator of length L can be represented in the form

$$2k(\omega_j)L = 2\pi j. \tag{5.37}$$

Using the expressions (5.36) and (5.37), we can represent the frequency of the j-th longitudinal mode in the form

$$\omega_j = j\frac{2\pi}{2L}V_{gr} + \omega_0\left(1 - \frac{V_{gr}}{V_{ph}}\right). \tag{5.38}$$

Thus, the frequency difference between two adjacent modes is equal to

$$\omega_{j+1} - \omega_j = \frac{2\pi}{2L}V_{gr} \equiv \omega_r. \tag{5.39}$$

The second term in equation (5.38) corresponds to the shear frequency denoted by

$$\omega_{CE} \equiv \omega_0\left(1 - \frac{V_{gr}}{V_{ph}}\right). \tag{5.40}$$

Using the equations (5.38)–(5.40) gives the expression for the frequency of the longitudinal mode with number j:

$$\omega_j = j\omega_r + \omega_{CE}. \tag{5.41}$$

The radiation field in the resonator is represented by expansion to a Fourier series

$$E(t) = \sum_{j=-\infty}^{\infty} S(\omega_j)\exp\left[-i(j\omega_r + \omega_{CE})t\right], \tag{5.42}$$

where $S(\omega_j)$ is the spectral amplitude of the mode.

Then, after a complete cycle through the resonator for a period $T = 2\pi/\omega_r$, the field strength in the resonator will have the form

$$E(t+T) = \sum_{j=-\infty}^{\infty} S(\omega_j) \exp[-i(j\omega_r + \omega_{CE})(t+T)]$$

$$= E(t)\exp\left(-i\left(2\pi j + \frac{2\pi\omega_{CE}}{\omega_r}\right)\right) == E(t)\exp(-i\Delta\varphi),$$

(5.43)

since $\exp(-i2\pi j) = 1$, and

$$\Delta\varphi \equiv 2\pi\frac{\omega_{\tilde{N}E}}{\omega_r}.$$

(5.44)

As can be seen from (5.43), the electric field of each pulse generated by the laser is a copy of the electric field of the previous pulse, the phase of which is shifted by an amount determined by the relation (5.44).

Thus, the radiation of the pulse formed as a result of longitudinal mode-locking of the radiation from a continuous laser is characterized by carrier frequency ω_0 at the centre of the gain band of the active medium. On either side of the frequency ω_0, a periodic comb of cavity modes is formed, separated by a frequency interval ω_r, which specifies the period of the pulse envelope (Fig. 38), which, due to dispersion, shifts along the frequency axis relative to the origin by the value ω_{CE}.

The pulse repetition rate is determined by the time of the bypass of the resonator. The carrier frequency can be shifted relative to the maximum of the envelope of the pulse by an amount ω_{CE}, which leads to a corresponding shift in the phase of the pulse maximum relative to the maximum of its envelope by an amount $\Delta\varphi$. When the pulse duration approaches the period of the carrier frequency (see (5.40) and (5.44)), the dependence of the electric field on time begins to be determined to a large extent by the magnitude of this phase shift. Figure 39 shows two such dependences for pulses of the same shape and duration (5fs), but with different phase shifts between the envelope maximum and the carrier frequency [38]. As can be seen, the dependences of the shape of the envelope pulses are significantly different, in spite of the fact that their energy does not change. This circumstance is not only of fundamental interest, it, as a rule, becomes significant when the pulse duration becomes

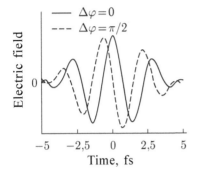

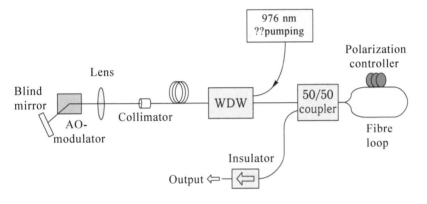

Fig. 39. Dependence of the electric field of the light wave of a pulse on time for a pulse of duration 5 fs at a wavelength of 750 nm with different phase shifts between the carrier and the envelope [38].

Fig. 40. Scheme of a Yb-fibre laser with active mode-locking [43].

very small and approaches 10 fs, since it begins to manifest itself in nonlinear effects.

Figure 40 shows a scheme of a Yb-fibre laser with active mode-locking [43]. Pumping is carried out by a continuous semiconductor laser diode with a wavelength of 976 nm, the radiation of which is introduced through the frequency divider (WDM) into the active fibre. The laser resonator is formed by a blind mirror and a fibre loop 3.55 cm long, containing a 50/50 divider and a polarization controller. The Q-switching of the resonator was carried out by an acousto-optical modulator with a maximum modulation frequency of 40 MHz.

Figure 41 a shows the experimental dependences of the radiation intensity at the exit of the fibre laser, obtained with a pumping power of 188 mW and an acousto-optical modulation frequency of 100 kHz. The pulse repetition period is ~10 µs. As the frequency of the acoustooptic modulator increases to 40 MHz, the fibre laser begins to emit pulses with a duration of ~25 ns (Fig. 41 b). The inset in Fig.

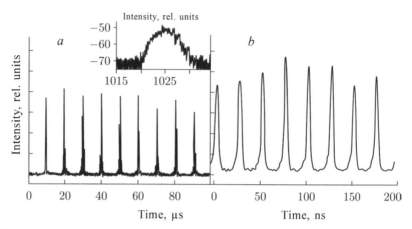

Fig. 41. The time dependences of the radiation intensity at the exit of the fibre laser: *a* – sequence of pulses generated by the laser at a frequency of the AO modulator of 100 kHz; *b* – sequence of pulses at the frequency of the AO modulator of 40 MHz [43].

41 *a* shows the dependence of the emission intensity of the fibre laser on the wavelength. As can be seen, the laser emits at a central wavelength of 1025 nm with a radiation bandwidth of ~3.68 nm.

Passive mode-locking of fibre lasers. Passive mode-locking is carried out in a different scenario [38]. For this purpose, a device is introduced into the laser cavity, which is capable of changing the transmission for the radiation circulating in the resonator, depending on its intensity. The material for such a device is a substance that has a nonlinear transmission in the frequency band coinciding with the gain band of the active medium of the laser. Since the action of such devices does not require external control, this method is called the passive mode-locking. To implement the passive mode-locking, it is important that generation starts immediately on a large number of modes with a fluctuation distribution of radiation intensity characteristic for multimode radiation in time (Fig. 42) [38].

As can be seen from Fig. 42, if the phases of the electromagnetic waves of all modes are not connected in any way, that is, the phase differences of neighbouring modes are randomly distributed, then their interference leads to the fact that the radiation intensity of the laser represents a disorderly set of fluctuation peaks in the interval *T* equal to the time of the complete passage of the laser ray between mirrors (5.28). With successive passages between mirrors, this set of fluctuation pulses has the character of thermal noise, amplified at each passage through an active medium. The average duration of an individual fluctuation of peak intensity τ_f is associated with the width

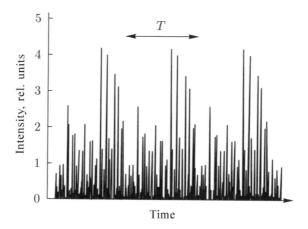

Fig. 42. Dependence of the intensity of multimode laser radiation on time in a disorderly set of excited modes [38].

of the laser radiation spectrum Δv by the relation $\tau_f \sim 1/\Delta v$. The width of the spectrum is determined by the number of longitudinal modes N, the frequency interval between which is $\delta v = 1/T$, i.e. $\Delta v = N\delta v$.

Thus, multimode lasing with a wide spectrum already contains short pulses of light which, because of the random distribution of phases, randomly fill the entire period T, and their intensity is small.

Thus, multimode laser generation with a wide spectrum already contains short pulses of light, which, because of the random distribution of phases, randomly fill the entire period T, and their intensity is small.

Figure 43 shows the scenario for the generation of ultrashort pulses in the case of passive mode-locking. First, the generation of broadband multimode radiation occurs. When multimode laser radiation passes through an active medium and a device with nonlinear transmission, its temporal intensity profile changes. Weaker fluctuation pulses are weakened by a nonlinear absorber more strongly than the more intense ones. As a result of the combined action of the active medium and the nonlinear absorber, successive passages of radiation between the resonator mirrors discriminate fluctuational pulses in intensity. In the end, a single ultrashort pulse remains in the resonator in the period T. Its shape will change due to dispersion and nonlinear effects in the interaction of laser radiation with matter inside the resonator. Thus, in the generation of ultrashort pulses with the use of the passive mode-locking method, two stages of the process can be distinguished: the formation of a single pulse

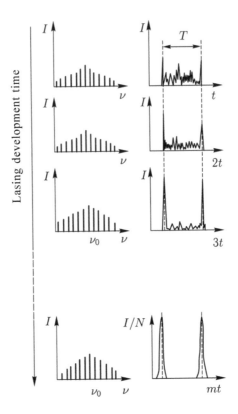

Fig. 43. The scenario for the development of USP generation in time (in terms of the number of passes in resonator *m*), when lasing starts immediately on all modes with arbitrary phases, and then, as it develops, the phases of all modes are synchronized.

from fluctuation pulses due to their nonlinear discrimination and the acquisition of the final form and duration.

It should be noted that the mechanism for the formation of ultrashort pulses in a passive mode-locked laser is very sensitive to the initial operating conditions of the laser and requires careful removal from the resonator of any elements capable of discriminating longitudinal modes.

Thus, the main part of the laser with passive mode-locking is a nonlinear optical element capable of decreasing the losses with increasing intensity of the radiation passing through it – the saturable absorbers (SA). Established inside the laser resonator, the SA produces a sufficiently rapid amplitude self-modulation of the radiation losses the magnitude of which depends on the intensity of the radiation passing through the SA. The dependence of the radiation

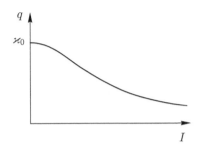

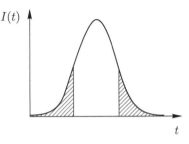

Fig. 44. The type of the dependence of the loss factor of the SA on the intensity of the radiation incident on it.

Fig. 45. Illustration of the action of the SA, leading to a cutoff of the leading and trailing edges of the pulse.

loss coefficient on the radiation intensity $q(I)$ can be approximately described by the equation [44]

$$q(I) = \frac{\varkappa_0}{1 + \beta I},\qquad (5.45)$$

where $\varkappa_0$ is the coefficient of SA losses in the absence of radiation; β is the coefficient of proportionality.

A graph illustrating the form of the dependence of the SA loss factor is given in Fig. 44.

As a result, a 'window' of positive amplification is formed on the time dependence of the total laser gain, which coincides with the most intense pulse. The width of this window depends not only on the duration of the pulse forming it, but also on the characteristic relaxation times of the amplification of the active medium and the state of the amplifier [45].

The role of a saturated absorber in passive mode-locking devices is always twofold [38]. In the initial phase of the generation of light pulses it plays the role of a discriminator, providing the isolation of a single pulse. Later, the shape of the pulse changes, since in successive passes through the saturable absorber the front and rear edges of the pulse are cut (Fig. 45), which causes a decrease in its duration.

A significant role in the change in the shape of the pulse is played by the temporal characteristics of the saturable absorber. It is customary to distinguish between fast (dozens of femtoseconds) and slow (tens of picoseconds) SAs, for which there is a different nature of their effect on the duration of the generated pulses 0 from the losses in the SA [44]:

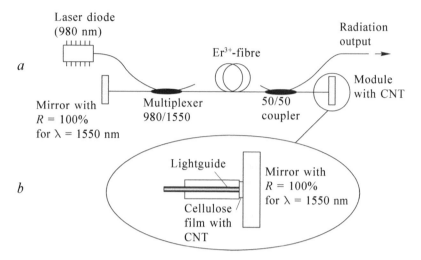

Fig. 46. *a* – diagram of an erbium fibre laser; *b* – scheme of a saturable absorber [46].

$$\text{for fast SA: } \tau \sim \left(\varkappa_0\right)^{-1/2}$$
$$\text{for slow SA: } \tau \sim \left(\varkappa_0\right)^{-1/4}. \tag{5.46}$$

When the pulse duration becomes sufficiently small (less than one picosecond), the dispersion of the group velocity and the effects of nonlinear-optical self-action of the pulse begin to affect the shape and duration of the pulse: self-focusing and phase self-modulation [38, 44, 45].

For illustration, Fig. 46 shows a scheme of an erbium fibre laser with a saturable absorber [46]. The pumping of the erbium active light guide was carried out with a laser diode with a wavelength of 980 nm the radiation of which was introduced into the light guide by means of a spectral multiplexer. The radiation emitted from the erbium fibre passed through a coupler, which provided a 50% radiation yield; The rest of it fell on a saturable absorber.

As a saturable absorber, films of optical quality based on carboxymethyl cellulose (CMC) and single-walled carbon nanotubes embedded in them were used. Single-walled carbon nanotubes (CNT) are a new nanomaterial with remarkable physical properties, the list of which has recently been supplemented by the detected ability of the CNT to saturate the absorption of laser radiation with wavelengths in the near-IR range.

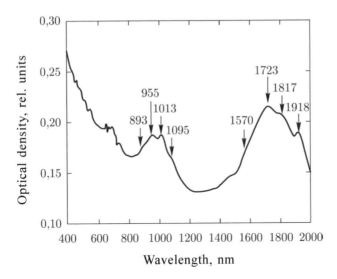

Fig. 47. Absorption spectrum of a film of carboxymethyl cellulose containing CNT [46].

Figure 47 shows the spectrum of optical absorption in a wide spectral range of a CMS film with CNT dispersed in it. The absorption spectrum corresponding to the transitions between the first, second, etc. symmetric van Hove singularities in the density of electronic states of the nanotube ensemble is clearly visible on the spectrum. The width of the bands is determined by the dispersion of the nanotube distribution along the diameters. The saturation absorption was 2.6% at a peak radiation intensity of 2 MW/cm^2. It was these films that were used to realize a continuous mode of mode-locking in erbium fibre lasers (Fig. 46). Figure 46 *b* shows the design of a device that combines a resonator mirror and a saturable absorber.

The emission spectrum of an erbium fibre laser with an active lightguide length of 90 cm is shown in Fig. 48. One end of the fibre was pressed against a mirror with 100% reflection at $\lambda = 1550$ nm, and the other to the same mirror, but through a CMC film containing CNT.

With a pumping power of 40 mW, the output power of the fibre laser reached 4 mW with a repetition rate of 28.5 MHz. The pulse duration was 470 fs with a spectral width of 11 nm.

If the blind mirror is replaced with a Bragg diffraction grating with an 85% reflection coefficient (Fig. 49), then it becomes possible

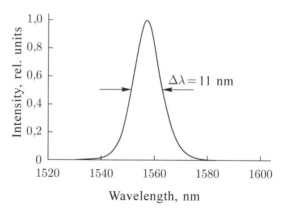

Fig. 48. The emission spectrum of an erbium-doped fibre laser [46].

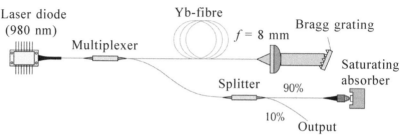

Fig. 49. Scheme of a Yb-fibre laser with tuning of the emission wavelength [46].

to reconstruct the laser radiation frequency within the gain contour of the active fibre by simply scanning the grating angle.

As shown in Ref. [46], 13% nonresonant absorption modulation provides synchronization of the modes of the fibre laser, which generates a sequence of pulses of 15 ps duration with a frequency of 16 MHz (Fig. 50).

The rotation of the grating allowed obtaining a stable passive synchronization of the laser modes and providing a smooth tuning of the radiation wavelength in the range 1045–1065 nm (Fig. 51).

The Yb-fibre laser was pumped by the continuous emission of a 30 mW semiconductor laser with a wavelength of 0.75 nm, and the power of the generated pulses were 0.5 mW.

Passive mode-locking of a fibre laser using a semiconductor saturable absorber mirror (SESAM). Since 1992 the passive mode-locking in lasers has been carried out using semiconductor saturable absorbers in combination with Bragg mirrors that have a high reflection coefficient [47]. In the foreign and domestic literature, the abbreviation SESAM is used to denote these devices. The use

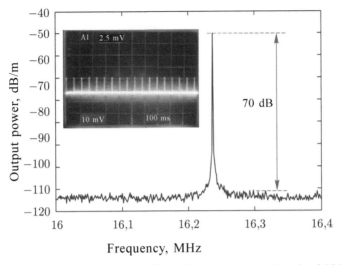

Fig. 50. The spectrum of a single pulse with a central wavelength of 1064 nm and a photograph of the oscillogram of a pulse train generated by a Yb-fibre laser [46].

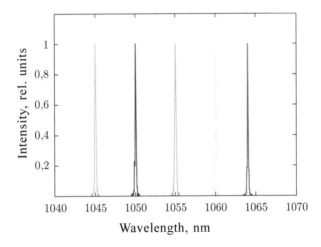

Fig. 51. Tuning of the emission wavelength of an Yb-fibre laser upon rotation of the diffraction grating [46]

of SESAM allows generating both pulses of nanosecond duration in the Q-switching mode and pulses with a duration from several picoseconds to tens of femtoseconds in the passive locking mode. As saturable absorbers, SESAM uses semiconductor materials with a wide absorption band: from visible to middle infrared wavelengths. In the process of manufacturing the SESAM, due to the variation in the structure of the semiconductor and the parameters of its growth,

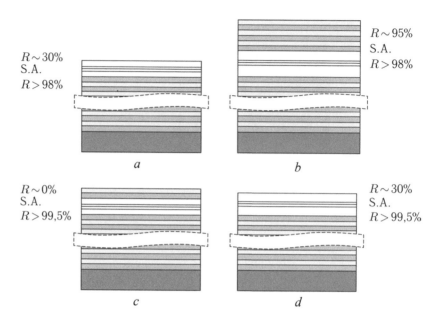

$R \sim 30\%$
S.A.
$R > 98\%$

$R \sim 95\%$
S.A.
$R > 98\%$

$R \sim 0\%$
S.A.
$R > 99,5\%$

$R \sim 30\%$
S.A.
$R > 99,5\%$

a   *b*

c   *d*

Fig. 52. Types of SESAM with semiconductor Bragg mirrors and a saturable absorber [44].

it is possible to control the relaxation time, the range of operating wavelengths, the nonlinear modulation of the reflection coefficient, and the saturation energy density [48].

The SESAM consists of a semiconductor Bragg mirror, and several layers of a saturable absorber, using different combinations of semiconductor layers, as shown in Fig. 52. The material from which the Bragg mirror is made has a wide forbidden zone, as a result of which absorption is not observed in it. As a rule, the Bragg mirror consists of alternating layers of AlAs and GaAs or GaInAs and InP [44]. The saturable absorber (SA) is a multilayer structure composed of nanometer thicknesses of GaInAs and GaInNAs semiconductor layers, for which the effect of quantization of energy levels is observed, as in a one-dimensional potential well, and located at the maximum of the electric field of the reflected laser radiation.

A Fresnel reflection is observed at the boundary of the semiconductor structure with air which together with the reflection from the Bragg mirror creates a resemblance of the Fabry–Perot resonator. Such a design has a low quality factor, since the typical Fresnel reflection is of the order of 30% (Fig. 52 *a*). As an example, Fig. 53 shows the spectral dependence of the reflection coefficient for the SESAM of this type, the absorber in which is a quantum well

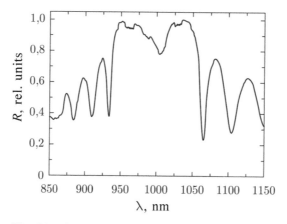

Fig. 53. The reflection spectrum of SESAM [44].

composed of five GaInAs and GaInNAs layers deposited on a Bragg mirror consisting of 26 pairs of AlAs/GaAs quarter-wave layers [44]. As can be seen, the working region of such an absorber lies in a rather wide range of wavelengths from 950 to 1050 nm.

To increase the quality factor, a semiconductor or dielectric protective layer can be applied to the surface of the absorbing layer, which also plays the role of a second mirror in the resonator with a reflection coefficient of 95%, as shown in Fig. 52 *b*.

With this design, the SESAM can work in two modes: resonant and nonresonant. Working in a resonance mode, the SESAM has a lower saturation energy density and a higher modulation depth at a narrower band of operating wavelengths.

Figure 52 *c* shows the SESAM with a non-reflective coating (Anti-reflection Coated Fabry-Perot Saturable Absorber: AR-FPSA) [88]. Such a saturable absorber operates in the nonresonant mode, while the non-reflective coating increases the depth of modulation, and also allows to reduce the saturation threshold.

To compensate for dispersion, SESAM saturable absorbers have a chirped Bragg mirror (see section 5.4.2.4) (see Fig. 52).

The absorption saturation in SESAM is due to the limited lifetime of the excited non-equilibrium carriers, as well as a certain time of thermalization, i.e., the establishment of an equilibrium temperature distribution in different zones of the absorbing layer. The presence in the SESAM of two different components of the relaxation time due to intraband thermalization and carrier capture and/or their recombination positively influences the establishment of the passive mode-locking. Due to the slow relaxation time, the value of the saturation energy flux decreases, thus improving the ability of the SESAM to initiate the passive mode-locking. The presence of a fast relaxation time allows the formation of high-quality ultrasonic pulses.

When choosing a saturable absorber, it is necessary to choose a compromise relaxation time, since the use of a fast saturable absorber can lead to an increase in the threshold or even the impossibility of initiating the passive mode-locking.

To effectively operate a slow saturable absorber, its relaxation time should be shorter than the lifetime of the metastable level of the active medium. This condition is satisfied in the case of optical fibres doped with rare-earth elements. The duration of the pulses generated in a passive mode-locked laser using a relatively slow saturable SESAM absorber is determined from equation [49]:

$$\tau \approx \frac{2\pi C}{\Delta\omega} \sqrt{\frac{g}{\Delta R}}, \tag{5.47}$$

where $\Delta\omega$ is the width of the gain band, g is the gain factor of the medium (which is equal to the total loss of the resonator), ΔR is the modulation depth of the saturable absorber, and C is the empirical coefficient that depends on the operating mode of the SESAM.

It should be noted that when a slow saturable absorber is used, pulses of a duration much shorter than the relaxation time can be generated. This is due to the compression of the trailing edge of the pulse due to the saturation of the active medium and the establishment of an optimal balance of losses and amplification in the laser cavity.

The absorption saturation effect in the SESAM due to the third-order nonlinear susceptibility leads to an insignificant modulation of the reflection coefficient, which is sufficient to trigger the passive Q-switching of the laser. Figure 54 shows a typical dependence of the reflection coefficient of the SESAM on the energy density of the incident radiation [44].

As can be seen, the main parameters characterizing the work of the SESAM are: the saturation energy density F_{sat}; amplitude of modulation of the reflection coefficient ΔR; the value of unsaturated losses ΔR_{ns} and the value of the saturating reflection coefficient R_{ns}. The nonlinear dependence of the reflection coefficient of the SESAM can be represented by the expression [44]

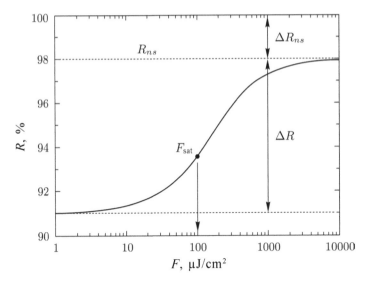

Fig. 54. Dependence of the reflection coefficient of the SESAM on the energy density of the laser pulse.

$$R(F) = 1 - R_{ns} - \Delta R \left[\frac{1 - \exp(-F / F_{sat})}{F / F_{sat}} \right]. \tag{5.48}$$

With low energy fluxes, SESAM operates in the radiation absorption mode, and with an increase in the energy density of the laser pulse a saturation of the absorption occurs in the resonator, which leads to a decrease in losses and an increase in the reflection coefficient of the radiation. Further rapid relaxation of the excited state of the absorber returns the SESAM to its original state. This effect leads to the launch of the mode of passive synchronization of laser modes.

The work of a fibre laser using the SESAM is illustrated using an example of an ytterbium fibre laser [50].

The scheme of the ytterbium fibre laser with the SESAM modulator of the quality factor is shown in Fig. 55 [50]. The laser is pumped by a laser diode at a wavelength of 915 nm. The fibre laser resonator is formed by a mirror with 100% reflectance and a semiconductor saturable mirror (SESAM). Based on GaInNAs, the SESAM operates in the spectral range of 940–1050 nm and consists of 26 pairs of AlAs and GaAs quarter-wave layers deposited on an *n*-type GaAs (001) substrate that provides Bragg reflection with a central wavelength of 1000 nm. The SESAM has a saturation energy

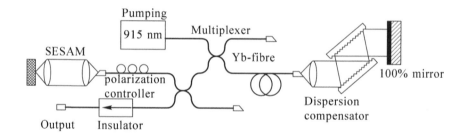

Fig. 55. Scheme of a Yb-fibre laser with an SESAM Q-switch [50].

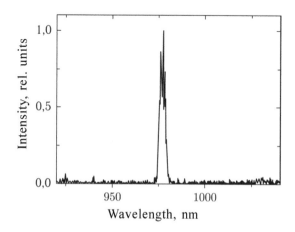

Fig. 56. The emission spectrum of a Yb-fibre laser with an SESAM Q-switch [50].

density of 3 μJ/cm² and a modulation amplitude of the reflection coefficient $\Delta R \sim 12\%$. A two-grid pulse dispersion compensator is also placed in the laser cavity (see section 5.4.2.2.).

Figure 56 shows the emission spectrum of a Yb-laser with an active lightguide length of 40 cm. As can be seen, the laser emits electromagnetic waves in a fairly narrow spectral range with a central frequency of 980 nm.

Mode-locking in the laser begins when the pump radiation power reaches 40 mW. With a pumping power of 50 mW, the laser emits pulses at a central wavelength of 980 nm with a frequency of 30 MHz, a duration of 2.3 ps and an energy of ~0.1 nJ.

5.4.2. Compensation of dispersion spreading pulses

The phenomenon of dispersion interferes with the production of extremely short pulses. Dispersion leads to a different speed of

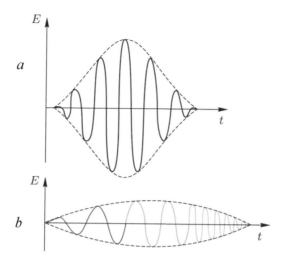

Fig. 57. Illustration of the phenomenon of dispersion 'spreading' of a laser pulse in a medium: a – initial pulse shape; b – pulse shape after propagation in a medium with dispersion.

propagation of light waves with different wavelengths (frequencies) of the wave. Due to the difference in the propagation velocities, the shape of the pulse changes and its duration increases (Fig. 57).

In addition, those wavelengths (frequencies) will appear at the leading edge of the pulse for which the propagation velocity in the medium will be greater, and on the trailing front – waves for which the speed of light is smaller. Therefore, a linear change in frequency from time to time occurs within a pulse with an increased duration due to phase self-modulation (PSM). This phenomenon of linear frequency modulation is called the chirp [38].

Two approaches to describing the process of dispersion, which leads to a broadening of the pulses in the optical fibres, are usually distinguished. The group velocity V_g of a mode propagating in a fibre with a propagation constant is defined as [43]

$$V_g = \left(\frac{d\beta}{d\omega}\right)^{-1} = c\left(\frac{d\beta}{dk}\right)^{-1}, \qquad (5.49)$$

where ω is the cyclic radiation frequency; k is the wave number, and c is the speed of light in a vacuum.

On the basis of this, when propagating with an optical fibre with a length L, the group delay is

$$\tau_g = \frac{L}{V_g} = \frac{L}{c}\frac{d\beta}{dk} = -\frac{L\lambda^2}{2\pi c}\frac{d\beta}{d\lambda}. \tag{5.50}$$

where λ is the wavelength of the radiation in a vacuum.

Suppose that the light pulse at the entrance to the optical fibre has a spectral width $\Delta\omega$, then the broadening of the pulse during propagation in the fibre will be

$$\Delta\tau = \left|\frac{d\tau_g}{d\omega}\right|\Delta\omega = \frac{d}{d\omega}\left(\frac{L}{V_g}\right)\Delta\omega = L\frac{d^2\beta}{d\omega^2}d\omega = L\beta_2\Delta\omega, \tag{5.51}$$

where the coefficient β_2 is called the group velocity dispersion (GVD), defined as

$$\beta_2 = \frac{d^2\beta}{d\omega^2}. \tag{5.52}$$

This coefficient is usually measured in [ps^2/km].

Expression (5.52) for the dispersive spreading of the pulse can be represented differently, using the concept of the wave interval of the light pulse $\Delta\lambda$:

$$\Delta\tau = \left|\frac{d\tau_g}{d\lambda}\right|\Delta\lambda = \frac{d}{d\lambda}\left(\frac{L}{V_g}\right)\Delta\lambda = LD\Delta\lambda, \tag{5.53}$$

where D is the dispersion group delay (GDD), determined by

$$D = \frac{d}{d\lambda}\left(\frac{1}{V_g}\right). \tag{5.54}$$

This coefficient is usually measured in [ps/nm·km].

The relationship between the dispersion group velocity and the dispersion group delay is given by

$$D = -\frac{2\pi c}{\lambda^2}\beta_2. \tag{5.55}$$

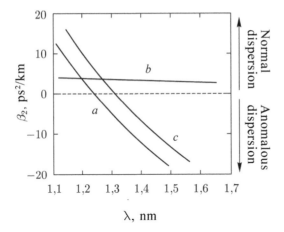

Fig. 58. Contribution of material (*a*) and waveguide (*b*) dispersion to the resulting (*c*) group velocity dispersion in a single-mode optical fibre [44].

The normal dispersion of the medium corresponds to the case when $D < 0$, that is, when $\beta_2 > 0$. Therefore, such a dispersion of the medium, when the GVD is positive, is also called the positive or normal dispersion. When $D < 0$, which corresponds to a negative GVD value: $\beta_2 < 0$ and we are talking about a medium with the negative or anomalous dispersion.

Along with material dispersion, optical waveguides have a waveguide or intramode dispersion of radiation due to the difference in the phase velocities of waves propagating in the fibre within the same mode [43].

Figure 58 shows the dependences of the GVD for single-mode quartz fibre waveguides due to the material and waveguide dispersion, and the resulting dependence on the emission wavelength [44]. As can be seen, the GVD, due to the material dispersion, is positive for the spectral region with wavelengths less than 1.2 µm. The GVD associated with the waveguide dispersion remains positive for all wavelengths. As a result, the total GVD is positive for wavelengths less than 1.3 µm, and turns negative for wavelengths exceeding 1.3 µm. Thus, when optic fibres are used as resonators of fibre lasers, ultrashort radiation pulses must be either provided in the region of small values of the anomalous dispersion or be able to control the dispersion of the pulse.

The decrease in the duration of the ultrashort pulse requires a transition to the region of higher negative GVD values. According to (5.53), if $D > 0$ then the pulse broadens. Therefore, in order to control the duration of the ultrashort pulse, it is necessary to have devices to control the group velocity dispersion of the pulses. Due to the dispersion, the pulse spectrum is divided into fragments that propagate with different group velocities, which results in the spreading of the pulse. In the region of normal dispersion ($\beta_2 > 0$), the long-wave components move faster than the short-wavelength components (this corresponds to a positive frequency modulation of the pulse) (Fig. 58 *a*), and for anomalous dispersion ($\beta_2 > 0$) the components change places.

To perform compression, the light pulse, it is necessary to inform the linear frequency modulation with the opposite sign, as in the case of a spreading. Indeed, if the instantaneous pulse frequency decreases from the beginning to the end of the pulse, then in a medium with normal dispersion ($\beta_2 > 0$), the head part of the pulse containing high-frequency spectral components will propagate more slowly than the tail part. As a result, the head and tail parts of the pulse begin to approach and the compression of the frequency-modulated pulse occurs in the dispersive medium. A simplified mathematical description of this process is given in [51].

If a laser pulse with an initial duration T_0 propagates in a dispersive medium with a quadratic phase-time dependence

$$\varphi_0 = \frac{\epsilon \omega_0 t^2}{2 \dot{Q}_0^2},$$ (5.56)

where $\epsilon = \pm 1$ determines the sign of the frequency modulation, $\Delta \omega_0$ is the full change of the frequency within the pulse, then the pulse duration changes with the distance z travelled according to the expression [51]

$$T(z) = T_0 \left[\left(1 + \text{sign}\left(\epsilon \beta_2 \right) \frac{\Delta \omega_0 T_0 z}{L_d} \right)^2 + \frac{z^2}{L_d^2} \right]^{1/2},$$ (5.57)

where L_d is the length of dispersion spreading of the pulse:

$$L_d = \frac{T_0^2}{2|\beta_2|}.$$ (5.58)

As can be seen from (5.57), the compression or decrease of the pulse duration in the dispersion medium occurs under the condition $\epsilon\beta_2$ <0.

This is in good agreement with the above qualitative arguments on the mechanism of dispersion compression of a laser pulse. One of the most important results achieved in the field of ultrashort pulse generation was the creation of devices capable of purposefully changing the GVD and controlling the frequency chirp. To this end, it has been proposed to use prisms, diffraction gratings, special interferometers and interference dielectric mirrors.

5.4.2.1. Prism compensators of group velocity dispersion

Figure 59 shows the diagram of a prism regulator of the GVD. The glass prism as a dispersive medium exerts a different retarding effect on the light of different wavelengths, since the refractive indices characterizing the degree of this deceleration and the decrease in the speed of light are different for radiation of different frequencies, i.e. different colours. Due to this, the glass prism, used as a dispersing element, decomposes white light into the iridescent strip.

Hence, if a pulse in which the frequency will change linearly is artificially created (say, the leading front will be 'painted' in blue, and the rear in red) (Figs. 57 and 59 *a*), then when passing through the prism delay line the optical pulse will contract, since the head 'blue' part of the pulse in a medium with a normal dispersion moves with a smaller group velocity than the tail 'red', which, as it moves, more and more catches the front, which leads to a shortening of the. If, on the contrary, the carrier frequency in the pulse is linearly increased (the head part of the pulse contains low frequencies, and the tail part is high), then, when a pulse is passed through a delay line with an anomalous dispersion, where the blue waves move faster

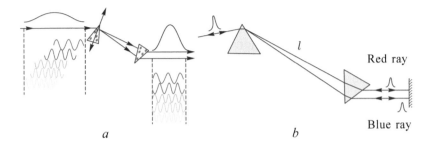

a   *b*

Fig. 59. Diagram of prism regulator of the GVD in the laser. *a* – a prism pair system to compensate the GVD; *b* – compensation of the transverse shift of the rays in the prism device.

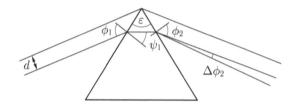

Fig. 60. The path of beams in the prism.

than the red ones, the tail high-frequency part of the pulse will catch up with the front high frequency, causing a pulse shortening. When pulse radiation passes through the prism (Fig. 60), the laser beam is oriented so that it passes as close as possible to the region of the top of the prism, in order to reduce the influence of the intra-prism dispersion of the prism material. The dispersion properties of the prism were calculated using the technique described in [51]. Using (5.53), the expression for the dispersion of group delay (GDD) can be represented as:

$$D = -\frac{d\tau_g}{d\lambda} = -\frac{\lambda}{cL}\frac{d^2 L_n}{d\lambda^2},$$ (5.59)

where L is the geometric length of the light path in the medium; L_n is the optical length of the light path in the medium; c is the speed of light in a vacuum.

As can be seen from Fig. 60, the optical path length of the radiation in the prism can be expressed by the following relationship:

$$L_n(\lambda) = \frac{dn_\lambda \sin \varepsilon}{\cos\phi_1 \cos(\varepsilon - \psi_1)},$$ (5.60)

where d is the beam width; n_λ is the refractive index of the prism material for a given radiation wavelength; The refracting angle of the prism; ϕ_1 is the angle of incidence of radiation on the prism; ψ_1 is the angle of refraction of the radiation in the prism.

In expression (5.60), the quantities n_λ and ψ_1 are variables, and

$$\psi_1 = \arcsin\left(\frac{\sin\phi_1}{n_\lambda}\right),$$ (5.61)

and all other values are constants.

Differentiation of (5.60) twice on λ gives the following expression [52]:

$$\frac{d^2 L_n}{d\lambda^2} = \frac{d\sin\varepsilon}{\cos\phi_1}\{\frac{1}{\cos(\varepsilon-\psi_1)}\left[1+\frac{\mathrm{tg}(\varepsilon-\psi_1)\sin\phi_1}{\sqrt{n_\lambda^2-\sin^2\phi_1}}\right]\frac{d^2 n_\lambda}{d\lambda^2}$$

$$+\frac{\sin\phi_1^2}{n_\lambda\cos(\varepsilon-\psi_1)\left(n_\lambda^2-\sin^2\phi_1\right)}\times \qquad (5.62)$$

$$\times\left[\frac{1}{\cos^2(\varepsilon-\psi_1)}+\mathrm{tg}^2(\varepsilon-\psi_1)-\frac{\mathrm{tg}(\varepsilon-\psi_1)\sin\phi_1}{\sqrt{n_\lambda^2-\sin^2\phi_1}}\right]\left(\frac{dn_\lambda}{d\lambda^2}\right)^2\}.$$

For the case of a radiation drop on a prism at the Brewster angle, expression (5.62) is simplified, since $\mathrm{tg}\phi_1 = n_\lambda$, and $\psi_1 = \varepsilon/2$:

$$\frac{d^2 L_n}{d\lambda^2} = \frac{2d\left(n_\lambda^2+1\right)}{n_\lambda^2}\left[\frac{d^2 n_\lambda}{d\lambda^2}+\frac{1}{n_\lambda^3}\left(\frac{dn_\lambda}{d\lambda^2}\right)^2\right]. \qquad (5.63)$$

As can be seen from Fig. 59 *a*, in a two-prism compensator the first prism not only introduces dispersive blurring into the pulse, but also carries out the angular separation of its frequency components. At the output of the first prism, the radiation emerges diverging at an angle $\Delta\phi_2$, since it undergoes a wave dispersion in the prism (Fig. 60). According to [51], the radiation characterized by an angular divergence of $\Delta\gamma_1$ and a spectral range $\Delta\lambda$, after passing through the first prism has an angle of deviation from the axis, given by the expression:

$$\Delta\phi_2 = K_1\Delta\gamma_1 + K_2\Delta\lambda, \qquad (5.64)$$

where K_1 and K_2 are the coefficients.

For a paraxial approximation, $\Delta\gamma_1 \approx 0$, so after exiting the first prism, the divergence angle of the radiation will be equal to

$$\Delta\phi_2 \approx K_2\Delta\lambda. \qquad (5.65)$$

After passing through the second prism, the deflection angle will be:

$$\Delta\phi_2' = K_1\Delta\gamma_2 + K_2\Delta\lambda, \tag{5.66}$$

but since $\Delta\gamma_2 = -\Delta\phi_2 = -K_2\Delta\lambda$, then $\Delta\phi_2' = 0$. That is, the dispersion divergence of the rays is compensated. Consequently, the radiation from a laser pulse undergoes a GVD only when passing between the spaces between prisms spaced apart by a distance l (Fig. 59 b). As shown in [51], when the radiation falls on the prism at the Brewster angle, the expression for the second derivative of the optical path of radiation as it passes between the prisms has the form [52]:

$$\frac{d^2L_n}{d\lambda^2} = 2l$$

$$\left\{ \left[\frac{d^2n_\lambda}{d\lambda^2} + \left(2n_\lambda - \frac{1}{n_\lambda^3} \right) \left(\frac{dn_\lambda}{d\lambda} \right)^2 \right] \times \sin\phi_2 - 2 \left(\frac{dn_\lambda}{d\lambda} \right)^2 \cos\Delta\phi_2 \right\}. \tag{5.67}$$

Since the angle $\Delta\phi_2$ is vanishingly small, then $\sin\Delta\phi_2 \sim 0$, and $\cos\Delta\phi_2 \sim 1$; hence the expression (5.67) reduces to the form

$$\frac{d^2L_n}{d\lambda^2} \approx -4l \left(\frac{dn_\lambda}{d\lambda} \right)^2. \tag{5.68}$$

It follows from the expressions (5.55), (5.56), and (5.68) that the GVD introduced by the propagation of the pulses between the two prisms is always negative.

Thus, two mechanisms operate in the prismatic compensator of the GVD: a positive GVD pulse in the prism material and a negative GVD in the space between the prisms. This makes it possible to smoothly regulate the GVD by changing the distance between the prisms and performing their parallel shift with respect to each other.

In the prism compensator, a spatial shift occurs in the different spectral regions of the pulse. To compensate for it, the reverse reflection from the mirror placed behind the second prism is used, as shown in Fig. 59 b.

5.4.2.2. Grating compensator of the group velocity dispersion

The prism regulator of the GVD value has a limited field of use since it requires very large prisms to compress highly stretched pulses [38]. More effective for this purpose is the GVD regulator based on a pair of reflective diffraction gratings (Fig. 61) [51].

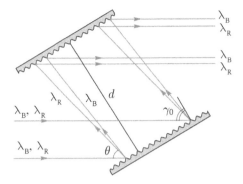

Fig. 61. The regulator of the group velocity dispersion [44].

The light pulse incident on the diffraction grating splits into a spectrum, and the spectral components propagate at different diffraction angles. Then these components fall on the second diffraction grating and, being reflected from it, go parallel to the initial impulse.

As can be seen from Fig. 61, the spectral components with a shorter (λ_B – blue) wavelength pass in the compensator a smaller path than the spectral components with a longer (λ_R – red) wavelength. Therefore, different frequency components have different time delays. The path length for radiation with a wavelength between two gratings can be calculated as follows [44]:

$$\Delta L(\lambda) = d \frac{1 + \cos\theta}{\cos(\gamma_0 - \theta)}, \qquad (5.69)$$

where d is the distance between the lattices; γ_0 is the angle of incidence, and θ is the diffraction angle

$$\theta(\lambda) = \gamma_0 - \arcsin\left(\frac{\lambda}{\Lambda} - \sin\gamma_0\right), \qquad (5.70)$$

where Λ is the period of the diffraction grating.

Using (5.50), (5.51), (5.69), and (5.70), we obtain the following expression [44] for the GVD coefficient:

$$\beta_2 = -\frac{\lambda^3 d}{\pi c^2 \Lambda^2}\left[1 - \left(\frac{\lambda}{\Lambda} - \sin\gamma_0\right)^2\right]^{-3/2}, \qquad (5.71)$$

where c is the wavelength of light in a vacuum.

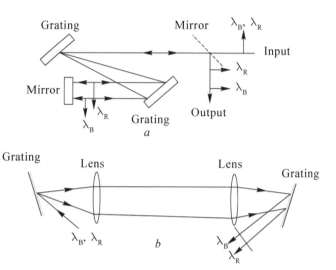

Fig. 62. *a* – compensation of transverse ray shift in the grating compensator of GVD; *b* – scheme of a grating with a positive GVD.

As follows from (5.71), a pair of diffraction gratings introduces an anomalous dispersion the value of which linearly depends on the distance between them and therefore can easily be controlled by their simple displacement. Modern technologies allow the production of gratings with linear distances between them up to 1 m.

As can be seen from Fig. 61, the various spectral components of the pulse shift in the transverse direction. This defect is eliminated in a two-pass scheme in which the laser pulse passes through the compensator twice (Fig. 62 *a*).

The prism regulators of the GVD allow to realize both negative and positive dispersion. To obtain a positive GVD a telescopic system consisting of two lenses is placed between the gratings (Fig. 62 *b*). The telescopic system turns the image so that as a result the path for red waves becomes shorter than for the blue ones.

In practice, a system of gratings with a telescope is used to stretch the ultrashort pulse, and the usual pair of gratings is used to compress the chirped pulse.

5.4.2.3. Compensator for group velocity dispersion based on the Gires–Tournois interferometer

The Gires–Tournois interferometer is a plane-parallel transparent plate, one of the faces of which has a high reflection coefficient ($r_0 = 1$), and the other has a lower reflection coefficient ($|r_1| < 1$) (Fig. 63).

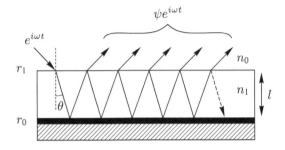

Fig. 63. Diagram of the Gires–Tournois interferometer.

The laser pulse introduced into the interferometer is reflected from the structure with the amplitude ψ because of the partial transmission of the upper surface of the interferometer.

The complex reflection coefficient for an interferometer is

$$r = -\frac{r_1 - e^{-i\delta}}{1 - r_1 e^{-i\delta}},$$ (5.72)

where r_1 is the complex amplitude of the reflection coefficient from the first surface,

$$\delta = \frac{4\pi}{\lambda} n_1 l \cos\theta,$$ (5.73)

where n_0 is the refractive index of the environment; n_1 is the refractive index of the plate material; l is the thickness of the plate; θ is the angle of refraction; λ is the wavelength of the radiation in a vacuum.

As noted in [44], the group delay for a propagating pulse successively reflected from the planes of the Gires–Tournois interferometer is

$$\tau_g = \frac{t_0 \left(1 - r_1^2\right)}{\left(1 + r_1^2\right) - 2 r_1 \cos\omega t_0} = t_0 \frac{1 + r_1}{1 - r_1} \frac{1}{1 + \frac{4 r_1}{\left(1 - r_1\right)^2} \sin^2\left(\frac{\omega t_0}{2}\right)},$$ (5.74)

t_0 is the time required for the pulse to fully travel between the upper and lower planes of the interferometer:

$$t_0 = \frac{2 n_1 l \cos\theta}{c}.$$ (5.75)

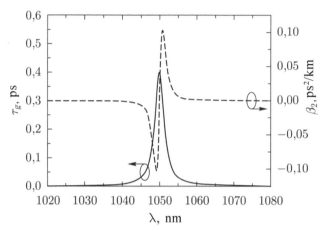

Fig. 64. Calculated dependences for group delay and GVD of the Gires–Tournois interferometer [44].

Using (5.51), (5.74), and (5.75), we can calculate the GVD for the laser pulse. Figure 64 shows the results of calculating the group delay and GVD for the Gires–Tournois interferometer with the following parameters: $r_1 = 0.9$; $n_1 = 3$ and $l = 2.1$ μm, demonstrating the presence of regions of both positive and negative GVDs, which can be used to compensate for GVD of the laser pulses.

Apparently, the feature of the Gires–Tournois interferometer is the presence of a strong dependence of the GVD on the wavelength of the radiation. This dependence is also periodic, as is the case for multi-beam interference.

5.4.2.4. Compensators of group velocity dispersion based on chirped Bragg mirrors

The Bragg mirrors, also called distributed Bragg reflectors or Bragg gratings, are a structure that consists of an alternating sequence of layers of two different optical materials (Fig. 65). The most commonly used design is a quarter-wave mirror, where the thickness of each optical layer corresponds to a quarter of the wavelength for which a mirror is designed. The last condition works for normal angles of incidence (90°). The principle of the Bragg mirror is based on the fact that a Fresnel reflection occurs at each boundary between two transparent materials with different refractive indices. For the wavelength for which a mirror is designed, the difference in the length of the optical path for the rays reflected in each successive layer differs by half the wavelength. Therefore, all waves reflected from the surfaces interfere, resulting in a strong reflection. The achievable reflectivity depends on the number of paired layers and on the refractive indices of neighboring layers. The width of the reflection band depends mainly on the contrast.

Figure 65 shows the distribution of layers in a Bragg grating consisting of 8 pairs of alternating layers of titanium oxide and silicon oxide. The dependences *2* and *3* show the distribution of the intensity of the radiation incident on the Bragg grating with wavelengths of 1000 and 800 nm, respectively. As can be seen, radiation with different wavelengths penetrates the Bragg mirror in different ways.

It is noteworthy that the intensity of the field outside and inside the mirror oscillates because of the interference of incident and reflected waves propagating toward them. In this case, the radiation with a wavelength of 800 nm is mainly concentrated in the depth of the Bragg grating and can freely pass through the layers of the mirror, and the radiation with a wavelength of 1000 nm practically does not penetrate into the depth of the grating.

Figure 66 shows reflection and chromatic dispersion as a function of the wavelength of the radiation incident on the Bragg mirror. The reflectance of the Bragg mirror has a sufficiently extended maximum, and the dispersion is small in the region of the maximum reflection

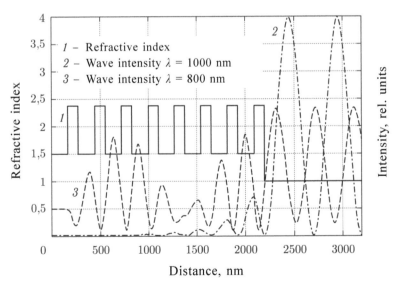

Fig. 65. *1* – distribution of the refractive index in a Bragg mirror; *2, 3*-spatial distribution of the radiation intensity with wavelengths of 1000 and 800 nm in a Bragg mirror.

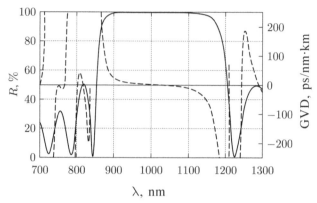

Fig. 66. Spectral dependences of the reflection coefficient and group velocity dispersion for a Bragg mirror.

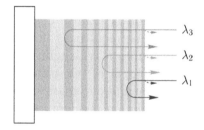

Fig. 67. Schematic representation of a chirped Bragg mirror [53].

of the lattice, increases rapidly, approaching the edges of the spectral range, and has regions of positive and negative dispersion.

It is these above features of the functioning of Bragg mirrors that formed the basis for the creation of GVD compensators based on chirped Bragg mirrors (Fig. 67). The chirped Bragg mirror (CBM) consists of alternating layers with high (n_H) and low (n_L) refractive indices and variable thicknesses multiples of $\lambda/4$.

As can be seen from Fig. 67, the reflection of long-wave (red) radiation occurs in the depth of the Bragg mirror, while short-wave (blue) radiation is reflected by layers that are closer to the surface. As a result, the different frequency components of the reflected laser pulse have different optical path values and different time delays. Such chirped Bragg mirrors make it possible to perform compression and decompression of femtosecond pulses very efficiently and with minimal losses and are used to compensate the dispersion in the cavity of the master oscillator of a femtosecond fibre laser.

A detailed description of the functioning and calculation of chirped Bragg mirrors is given in [53–55].

The simplest way to produce a CBM is to compose it from a sequence of multilayer periodic Bragg gratings tuned strictly to a certain frequency. In [56], the results of a study of a CBM consisting of three sections of different-period multilayer Bragg mirrors are presented, the total number of consecutive layers in succession was 91. This construction had a high reflection coefficient (~99.7%) in the spectral range from 600 to 1200 nm. Structurally, the arrangement of the layers in the CBM is described by formula

$$S(0.5H_0, L_0, 0.5H_0)15(0.5H_1, L_1, 0.5H_1)15(0.5H_2, L_2, 0.5H_2)15S_0, \quad (5.76)$$

where the following designations are used: S_0 is the SiO_2 substrate (n_S = 1.4656 is the refractive index at the wavelength λ_0 = 1100 nm), S is the ambient air (n_0 = 1.0, H_0, H_1, H_2 and L_0, L_1, L_2 are Nb_2O_5 layers (n_H = 2.2393 at a wavelength of λ_0 = 1100 nm) and SiO_2 layers (n_L = 1.46566 at a wavelength of λ_0 = 1100 nm), respectively. Optical layer thicknesses were selected as follows:

$$n_{H_0}d_{H_0} = n_{L_0}d_{L_0} = 0.77273; \ n_{H_1}d_{H_1} = 0.77273 n_{L_1}d_{L_1} = 0.59091;$$

$$n_{H_2}d_{H_2} = 0.59091 n_{L_2}d_{L_2} = \frac{\lambda_0}{4} = 275\,\text{nm}. \tag{5.77}$$

Figure 68 shows the spectral dependences for the reflection coefficient, the group delay time and the group delay dispersion for the CBM consisting of 91 alternating layers of Nb_2O_5 and SiO_2 (the first and last layers are made of Nb_2O_5) deposited on a fused quartz substrate. The optical thickness of the layers increases from $0.5\lambda_0/4$ to $1.15\lambda_0/4$, starting from the substrate (curves 2, 4, 5) and from the air side (curves 1, 3, 6). As can be seen, for the reflection coefficient there are three different ranges (I–III) lying in the range from 780 to 1000 nm. The dependences of the group delay time and the group delay dispersion are modulated by the sharp peaks observed in the strong reflection region of the CBM, whose presence is due to the interference of waves, with high efficiency reflected from each individual mirror and interfaces. Therefore, the smoothness of the dependences increases if the direction of the light wave incident to the CBM is changed.

To ensure the compression of laser pulses, the dispersion of the group delay in the CBM should be negative, and the group delay should increase linearly with increasing wavelength. To this end, as

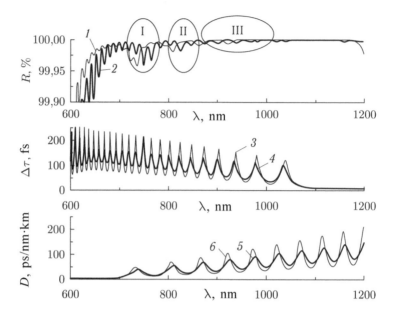

Fig. 68. Spectral dependencees for the reflection coefficient (R) – *1, 2*; group delay ($\Delta\tau$) – *3, 6* and group delay dispersion (D) – *5, 6*, of the pulse for cases when the optical thickness of layers increases from the substrate (*2, 4, 5*) and from the air side (*1, 3, 6*).

shown in Ref. [54], the distribution of the refractive index in a CBM must satisfy the condition:

$$n(x) = \sqrt{n_H n_L}\, \exp\left\{ \ln\left(\sqrt{\frac{n_H}{n_L}} \right) \exp\left(-\frac{x^2}{2\sigma^2} \right) \sin[x(k_0 + c_1 k_0 x)] \right\}, \quad (5.78)$$

where

$$x = \int_0^z n(u)\,du, \quad (5.79)$$

but $n(x)_{max} = n_H$; $n(x)_{min} = n_L$; σ is the width of the Gaussian function enveloping the pulse; k_0 is the wave number at the central wavelength (λ_0), which satisfies the Bragg condition; $k = 2\pi/\lambda$; c_1 is the linear spatial parameter of the chirp for the CBM [54]. Depending on the choice of parameters σ and c_1, the CBM can provide both growth and decrease in the group delay time of the pulse. Figure 69 shows the dependences $\Delta\tau$ on the wavelength for $\sigma = 2^{1/2}$ $n_H = 2.3$ (TiO$_2$); $n_L = 1.45$ (SiO$_2$), showing increasing ($c_1 = +0.02$) (*1*) and decreasing ($c_1 = -0.02$) (*2*) dependences for the group delay time [55].

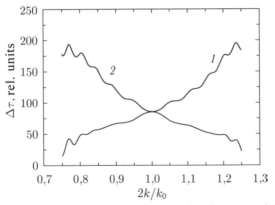

Fig. 69. Calculated dependencees for the group delay.

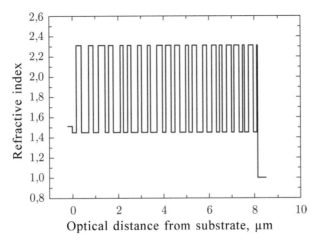

Fig. 70. Distribution of the optical thicknesses of layers in a CBM [55].

In practice, it is extremely difficult to ensure a smooth change in the refractive index in the CBM layers. Therefore, it is attempted to optimize the optical thickness of the layers by selecting their thickness. Figure 70 shows the spatial distribution of the optical thickness of the layers in the CBM, corresponding to the following formula:

$$S_0(0.87L_1, 14H_1, 58L_0, 98H_1, 18L_1, 45H_0, 75L_0, 96H_1, 57L_0,$$
$$85H_0, 73L_0, 84H_1, 45L_0, 85H_1, 31L_0, 69H_1, 30L_1, 29H_0, 69L_1,$$
$$30H_0, 81L_1, 07H_1, 25L_0, 67H_0, 81L_0, 96H_1, 35L_0, 88H_1, 03L1, \qquad (5.80)$$
$$09H_0, 62L_0, 66H_0, 87L_1, 12H_0, 62L_1, 21H_0, 63L_0, 43H_0,$$
$$93L_1, 07H_0, 87L_0, 16H)S,$$

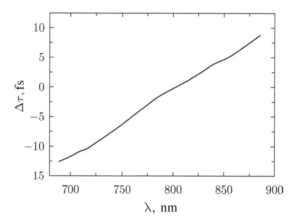

Fig. 71. Experimental dependence of the group delay time [55].

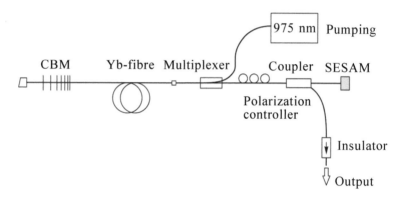

Fig. 72. Scheme of an ytterbium fibre laser with a CBM [57].

where S_0 is the substrate of fused quartz ($n_S = 1.51$), S is the air ($n_A = 1$); H and L are four-wave layers of TiO_2 ($n_H = 2.315$) and SiO_2 ($n_L = 1.45$) at the radiation wavelength $\lambda = 790$ nm.

Figure 71 shows the dependence of the group delay time for a pulsed titanium–sapphire femtosecond laser obtained using the CBM of the above-described structure.

The chirped Bragg mirrors (CBM) are very simply implemented on the basis of fibre Bragg gratings with a variable period.

The scheme of an ytterbium fibre laser in which a CBM is used is shown in Fig. 72 [57]. Pumping of a single-mode optical fibre with a length of 2 m doped with Yb^{3+} ions is provided by a laser diode with a wavelength of 976 nm. The SESAM consistent with the fibre lightguide whose saturable absorber is fabricated from quantum wells of InGaAs, and the mirror represents a multilayer Bragg structure

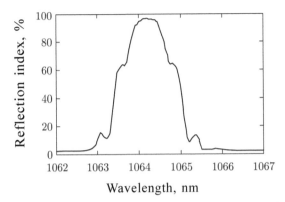

Fig. 73. Spectral reflection of a fibre CBM [57].

of GaAs/AlGaAs layers deposited on a GaAs substrate, has the following parameters: a reflectance of more than 90% in the spectral range of 1040–1100 nm, the relaxation time constant is ~500 fs, the saturation energy density is 60 $\mu J/cm^2$ and the modulation amplitude of the reflection coefficient is 6%. The second reflector in the laser cavity is a fully fibre CBM, which has a reflection coefficient close to 97%.

The fibre chirped Bragg mirror is a fibre Bragg grating with a linear spatial chirp; the spectral dependence of its reflection coefficient is shown in Fig. 73. The half-width of the spectral dependence of the CBM is ~1.56 nm.

The self-mode-locking mode in a fibre laser is triggered when the pump radiation power exceeds a threshold value of 14 mW. Starting from this power value, the fibre laser generates pulses with a repetition rate of 50 MHz. In this case, the duration of the generated pulses and their power depend on the power of the pump radiation. Thus, at a pumping power of 14 mW, the duration of the generated pulses is ~2.9 ps and their power is ~1 mW. An increase in the pumping power to 22 mW leads to a shortening of the pulse duration to ~2.6 ps, and their average power increases to ~2 mW.

Figure 74 shows the experimental and calculated dependences for the power spectrum of pulses generated by the laser obtained for the above laser pumping powers. As can be seen, the increase in the pump power leads to a broadening of the laser emission spectrum, which is due to the shortening of their duration.

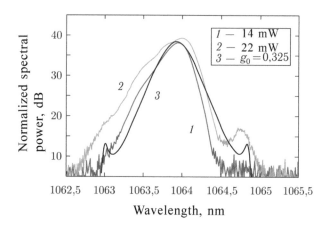

Fig. 74. Spectral dependences of the radiation power of a Yb-fibre laser with a CBM: *1* – pumping power 14 mW; *2* – pumping power 22 mW; *3* – computed dependence at the laser gain factor $g_0 = 0.325$ [57].

5.4.3. Amplification of ultrashort pulse in fibre lasers

The compression of light pulses reduces the pulse duration and increases its peak power. In this case, if there are no losses in the system, the energy of the pulse does not change. As a rule, the energy of a single pulse in a sequence of ultrashort pulses generated by a fibre laser does not exceed tens of nJ. Therefore, to increase the energy of the pulse, its amplification is necessary. To do this, the laser pulse is transmitted through the amplifying medium.

However, a number of problems arise in amplification of the USP. First, because of the dispersion of the active medium of the amplifier, the duration of the ultrashort pulse increases. Secondly, when the radiation intensity reaches 10^{11}–10^{12} W/cm^2 in amplifying media, self-focusing takes place, leading to optical breakdown and damage to the medium. Since due to the short pulse duration the radiation intensity reaches these values even at relatively low energies, this does not allow direct, multipass methods to obtain a noticeable increase in the USP energy. To overcome this problem, a special technique for amplifying the ultrashort pulse (Fig. 75) is used: the method of amplification of chirped pulses [58].

The essence of the method is that the pulse to be amplified is passed through a delay line (including fibre) with a large positive dispersion of the group velocity. Dispersion decomposes the pulse into colours in time, colouring one end of it in red and the other in blue. Since the width of the pulse spectrum is small and the

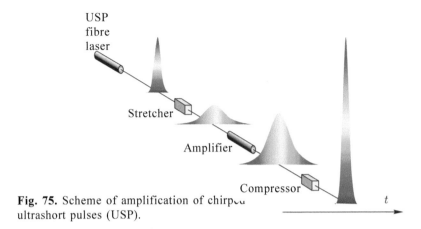

Fig. 75. Scheme of amplification of chirped ultrashort pulses (USP).

group velocities within this frequency range vary approximately linearly, the pulse moves uniformly in the medium with the GVD, and the frequency will vary linearly along the pulse, that is, the pulse becomes chirped, and its duration strongly increases up to 10^4 times. Accordingly, the peak power of the pulse decreases. Such a chirped pulse can be effectively amplified by safely passing it through an amplifier, thereby increasing its energy by a factor of 10^3–10^4. The device that carries out the chirping of the USP is called a stretcher or an expander.

After amplification, the pulse is transmitted through a delay line in the form of an environment with an anomalous dispersion-compressor. As a result of selecting the compressor parameters, the chirp is compensated to zero and the pulse acquires the original ultrashort duration. Thus, to implement the process of pulse compression, it is necessary to have a medium with an adjustable negative dispersion. Today, these devices with adjustable dispersion used most often as described in section 5.4.2 include prism and lattice compensators of group velocity dispersion. Note that in the circuit for amplification of chirped pulses shown in Fig. 75 the duration of the output pulse can not be shorter than the pulse duration at the input.

Nevertheless, the principle underlying the method of amplification of chirped pulses can be used to reduce the duration of the ultrashort pulse. For this purpose, the self-modulation effect of the pulse is used, which occurs when the ultrashort pulse is passed through optical fibres or capillaries and leads to a broadening of the spectrum, accompanied by the appearance of a chirp (see Chapter 1). The resulting broadening of the pulse spectrum can be so great that the monochromatic light of the laser is converted into white light containing all the colours of the rainbow, i.e., covering all or almost the entire range of visible light. Such radiation with an ultrawide spectrum is called a supercontinuum (see Chapter 2). Further, the resulting chirped pulse with a large spectral width is compressed by a compensator with a suitable dispersion value, but not to the original duration, but to a duration corresponding to its increased spectral width. This method allows one to obtain USPs lasting several femtoseconds.

An important problem for femtosecond lasers is the removal of heat from the active element, since they operate in a continuous mode. For this reason for active elements it is desirable to use materials with high thermal conductivity. The Ti-sapphire has the best results; its thermal conductivity is comparable with the thermal conductivity of metals. In fibre lasers, heat is released from the material of the active core into a fused quartz cladding. Because of the considerable length of the optical fibre in these lasers, the heat flux removed is lower than in lasers with ordinary rod active elements, and the volume of the outer cladding is much larger than the core volume. Thus, fibre lasers do not require special cooling. However, in amplifying systems, due to the high power values of amplified ultrashort pulses, it is necessary to reduce the pulse repetition rate, so that the released heat can dissipate into the surrounding space.

References

1. Snitzer E., Phys. Rev. Lett. 1961. V. 7. P. 444–446.
2. Snitzer E., Appl. Op., 1966. V. 5, No. 10. P. 1487–1499.
3. Kao K.C., Hockham G.A., Proc. IEE, 1966. V. IEE 133. P. 1151–1158.
4. Dianov E.M., Usp. Fiz. Nauk. 2013. Vol. 183, No. 5, P. 511–518.
5. Dianov E.M., Usp. Fiz. Nauk. 2004. V. 174, No. 10. P. 1139–1142.
6. Dianov E.M.,Vestnik RAN. 2007. 77, No. 8. P. 714–718.
7. Kurkov A.S., Dianov E.M., Kvant. elektronika. 2004. ~V. 34, No. 10. P. 881–900.
8. Kurkov A.S., Vestnik Nizhegorod. Univ. 2008, No. 2. Pp. 32–38.
9. Dawson J.W., Fibre Lasers: Technology, Applications and Associated Laser Safety, Presentation to LSO Workshop, September 11, 2012.
10. Sahu J.K., et al., Electron. Lett. 2001. V. 37. P. 1118–1119.
11. Furusawa K., Optics Express. 2001. V. 9, No. 13. P. 714–720.
12. Malykin G.B., Usp. Fiz. Nauk. 2000. V. 170. P. 1325–1349.
13. Kulchin Yu.N., Distributed fibre-optic measuring systems, Moscow, Fizmatlit, 2001..
14. Rodica S.A., in: Advance in Optical Amplifiers, ed. P. Urquhart,I n Tech, Croatia, 2011. P. 255–278.
15. Bespalov V.G., Makarov N.S., Izv. RAN. Ser. Fiz.. 2002. Vol. 66, No. 3. P. 550–352.
16. Nilsson J., et al., Optical Fibre Technology. 2004, No. 10. P. 5–30.

17. Dianov E.M., et al., Kvant. elektronika. 1997. V. 24, No. 1. P. 3–4.
18. Jackson S.D., Li Y., IEEE J. Quantum Electron. 2003. V. 39. P. 1118–1122.
19. Kurkov A.S., et al., Kvant. elektronika. 1999. Vol. 27, No. 3. P. 239–240.
20. Townsend J.E., et al. Electron. Lett. 1991. V. 27. P. 1858–1860.
21. Hanna D.C., et al. Opt. Commun., 1989. V. 75. P. 283–286.
22. Hayward A., et al. Electron. Lett. 2000. V. 36. P. 711–712.
23. Clarkson A., et al. Opt. Lett. 2002. V. 27. P. 1989–1992.
24. Hanna D.C., Percival R.M., et al. Continuous–wave oscillation of a holmium–doped silica fibre laser, Electron. Lett. 1989. V. 25. P. 593.
25. Jackson S.D., Mossman S., Appl. Phys. B. 2003. V. 77. P. 489.
26. Jackson S.D., Mossman S., Appl. Opt. 2003. V. 42. P. 3546
27. Evtikhiev N.N., et al., Information optics. Moscow, Izd. MEI, 2000.
28. Raman Amplifiers for Telecommunications. Physical Principles, Ed. Islam M.N. Springer: Series in Optical Sciences, 2004. V. 90/1. P. 300.
29. Stolen R.H., Ippen E.P., Appl. Phys. Lett. 1973. V. 22. P. 6–8.
30. Rini M., et al. Opt. Commun. 2002. V. 203, No. 1, 2. P. 139–144.
31. Kurkov A.S., et al., Kvant. elektronika. 2000. V. 30, No. 9. P. 791–793.
32. Dianov E.M.,et al., Kvant. elektronika. 1999. Vol. 29, No. 2. P. 97–100.
33. Dianov E.M., et al., Opt. Lett. 2000. V. 25, No. 6. P. 402–404.
34. Dianov E.M., et al., Laser Phys. 2003. V. 13. P. 397–402.
35. Turitsyn S.K., et al., Phys. Rev. Lett. 2009. V. 103, No. 13. 133901.
36. Turitsyn S.K., et al., Nature Photonics. V. 4, No. 4. P. 231–235.
37. Babin S.A., Batkin I.D., Avtometriya. 2013. Vol. 49, No. 4. P. 3–29.
38. Kryukov P.G., Kvant. elektronika. 2001. V. 31, No. 2. P. 95–119.
39. Ochotnikov O., et al., New J. of Physics. 2004. V. 6. P. 177–199.
40. Hayrapetyan V.S., Ushakov O.K., Physics of lasers. Novosibirsk, SSGA, 2012.
41. Pupeza I., Power Scaling of Enhancement Cavities for Nonlinear Optics. Max–Plank Inst., Munchen. 2011.
42. Zakharov V.P., Shakhmatov E.V., Laser technology: Proc. allowance. Samara: Samara Publishing House of State Aerospace. University, 2006.
43. Yen T.H., et al., Laser Physics. 2012. V. 22, No. 2. P. 441–446.
44. Gomes L., Ultrafast Ytterbium Fibre Sources Based on SESAM technology. Tampere Univ. Technol., 2006.
45. Haus H.A., IEEE J. Sel. Top. of Quant. Electron. 2000. V. 6. P. 1173–1185.
46. Tausenev A.V., et al., Kvant. elektronika. 2007. V. 37, No. 3. P. 205–208.
47. Keller U., et al., Optics Letters. 1992. V. 17, No. 7. P. 505–512.
48. Maas D.J.H.C., et al., Optics Express. 2008. V. 16, No. 23. P. 18646.
49. Paschotta R., Keller U., Applied Physics. 2001. V. B73, No. 7. P. 653–660.
50. Okhotnikov O.G., et al., IEEE Photonics Technology Lett. 2003. V. 15, No. 11. P. 1519–1521.
51. Sukhorukov A.P., Soros Journal. 1997. No. 7. P. 81–86.
52. Zaitsev V.P., et al., Vopr. Atom. Nauki i Tekhniki. 2010. No. 3. P. 107–110.
53. Dombi P., et al., Optics Express. 2005. V.13, No. 26. P. 888–894.
54. Szipocs R., K'ohazi–Kis A., Appl. Phys. 1997. V. B65. P. 115–135.
55. Szipocs R., et al., Appl. Phys. 2000. V. B70. P. 51–57.
56. Pervak V.Yu., et al., Quantum Electronics & Optoelectronics. 2008. V. 11, No. 2. P. 154–158.
57. Katz O., Sintov Y., Opt. Comm. 2008. V. 281, No. 10. P. 2874–2878.
58. Nelson L.E., et al., Phys. B. 1977. V. 65. P. 275–294.

6

Photonics of Nanostructured Biomineral Objects and Their Biomimetic Analogues

Introduction

Nature has always been an invaluable source of inspiration for scientists and engineers. The great scientific and technological revolutions that led to the rapid development of mankind were begun with the study and understanding of the processes taking place in the living nature. In the process of researching original solutions obtained during the evolution of living nature, new approaches to the development of engineering structures and technological processes were found, which served as a prerequisite for the formation of a new independent field of research, called biomimetics [1]. In connection with the latest achievements in the field of nanotechnology, which opened the prospects for the atomic design of various objects, the possibilities for an interdisciplinary partnership of physicists, chemists, engineers, technologists and biologists have expanded greatly, significantly pushing the horizons of biomimetic science, ensuring its penetration into such a seemingly very a distant field of study, like photonics.

The systems of optical transmission and information processing developed in the 20th century not only allowed us to connect distant corners of different continents, but also significantly changed the way of the life of the mankind. Open communication opportunities associated with the use of optical radiation, stimulated in the 21st century the search for fundamentally new technologies for

Fig. 1. Examples of photonic crystals created by wildlife: *a* – pollen of the wings of butterflies; *b* – change in the colour of the wing of a butterfly when it is immersed in a medium with a refractive index different from the refractive index of air; *c* – pearl shell coating; *d* – chitinous shell of beetles.

the development of the elemental base of communication systems, generation and detection of radiation, optical sensors and quantum computers. To solve these problems, more and more attention is being paid to such nanophotonics objects as photonic crystals [1].

As is known, the living nature has already created a variety of materials with photonic crystal properties, among which there are a variety of: pollen of butterfly wings, chitinous cover of beetles, mother of pearl shells, etc. (Fig. 1), the growth of which is based on one of the most promising nanotechnological directions – self-organization.

The basic structural components of living systems almost entirely consist of ordered arrays of protein and hydrocarbon molecules. This feature of living systems is due to the ability of biological macromolecules to self-organize in solution [2]. The latter property allows the production of uniquely complex nanostructures, providing high productivity per unit of mass without excessive requirements to the raw materials and energy. Unfortunately, most natural protein complexes do not possess resistance to temperature and chemical influences, and are also susceptible to the destructive effects of bacteria. In this connection, the use of direct analogs of biological systems to create ordered nanostructures has not found wide application.

As is known, the overwhelming majority of the locomotor and protective structures of living organisms is built on the basis of biomineral materials, which are complex composite substances. They consist of two main components: organic and mineral, the interrelation of which determines the structuredness of biological composites at the nano-, micro- and macrolevels, which provides unique characteristics of living systems based on them and is of considerable interest for their modeling to create new materials. In [2], the main building blocks that make up the micro- and nano-level of biological materials are systematized and isolated. As shown, these are molecular units (amino acids), proteins (collagen, keratin, elastin, etc.), polysaccharides (chitosan, Na-alginate, Na-hyaluronate, etc.) and minerals associated with the organic matrix in biomineral composite structures. The latter are the subject of active research in the field of biomimetics [3]. The greatest number of studies of biomineral structures is associated with the study of their chemical, biochemical and mechanical properties. Relatively recently discovered unique optical properties of natural biosilicates [4] initiated the increased interest of researchers in the field of photonics to these structures.

A vivid example of organisms based on the self-organizing process of biomineralization is the deep-sea glass sponge (DSGS), which has a cellular mechanism for the selective accumulation of silicon from water and a complex mechanism of protein functioning that allows them to create an openwork skeleton system, the main elements of which are spicules – ordered nanostructured formations from hydrated silicon dioxide [2, 3]. In works [4–6] it was shown that the spicules of some species of marine glass sponges transmit light radiation in the same way as optical fibres do.

In connection with this, the study of the morphology, physical and chemical properties of the elements of the construction of the biomineral skeleton of DSGS, as well as the very process of biosilification in living nature, are of considerable interest for the development of biomimetic nanotechnologies.

The basis of the principles of building biomineral structures: hierarchy of structural organization, multifunctionality and self-organization, can form two main directions of biomimetic nanotechnologies: biochemical and chemical. Biochemical modelling is based on the search for and isolation of organic matrix proteins or their active fragments, and the subsequent synthesis of biomineral structures on their basis. The chemical direction of biomimetics is based on the synthesis of materials that model specific processes in natural biomineral structures.

The present chapter is devoted to the presentation of the results of complex studies performed in the study of the morphology and physico-chemical characteristics of the material of the spicules of deep-sea marine glass sponges aimed at finding promising technologies for their biomimetic modeling and obtaining new nanostructured functional materials, including photonic devices and systems.

6.1. Morphology and physico-chemical characteristics of spicules of deep-sea glass sponges

About 5000 species of sea sponges are known in nature, of which about 500 species are classified as DSGS [4], which live at depths of 30–5000 m.

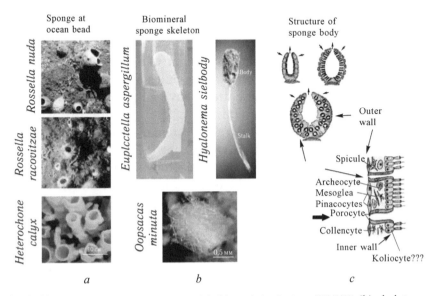

Fig. 2. Photographs of DSGS in natural habitats (a); photos of DSGS (b) skeletons; scheme of the body structure of the sponge (c).

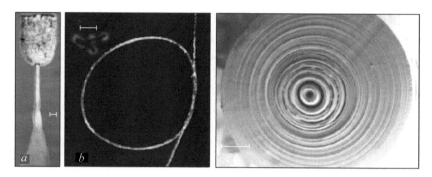

Fig. 3. Photograph of DSGS *Hyalonema sieboldy* (scale 1 cm) (*a*), a fragment of its basal spicule (scale 1 mm) (b) and SEM photographs of the treated end of the spicule (magnification 1157 times) (*c*) [13].

Deep-sea glass sea sponges are filter-type organisms (Fig. 2). They lead a sedentary way of life, attached to the ocean floor (Fig. 2 *a*). The strength of the sponge body and protection from external enemies is provided by a frame made of skeletal silicon-organic spicules, silicon-organic basal spicules are used to attach the sponge to the bottom (Fig. 1 *b*).

The length of the spicules, depending on their functional purpose, varies from 0.001 to 2 m, and the diameter varies from 5 μm to 3 cm [7, 31]. The life support of the sponges is illustrated in Fig. 2 *c*: the sponge through multiple channels in the walls of its body draws in itself seawater with the biological substrate contained in it, which is the food product of the sponge, and the remaining seawater with the products of the sponge metabolism is ejected through the outlet.

A photograph of a typical specimen of this kind of sponge is shown in Fig. 3. The basal (anchor) spicules of these sponges, which serve to attach DSGS to the bottom, have an amazing elasticity that allows them to be literally tied into a knot (Figure 3b), while the skeleton spicules have considerable toughness and brittleness.

A detailed scanning electron microscope (SEM) study of transverse sections of basal spicules shows that each of them consists of a large number of axial concentric layers, the number of which can vary from several tens to several hundreds depending on the type and age of the sponge (Fig. 3 *c*) . Each of the concentric layers of the spicule consists of tightly packed into the matrix of collagen-like nanofibrils and nanoparticles of hydrated silica with a size of 20 to 120 nm, separated by nanometer protein layers (Fig. 4) [8–11]. At the same time, regardless of the functional purpose, all spicules have a central core-axial protein filament 1–2 μm in diameter, the

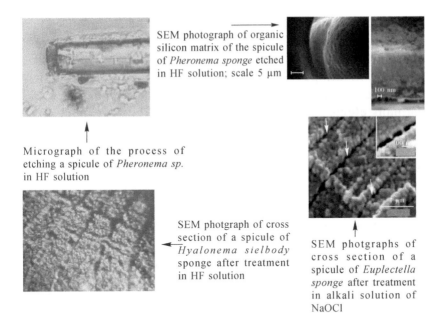

Fig. 4. Microscopic structure of the material of the spicules of DSGS [12].

localization region of which is clearly visible in the photographs in Figs. 3 and 4, and a set of surrounding layers of silicon dioxide and a protein component [12, 13]. The thickness of the layers containing nanoparticles from silica is more than 100 times greater than the thickness of the layers separating them from the protein.

An additional study of the DSGS spicules by methods of atomic force microscopy, X-ray phase analysis, and small-angle X-ray scattering indicates that most of the silicon dioxide contained in the material is amorphous and that the size of the nanoparticles is polydisperse.

In the course of the research it was established that the DSGS spicules are oriented and structured amorphous polymeric systems [4]. The orientation of polymer molecules in such systems is determined by the organic matrix [15], on which the polymer silicon oxides in the form of nanoparticles with an average size of 13–38 nm [15] precipitated during polycondensation. This agrees well with the model of nanostructural organization of sponge spicules proposed in [16].

The determination of the general regularity of the distribution of basic chemical elements in the nanocomposite material of the spicules of sea glass sponges was carried out using energy dispersive

X-ray spectrometry (EDS). The distribution of chemical elements along the cross section of the spicules of sponges *Pheronema sp., Ph. raphanus, H. sieboldy, H. populiferum and S. hawaiicus* from the centre to the periphery was studied. For all selected spicules of sponges, regardless of their species and functional purpose, the dominant elements of the composition are silicon (max ~33%), oxygen (max ~66%) and carbon (max ~9%), and also in small concentrations potassium and sodium [17]. It was also found that the chemical composition of the layers is not constant and depends on their location, the type of sponge and the type of spicule. As an

Table 1. Distribution of chemical elements along the cross-section of the skeletal spicule of the sponge Pheronema sp. [17]

Region studied	Chemical element, at.%				
	Si	C, H, N	O	K	Na
Centre of spicule, 1–2 μm	4.7±0.3	22.1±0.3	66.6±0.3	—	0.13±0.07
Central cylinder, 3–5 μm from centre	4.7±0.3	41.3±0.3	53.7 ±0.3	—	010±0.07
Layered region 7–10 μm from centre	3.6±0.3	29.6±0.3	66.6±0.3	—	0.08±0.07
Layered region 13–20 μm from centre	11.6±0.3	21.5±0.3	66.5±0.3	0.03±0.07	0.24±0.07
Layered region 50–100 μm from centre	22.4 ±0.3	10.8±0.3	66.6±0.3	0.17±0.07	—
Laminated shell	33.2±0.3	—	66.5±0.3	0.11±0.07	—

illustration, Table 1 shows the distribution of chemical elements in the skeletal spicule of the sponge *Pheronema sp.*

Studies of the degree of hydration of silicon oxide and the quantitative content of silanol groups in the spicules of deep-sea sponges were carried out by NMR ^{29}Si nuclear magnetic resonance [18]. Analysis of the data obtained showed that the ratio of the $Q2$: $Q3$:$Q4$ groupings corresponding to the $Si(OSi)_2(OH)_2/Si(OSi)_3OH$ / SiO_4 content in the test spicules is as follows: 57–68.6% SiO_4, 21–39% monohydrated $Si(OSi)_3OH$ and 3–9.7% Si hydrate $(OSi)_2 (OH)_2$, which together gives 30 to 40% of the silanol groups $Q2$ and $Q3$. The

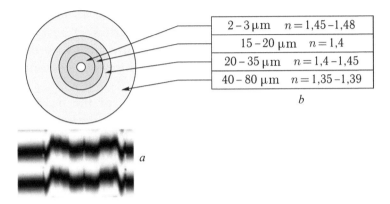

$2-3\,\mu m$	$n = 1,45-1,48$
$15-20\,\mu m$	$n = 1,4$
$20-35\,\mu m$	$n = 1,4-1,45$
$40-80\,\mu m$	$n = 1,35-1,39$

b

a

Fig. 5. Interferogram of the cross-section of the end of the spicule (*a*) and the distribution of the refractive index (*b*) [22].

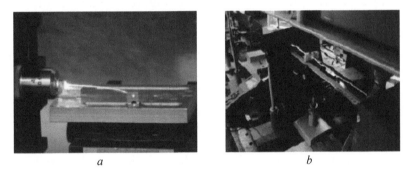

a   *b*

Fig. 6. Propagation of He–Ne laser radiation (*a*) and white light (*b*) in spicules [20].

presence of such a quantity of silanol groups indicates the developed surface of the samples under study [18].

A study of sponge material by differential thermogravimetric analysis showed that basal spicules, whose purpose is to retain the body of the sponge on the seabed and withstand currents, are flexible and have a silicon–carbon–water ratio of 3.7: 1.3: 1.0. This distinguishes them from the skeleton spicules, whose task is to ensure the rigidity of the body structure of the sponge and to protect against external threats having a silicon–carbon–water ratio of 4.4:1.4:1.0.

Measurements of the mechanical characteristics of basal spicules by the dynamic ultramicrohardness method [13, 19] showed that for the material of the basal spicules of sea sponges, the Young's modulus value is close to its corresponding values for natural opal. It was found that the distribution of Young's modulus over the cross-section of the basal spicule is non-uniform and varies from periphery to centre in the range from 33 000 to 40 000 Gpa. The observed behaviour of the mechanical properties of basal spicules is explained by the layering of their structure and the presence of an organic matrix [9, 20, 21], which allows them to have increased mechanical strength and flexibility.

The distribution of the refractive index of the basal spicule material along its cross-section measured using the interferometric technique [21] showed that the refractive index value for the central core which consists from the layers surrounding the axial filament ~(1.45–1.48) and its value falls to the edges of the spicule cross section to ~(1.39–1.41) (Fig. 5) [11, 12, 21, 22].

The significant difference in the refractive index of the material of the spicules in comparison with the refractive index of the surrounding medium (for air n = ~1, for water n = ~1.3) makes the structure of the spicule light-guiding. Figure 6 shows photographs illustrating the transmission of light through the spicules of the DSGS.

The spectral characteristics of the material of the spicules were studied in the spectral range from 200 to 1700 nm [20]. As an example, Fig. 7 shows the experimentally obtained spectral dependence of the transmission of the material of the basal spicule *Pheronema raphanus*, which is characteristic for all types of spicules of DSGS. From the above dependence it can be seen that the material of the spicules passes well enough radiation in the visible and near infrared ranges. In the vicinity of 770, 960 and 1150 nm, the

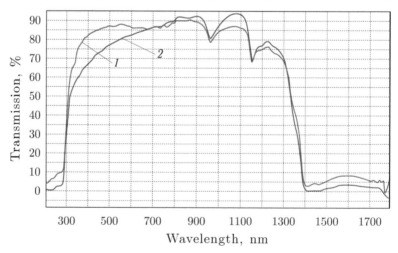

Fig. 7. Spectral transmission of the material of the spicules of the DSGS *Pheronema raphanus*: *1* – basal, *2* – skeletal spicules [22].

absorption regions due to the presence of hydrated silicates in the spicules are observed, which coincides with the results independently obtained in [21] for the basal spicules of the *Hyalonema sieboldi* sponge and other species of marine sponges [9]. In this case, for practically all the samples of the spicules chosen by us, the measured average level of light power loss for the waveguide propagation regime of laser radiation with a wavelength of $\lambda = 633$ nm is ~0.1 dB/m.

The layered structure of the spicules is associated with the periodicity of the change in the refractive index over the cross section with a nano- and micrometer scale. In such a layered structure it is possible to propagate guided radiation modes with an effective refractive index lower than the refractive index of silicon dioxide. These modes are strongly associated with light-guiding axial layers. Due to the large difference between the refractive indices of the alternating axial layers, the flux of light power in the radial direction undergoes a strong reflection. In this case, taking into account the large number of layers in the spicule, for certain values of the effective refractive indices of propagating modes, the condition of the phasing of the light fluxes propagating in the layers can be realized, which in turn, as for the photonic crystal structures (see Chapter 3), leads to a decrease in the losses on the radiation into the cladding [23].

As was shown in [24, 25], in the DSGS spicule light radiation can propagate only at certain angles to its axis. The resonant wavelength for such radiation directly depends on the thickness of the layers with a high refractive index. As a result, for spicules, by analogy with photonic crystals, there should exist photonic band gaps. By virtue of the smallness of the thicknesses of axial layers with a high refractive index, resonance conditions will be satisfied for waves propagating at angles close to $\pi/2$ with respect to the normal to the axis of the spicule. This means that for the light radiation guided in the basal spicules of DSGS the single-mode regime for guided modes with an increased size of the mode spot is most preferable. As the results of numerical simulation show, the size of the single-mode spot can reach ~20 wavelengths of guided light radiation [26]. With a variation in the thickness of the layers from 200 nm to 1 μm, the resonance values for the guided radiation wavelength will lie in the range from 300 to 1200 nm, which agrees well with the results of the experiments.

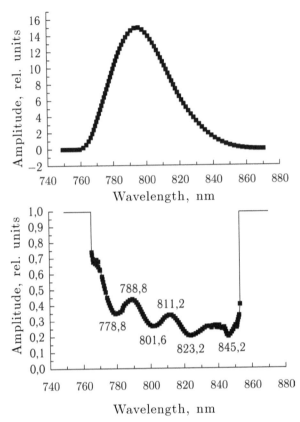

Fig. 8. Spectrum of ULP before (*a*) and after (*b*) passing through the spicule of DSGS *Hyalonema sieboldi* (length 40 mm, diameter 190 μm) [26].

A more detailed verification of the presence of the photonic-crystal properties of the DSGS spicules was based on a study of the process of transmission through them of femtosecond-duration pulses of a titanium–sapphire laser (40 fs) with an extremely low energy (~0.01 nJ) (ULP – ultrashort light pulses). The power spectrum of the laser radiation at the entrance to the spicule is shown in Fig. 8 *a*. For all the samples of the spicules, the presence of photonic-crystal band gaps was observed, as evidenced by the modulation of the spectrum of the transmitted radiation. This fact is illustrated by the dependence shown in Fig. 8 *b* of the transmission spectrum of the basal spicule 190 μm in diameter, 40 mm long, with a central channel (core) diameter ~2 μm and a thickness of axial layers with a high refractive index ~280 nm.

Thus, the results of the studies make it possible to conclude that the presence of periodic axial cylindrical layers of silica in the basal spicules of DSGS leads to the formation of photonic band gaps in them, as a result of which the spicules of deep-sea glass sponges are a new kind of natural photonic crystals.

6.2. The role of the photonic-crystal properties of the spicules of deep-sea sponges during their metabolism

Already the first studies of the transmission of the basal spicules of the sponges of *H. sieboldy*, *S. hawaiicus* and the spicule of *Ph. raphanus* showed the colour gradient from white to red along the length of the spicule (see Fig. 6 *b*), which allowed them to be classified as a differential spectral filter intensively scattering through the side surface of the spicule a part of the visible radiation spectrum in the wavelength range from 300 nm to 600 nm [11, 27].

A more detailed study of the intensity of scattering of optical radiation through the lateral surface of the basal and skeleton spicules of the sponges along the normal to their axes for different wavelengths (Fig. 9 *a*) showed that the layered organosilicon structure of the spicule, in combination with its conical shape (Fig. 9 *b*), leads to the appearance of periodic spatial modulation of the intensity of scattered light radiation with a modulation period that depends on the wavelength.

It can therefore be assumed that the spicules of marine glass sponges are one-dimensional natural photonic crystals with spatial chirping of the radial periodic structure (Fig. 9 *c*). It is the conical shape of the layers in such photonic crystals that varies in length and makes them poorly conductive from the thick end of the spicule to the thin as a result of strong dispersion of radiation into the surrounding medium through the cladding. Conversely, it allows the spicules to capture light through the cladding and transform it in the direction from the thin end of the spicule to the thick one with small losses [28, 29].

Analysis of the energy balance of DSGS showed 10 to 30% of the deficit required for the existence of sponges of energy coming from the processing of sponge phyto- and zooplankton [30]. A study of the absorption spectra of acetone extracts obtained from various fragments of the body of the *Ph. Raphanus* sponge (Fig. 10), the determination of the concentration of photosynthetic pigments and the composition of fatty acids included in them showed that the main symbionts in the marine glass sponge *Ph. raphanus* are cyanobacteria and they are found in structures localized on individual rays of the skeleton megasclerial spicules of DSGS [30]. The obtained results of the studies give serious arguments in favour of the hypothesis of the presence of the photoreceptor system in the *Ph. raphanus* sponge and the important role of photosynthetically active symbionts of deep-sea sponges covering spicules in their energy balance.

Apparently, a significant part of the symbionts and the resulting phytoplankton are photosynthetically active, i.e., participates in the process of photosynthesis and partially (10–30%) covers the energy costs of deep-sea glass sponges. The presence of an energy imbalance makes it possible to justify the presence of the photonic crystal properties of the DSGS spicules as a necessary element in the chain of organization of the process of photosynthesis by sponges. Figure 11 *a* shows a photograph of *Ph. Raphanus*, on which the holonosoic spicules emerging from its body are visible (Fig. 11 *b*). As can be seen from Fig. 11 *b*, the skeleton spicules have the form of a spindle, one part of which, inside the sponge body, has an inhomogeneous

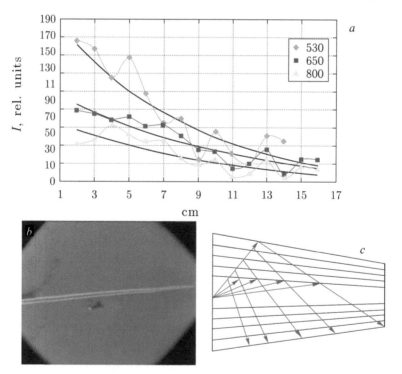

Fig. 9. Scattering of the radiation channeled in a skeleton spicule of DSGS *Hyalonema sielbody* along the normal to the axis (*a*) (the inset gives the notation for the wavelengths of channeled radiation, nm); SEM-photograph of the spicule (*b*);

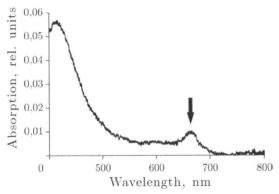

Fig. 10. Absorption spectrum of the acetone extract of the sponge (the arrow indicates the absorption peak characterizing the presence of chlorophyl A in the solution).

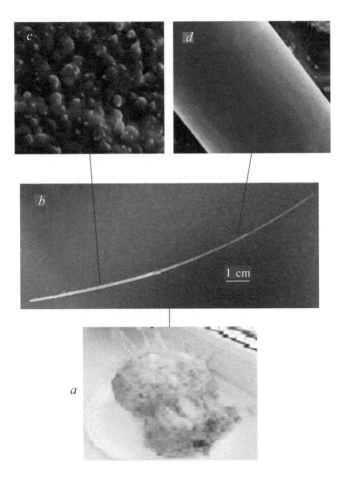

Fig. 11. *a* – picture of *Ph. Raphanus* sponge; *b* – picture of the skeleton spicule; *c, d* – SEM images of the surface of the skeleton spicule [30].

surface structure (Fig. 11 *c*), and the other part of the spicule facing the aqueous medium has a smooth laminated surface (Fig. 11 *d*). The surface of the spicule inside the sponge is covered with an organic substance containing cyanobacteria.

Thus, it can be assumed that the skeleton spicules in sponges, in addition to the protective functions and functions of stiffening the body, sponges act as a kind of optical antenna. The protruding part of the spicule-receiving antenna, which, due to the property of spatial chirp of the photonic crystal, receives the photon from the outside and transports it into the interior of the sponge body, where, due to the back narrowing of the spicule, it radiates it into an organic material containing cyanobacteria and, therefore, the photosynthesis process is supported.

Thus, the presence in skeleton spicules of the GSGS properties necessary to provide a system for receiving and transporting light radiation to photosynthetic objects allows us to take a fresh look at the functional purpose of sponges and on the questions of the spreading of sponge species in the ocean [31].

6.3. Nonlinear optical properties of the spicules of deep-sea glass sponges

In the case of waveguide excitation of basal spicules by pulses of the second harmonic of a YAG:Nd laser (λ = 532 nm) with a pulse duration of 12 ns, a repetition frequency of 10 Hz, and a pulse energy of ~30 mJ at the exit from the spicules, a significant increase in

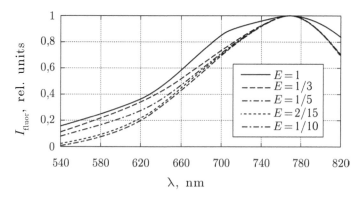

Fig. 12. Normalized experimental dependences of the fluorescence spectra of the spicule material obtained for different intensities of the input laser radiation. To excite the spicules, the second harmonic of the YAG:Nd laser was used: λ = 532 nm, pulse duration 12 ns, pulse frequency 10 Hz, maximum pulse energy E = 30 mJ [13].

the fluorescence intensity in the long-wave region (Fig. 12) [13]. The character of the dependence differs from the corresponding dependence of the fluorescence spectrum measured for a conventional multimode quartz fibre waveguide [31]. The different behaviour of the experimental dependences of the fluorescence spectra for the spicules of DSGS and quartz optical fibres is apparently related to the content of large organic molecule complexes in the spicules of the marine glass sponges and the nonlinear process of conversion of the light energy due to them.

The uniqueness of DSGS spicules as lightguides with a periodically changing profile of the transverse distribution of the refractive index lies in the complex character of the frequency profile of the dispersion of the light pulses propagating in them, which differs significantly from the case for ordinary quartz optical fibre waveguides. As a result, new nonlinear–optical phenomena and new regimes of the spectral–temporal transformation of ultrashort light pulses (ULP) can be observed in such 'lightguides'.

In this connection, a model has been developed that describes the propagation of ULPs in spicules, allowing one to take into account the combined manifestation action of the following nonlinear-optics dispersion effects, nonlinear polarization, and ionization nonlinearity, which leads to self-modulation of the light field phase transmitted through the spicule and effective transformation of the pulse spectrum into a supercontinuum. Numerical modelling and experimental study of the processes of the propagation of SRS in the spicules of the DSGS were also carried out [27]. The experiments were conducted using a titanium–sapphire laser complex Spitfire Pro (Spectra Physics, USA) with a light pulse duration of 40 fs, single laser pulse energy ~0.9 mJ, pulse repetition frequency of 100 Hz, central wavelength of ~800 nm and a width of the spectrum $\Delta\lambda_{1/2} = 35$ nm.

When an USP with an energy of ~5 nJ was input into various samples of DSGS spicules with a length of $L = 5$–15 mm, the phenomenon of self-focusing of radiation was observed with the formation of 'hot zones' in the transverse distribution of the intensity of the transmitted light beams. In these areas, a noticeable transformation of the emission spectra was observed [32]. The study of self-focusing processes of SRS in spicules made it possible to determine the value of the nonlinear optical coefficient for their material, which amounted to $n_2 \sim 8.8 \cdot 10^{-16}$ cm²/W, which is more

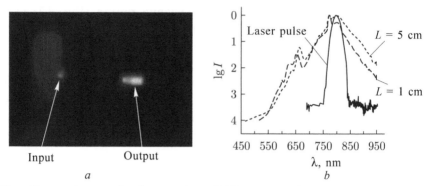

Input   Output

a   b

Fig. 13. *a* – photograph of generation of USP supercontinuum (λ = 800 nm; T = 40 fs; Q = 100 nJ) in the spicule of *Hyalonema sieboldi* spunge (diameter 200 µm, length 1.5 cm); *b* – spectrum of supercontinuum after passage through the spicule of the *Sericolophys hawaiicus* sponge with a diameter of 250 µm [32].

than three times the corresponding value for fused quartz and $n \sim 10^3$ times for air.

With an increase in the energy of the laser pulses, starting at ~20 nJ, a stable supercontinuum formation was observed in the spectrum of the light radiation transmitted through the spicules, and for energies of 0.1 mJ to 0.9 mJ, a prolonged multiple exposure led to an optical breakdown of the spicule material. Figure 13 *a* is a photograph illustrating the transformation of the spectrum of a femtosecond pulse with an energy of 100 nJ while passing through a segment of the spicule of the *Hyalonema sieboldi* sponge 200 µm in diameter and L = 1.5 cm in length to the spectrum of the supercontinuum that which stretches over the entire spectral area of visible radiation up to 400 nm. The experimental dependences of the USP spectrum after it has passed through the spicules of different length, shown in Fig. 13 *b*, demonstrate a considerable broadening of the USP spectrum in the Stokes and anti-Stokes regions with increasing spicule length.

Since the peak intensity of unfocused USP in the experiments could reach values of ~$7 \cdot 10^{10}$ W/cm², this led to a significant increase in the contribution of such nonstationary effects as nonstationary spatial self-focusing and multiphoton ionization of the spicule material, which, with increasing length, broaden the short-wavelength supercontinuum boundary up to 300 nm.

Simulation of the spectral distribution of the intensity of transmitted USP in the IR region of 750–850 nm and comparison of the results with experimental data showed that the group velocity dispersion of $\beta_2 = 2.8 \cdot 10^{-3}$ ps^2/m for the spicule is an order of magnitude smaller than the dispersion of an analogous optical fibre waveguide from fused quartz, which is explained by the substantial contribution of the waveguide part of the dispersion due to the multilayer quasiperiodic structure of the cladding [25] in comparison with the component due to the material of the spicule.

6.4. Biomimetic modelling of biosilicate nanocomposite material of DSGS spicules

6.4.1. Sol–gel technology of chemical modelling of biomineral materials and their optical characteristics

Along with studying and developing the processes of biomineralization in living nature, many laboratories make numerous attempts to carry out synthesis of biomineral nanocomposite materials with the help of available biopolymers – proteins and polysaccharides. In this case, the synthesis of biomimetic nanocomposite materials is carried out with the help of sol–gel chemistry methods, since it is the main technological device in the preparation of inorganic oxides (silicon, titanium, aluminum and other chemical elements) complimentary to chemical processes in living systems.

The methods of sol–gel chemistry are the main technological device in obtaining biomineralic oxides [33]. Synthesis of the substance is usually carried out using a two-step process of incorporating biopolymers into the silicate matrix at the sol–gel transition stage and does not lead to mineralization of bio-macromolecules.

A new one-stage approach to the synthesis of biomimetic hybrid nanocomposite materials based on the silicon-containing precursor tetrakis (2-hydroxyethyl) orthosilicate (THEOS) of 50% was proposed in [32, 34], and polysaccharides: Na-alginate (0.5–1%); Na-hyaluronate (0.1–2%); xanthan (0.5–2%), in which there is no stage of sol solution formation occurring in the traditional two-stage process.

In this case, the structure of the inorganic component of the resulting nanocomposite material is determined by the organic matrix–polysaccharide, similar to the formation of inorganic compounds in living organisms, resulting from their deposition (biomineralization) on biomacromolecules acting as templates. Consequently, the mechanism of formation of such biosilicates

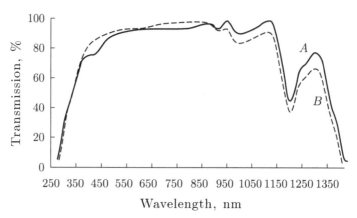

Fig. 14. Spectral transmission of biosilicate nanocomposite materials: *A* – biosilicate based on Na-alginate (1.0%); *B* – biosilicate based on Na-hyaluronate (1.0%) [32].

resembles one of those mechanisms that is used by the nature in the synthesis of spicules.

This method, proceeding at low temperature, was used to produce samples of optically transparent composite biosilicate materials the spectral transmission of which in the range from 350 to 1400 nm is practically the same as the spectral transmission of the material of the DSGS spicules [32].

Figure 14 shows the experimental dependences of the spectral transmission of synthesized nanocomposite biominerals based on Na-alginate and Na-hyaluronate, which are very close to the spectral transmission of the material of the DSGS spicules (Fig. 7). The refractive index of the synthesized materials was $n = 1.517$, which was slightly higher than for the material of the spicules of natural DSGS (Fig. 5).

Small concentrations of organic macromolecules in the resulting materials play the role of a morpho-forming matrix in the form of a complex intersection of fibrils that permeate all material (Fig. 15 *a*). The rest of the material is filled with silicate particles of spheroidal shape (average size ~60 nm) deposited on the organic matrix. Figure 15 *b* is a photograph of a transparent nanocomposite synthesized in an aqueous solution with 50% THEOS precursor content and 1% sodium hyaluronate. The length of the sample is 15 mm [32].

Investigation of the interaction processes of unfocused laser pulses of femtosecond duration (energy ~1 mJ, diameter 7 mm, repetition frequency 100 Hz) with synthesized nanocomposite biomimetic media demonstrated significantly higher levels of nonlinear optical

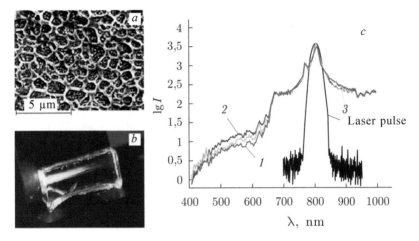

Fig. 15. *a* – SEM photograph of a section of a sample of a nanocomposite material based on Na-hyaluronate; *b* – photograph of a sample of nanocomposite material based on Na-hyaluronate and the filament formed therein; *c* – spectra of the supercontinuum generated in nanocomposite materials based on: *1* – (THEOS (50%) + xanthan (0.5%)); *2* – (THEOS (50%) + Na-hyaluronate (0.25%)); *3* – (THEOS (50%) + Na-alginate (0.25%)) [32].

features in comparison with the DSGS spicules [32]. The USP spectra in Fig. 15 passed through 10 mm long samples of composites based on xanthan, alginate and sodium hyaluronate show a much larger nonlinear character of the interaction of intense USP with these materials. Here the main role is played by the effect of phase self-modulation of the pulses due to the Kerr nonlinearity of the medium and self-focusing of the pulses, which leads in combination with other nonlinear optical processes to the formation of filaments in the samples (Fig. 15 *b*) and the formation of the supercontinuum spectrum. The best results on the efficiency of conversion of pulse energy to the supercontinuum spectrum were shown by samples with sodium hyaluronate for which higher energy levels in the supercontinuum spectrum were observed (Fig. 15 *c*). Samples with sodium alginate proved to be optically unstable, and samples with xanthan showed a high degree of absorption of USP.

A comparison of the total energy of the supercontinuum (P_{SC}) generated in the wavelength range 400–650 nm for samples of the same geometry with different bio-organic additives (up to 1% by weight) has shown that for samples with sodium hyaluronate P_{SC} is more than twice as high than for the samples with xanthan. A stable picture of the generation of the supercontinuum spectrum was observed in samples with sodium hyaluronate, even with a

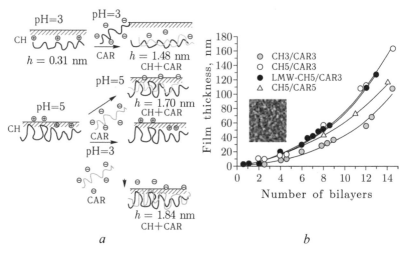

Fig. 16. *a* – scheme of formation of bilayer coatings of chitosan (CH)/carrageenan (CAR) obtained at different pH values of solutions, *h* is the thickness of the layer; *b* – dependence of the thickness of the film on the number of bilayers (the numbers in the notation of the layers correspond to the pH of the solutions upon adsorption); inset – AFM image of the surface of a multilayer coating with 10 bilayers of CH3/CAR3.

thickness of 1 mm. An approximate estimate of the value of the nonlinear-optical coefficient for a material based on sodium hyaluronate, performed using experimental data on the investigation of the filamentation process of USP in the samples yielded the value $n_2 \sim 29 \cdot 10^{-14}$ cm^2/W, which is more than two orders of magnitude higher than the corresponding value of n_2 for the material of the spicules of marine DSGS [32].

Another important feature of the generation of the supercontinuum spectrum in this material is the dependence of the total power of the supercontinuum P_{SC} spectrum on the concentration of sodium hyaluronate polysaccharide and the length of the sample [32], which makes it possible to control the nonlinear optical characteristics of both the material itself and the functional elements based on it.

6.4.2. 2-D and 3-D biomimetic nanocomposite biomineral structures for photonics, biomedicine, catalysis and sorption.

Interpolyelectrolyte reactions are an effective way of synthesizing hybrid structures. In essence, these are reactions with varying degrees of reversibility, which are primarily controlled by the activity and nature of low-molecular electrolytes in the reaction medium, through which self-organization processes are provided in such systems. The interaction of organic polyelectrolytes with colloidal particles makes it possible to create various composite materials with an ordered structure and new functional properties [33, 34].

Figure 16 shows a model for the formation of multilayer coatings on a flat surface by the layer-by-layer method using natural polyelectrolytes – polysaccharides of chitosan and carrageenan having opposite charges of molecules. It also shows an AFM image of a surface covered with such ordered multilayers of polymers.

The change in the pH, the molecular weight of chitosan, the carrageenan type and salt composition in the formation of multilayer chitosan/carrageenan coatings allowed us to establish that only the increment of the film thickness depends on the conditions for the formation of multilayer coatings, and the exponential nature of growth is preserved. This behaviour is a characteristic feature of rigid-chain natural polymers and is extremely rare for synthetic polyelectrolytes. At the same time this property ensures the formation of a thick coating thickness in a few hundred nanometers for a small amount of adsorption cycles.

Such multilayers can be used to create antimicrobial coatings on the surface of medical materials based on chitosan and carrageen [36]. As a result of a series of tests to assess the initial rate of adhesion and viability of bacteria on surfaces modified with chitosan/ carrageenan multilayers and covalently grafted with chitosan layers, it was found that both types of chitosan-containing coatings

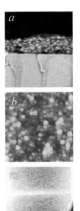

Fig. 17. Multilayer fibre-optic biomineral structure for photonic sensors: *a* – SEM-image of the transverse section of the structure on the surface of the planar fibre; *b* – AFM image of the structure surface; *c* – photograph of the track of the laser radiation channeled in the structure 'planar optical fibre–multilayer coating'. *d* – photo of a waveguide chemosensor.

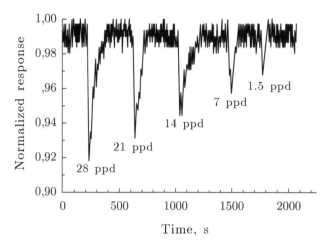

Fig. 18. Response of the waveguide sensor based on a biomimetic multilayered nanocomposite structure doped with a bromothymol blue dye, for the presence of ammonia vapour.

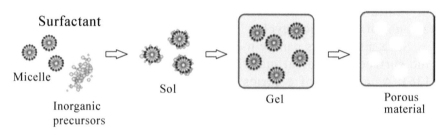

Fig. 19. Diagram of template synthesis of porous materials [34].

significantly reduce bacterial adhesion, with multilayer coatings reducing adhesion 50–100 times, and less hydrated covalently grafted coatings 10–20 times.

The use of multilayer coatings based on heterogeneously charged polymers makes it possible to create sensor coatings on plane waveguides (Fig. 17). Important advantages of such coatings for the formation of sensitive layers of optical sensors are high optical homogeneity, low scattering [37], controlled permeability for gases [38], and the possibility of introducing a strictly defined number of functionalizing additives, including dyes [38]. Multilayered touchscreens with a chitosan/carrageenan with bromothymol blue indicator provide an ammonia detection limit of less than 1 ppm with a response time of less than 30 s (Fig. 18) [35].

When creating three-dimensional ordered structures, an important role is played by the organization of polyelectrolytes in stable three-

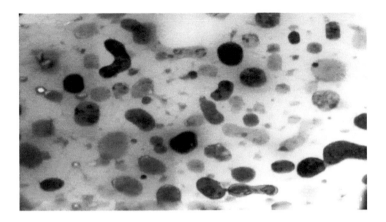

Fig. 20. TEM image of siloxane–acrylate microemulsion particles with gold nanoparticles immobilized therein [34].

dimensional forms in the form of micelles, emulsions, latexes. Self-organization of such structures in the presence of colloids with different charges of the surface makes it possible to obtain structured gels and materials based on them. An example of obtaining such three-dimensional structures is the template synthesis of porous materials, the scheme of which is shown in Fig. 19 [34].

When an organic template with a high surface charge density of colloidal inorganic components with an opposite charge is introduced into the solution, an ordered gel is formed. The resulting gel can be converted to a xerogel and, after removal of the template, forms uniform regular porous structures with pore sizes corresponding to the size of the template used.

Some organic templates can be used as nanoreactors for the synthesis of inorganic nanoparticles. For example, in the thermal reduction of $[AuCl_4]$ in a siloxane–acrylate microemulsion solution, which is a hydrophobic core with a hydrophilic shell, emulsion particles can be obtained in which gold nanoparticles are immobilized (Fig. 20).

Using such templates with immobilized nanoparticles, it is possible to obtain porous structures with nanoparticles distributed in the pores (for example, noble metal nanoparticles). Examples of such materials are given in the SEM images in Fig. 21.

It is obvious that many different ways arise for obtaining structures with different chemical nature of the solid phase and with different pore size distributions in each stage of this synthesis of

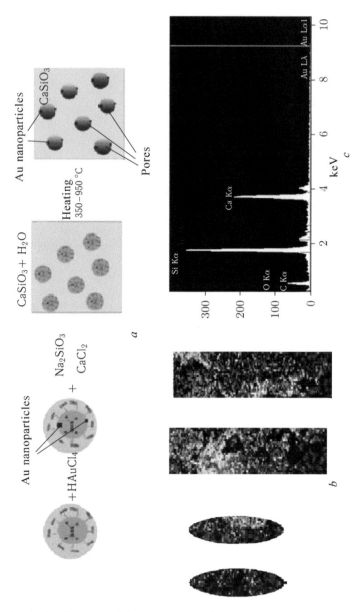

Fig. 21. Synthesis of porous calcium silicate with immobilized with gold nanoparticles in pores. *a* – scheme of synthesis; *b* – SEM image of the material; *c* – surface microanalysis of the surface.

composite porous materials. For example, in the synthesis of porous manganese oxides with immobilized gold nanoparticles, the structure and properties of the resulting materials are found to depend on the temperature of removal of the template, the rate of burnout of the template, and so on [39].

With the use of template synthesis methods, selective sorbents were obtained for sorption of strontium from sea water [40], hydrothermal oxidation catalysts [39], and a number of catalysts based on molybdenum and tungsten oxides for the catalysis of gas-phase reactions [41].

6.5. Biosilification in living systems using cloned proteins silicateins

The production of biosilification processes occurring in living nature is many gigatons per year, which is immeasurably greater than that produced by the entire world industry. At the same time, biological objects operate under environmental conditions and do not impose special requirements on the purity of the initial product and do not require rigid parameters of the synthesis process. Recently discovered silicate proteins from glass sponges are promising for use in nanobiotechnology due to their ability to polymerize silica monomers [42]. Silicateins perform a dual function-enzymatic and structure-forming, which makes it possible to obtain ordered silicate nanofibres *in vitro*. These proteins catalyze the synthesis of not only silicates, but also other materials based on metal oxides: TiO_2, ZrO_2 and Ga_2O_3 [43–45].

It was shown that silicateins are capable of catalyzing the formation of metal nanoparticles and bimetallic nanocrystals of barium fluorotitanate $BaTiOF_4$. With the help of silicates, nanotubes, nanofibres or nanofilms can be produced on the basis of both silica and other compounds, which makes it possible to use silicateins in the technologies of enzymatic formation of inorganic materials for use in semiconductor and optoelectronic devices [42]. The spicules of glass sponges are most interesting for nanotechnological studies, since only these sponges form long, highly ordered structures of polymer silicon. The genes of silicates $LoSilA_1$, $LoSilA_2$, $LoSilA_3$ (family of silicatein-α), $LoSilB$ (family of silicatein-β), and the gene of silicatein-like catcapsin LoCath from the sponge Latrunculia oparinae were cloned in the process of the performed studies [46]. For the first time, pacific ocean sponges were used for this purpose, the collection of which was carried out from depths of 300 to 600 m during the voyages of the Akademik Oparin research vessel (Far East Branch of the Russian Academy of Sciences) in the region of

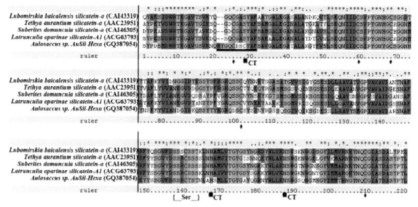

Fig. 22. Amino acid sequence of silicatein AuSil-Hexa from *Aulosaccus schulzei* glass sponge. The asterisks, dots and double dots indicate identical residues and amino acid substitutions with low and high similarity. Characteristic sites within silicateins are shown: amino acids forming the catalytic triad CT (Cys/Ser, His and Asn). Arrows indicate the cysteine residues involved in the formation of disulphode bonds. Ser – serine cluster [46].

the Middle Kurile Islands nd Vietnam. Also, for the first time in the search for genes, we used the technology of screening a large number of cDNA clones containing all possible isoforms of genes of interest to us.

The determination of the full-length sequences of cDNAs of silicatein genes was first carried out using a method based on the template switching effect and selective suppression of PCR polymerase chain reaction [44]. Further in the glass sponges we found a silicate gene, called AuSil-Hexa. The full-length gene was cloned by RACE-PCR and a phylogenetic analysis of the corresponding protein was performed. A detailed phylogenetic analysis showed that the AuSil-Hexa protein is a silicatein.

Thus, for the first time in the world, a silicatein gene was cloned from the glass sponges, the amino acid sequence of the corresponding protein was compiled (Fig. 22), from which it follows that AuSil-Hexa contains an unusual catalytic site in which the serine of the active site is replaced by cysteine [47]. The LoSilA$_1$ gene was expressed in various bacterial systems, as well as in plant cell cultures. The recombinant silicotene LoSilA$_1$ did not possess the ability to polymerize conventional silicon monomers under in vitro conditions, however, in the presence of an organic silicon compound, THEOS rumored hexagonal prisms from silica of 200–400 nm size (Fig. 23) [48].

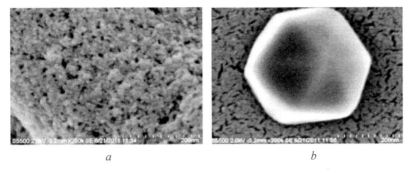

a   *b*

Fig. 23. Biological synthesis of SiO_2 silicon nanocrystals using recombinant silicatein protein, produced by the bacterial medium E-coli: *a* – negative control; *b* – crystal of silicon oxide [48].

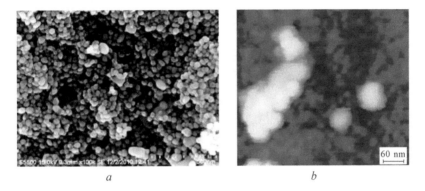

a   *b*

Fig. 24. Scanning electron microscopy (a) and atomic force microscopy (b) of silver nanoparticles obtained by phytosynthesis in the presence of recombinant silicotene LoSilA1.

Expression of the silicatein gene in plant cell cultures also showed an interesting result. It turned out that $LoSilA_1$ regulates the process of reduction of silver Ag(I) ions to nanoparticles containing Ag(0). This process, called Phytosynthesis or Green Synthesis of Metal Nanoparticles, began to develop after the publication of data on the successful production of nanoparticles of gold [48]. The presence of recombinant silicatein in plant cells made it possible to regulate the size of silver nanoparticles and triply increased the efficiency of phytosynthesis (Fig. 24) [48].

In recent studies, it has been shown that the use of extracts of plant cell cultures makes it possible to obtain more reproducible results in the production of metal nanoparticles. Since the cell culture is a simpler system than an entire plant, and culturing parameters are easily regulated, plant cell cultures can serve as a good model system for studying the formation of metal nanoparticles in plant cells based on the expression of silicatein proteins in them, and become the basis for a new biotechnological way of obtaining them.

References

1. Parker A.R., et al., Nature. 2003. V. 426. P. 786.
2. Meyers M.A., et al., Progress in Materials Science. 2008. V. 53. P. 1.
3. Yahya H., Biomimetics: Technology Imitates Nature. London, 2006.
4. Sundar V.C.. et al., Nature. 2003. V. 424. P. 899.
5. Aizenberg J., et al., Science. 2005. V. 309. P. 275.
6. Müller W.E.G., et al., Biosensors and Bioelectronics. 2007. V. 21. P. 1149.
7. Cattaneo-Vietti R., et al., Nature. 1996. No. 383. P. 397.
8. Leys S.P., et al., Advances in Marine Biology. 2007. V. 52. P. 1-145.
9. Müller W.E.G., et al., Biosensors and Bioelectronics. 2006. V. 21. P. 1149.
10. Leys S.P., et al., Advances in Marine Biology. 2007. V. 52. P. 1–145.
11. Kulchin Yu.N., et al., Vestnik DVO RAN. 2007. No. 1. P. 27.
12. Kulchin Yu.N., et al., Optical Memory and Neural Networks, 2007. v. 16. P. 189.
13. Kulchin Yu.N., Kvant. elektronika. 2008. V. 38. P. 51.
14. Small Angle X-ray scattering, ed. by Glatter O. and Kratky O., Academic press Inc. (London), 1982.
15. Kulchin Yu.N., et al. in: Perspective directions of development of nanotechnologies in the Far East Divison of the RAS, volume 2 (Ed. Yu.N. Kulchin), 2009. Vladivostok, IAPU DVO RAN. P.107.
16. Ehrlich H., et al., Journal of Nanomaterials. 2008. V. 10. P. 1155.
17. Kulchin Yu.N., Khim. Fiz. Mezoskopiya. 2009. V. 11, No. 3. P. 310.
18. Ernst R., et al., NMR in one and two dimensions, transl. from English., ed. K.M. Salikhov. Moscow, Mir, 1990..
19. Kulchin Yu.N., Rare Metals. 2009. V. 28. P. 66.
20. Ascension S.S., et al., Ross. Nanotekhnologii. 2010. V. 5. P. 126.
21. Aizenberg J., et al., Biological glass fibres: Correlation between optical and structural properties, Proc. Natl. Acad. Sci. USA, 2004. V. 101. P. 3358.
22. Kulchin Yu.N., et al., in: Biosilica in Evolution, Morphogenesis, and Nanobiotechnology, ed. W.E.G. Mülle and M.A. Grachev. Springer, 2009. P. 315.
23. Konorov S.O., Zh. Eksper. Teor. Fiz. 2003. V. 123. P. 975.
24. Voznesenskii S.S., et al., Ross. Nanotekhnologii. 2011. V. 6. No. 1, 2. P. 60.
25. Kulchin Yu.N., Optika Spektroskopiya. 2009. V. 107. P. 468.
26. Kulchin Yu.N., et al., Pis'ma Zh. Tekh. Fiz., 2008. V. 34. P. 1.
27. Kulchin Yu.N., et al., in: Perspective directions of development of nanotechnologies in the Far East Divison of the RAS, volume 3 (Ed. Yu.N. Kulchin), 2010. Vladivostok, IAPU DVO RAN. P. 57.
28. Kulchin Yu.N., Optika Spektroskopiya. 2011. V. 111. No. 5. P. 858.
29. Kulchin Yu.N., Vestnik RAN. 2013. P. 83, No. 2. P. 1–13.
30. Drozdov A.L., et al., DAN. 2008. V. 420, No. 4. P. 565.
31. Kulchin Yu.N., et al., Photonics of biomineral and biomimetic structures and materials. Moscow. Fizmatlit, 2011. P. 224.
32. Kulchin Yu.N., et al., Laser Physics. 2011. V. 21. P. 1.
33. Dunn B., Acta Mater. 1998. V. 46. P. 737.

34. Shchipunov Yu.A., in: Perspective directions of development of nanotechnologies in the Far East Divison of the RAS, volume 2 (Ed. Yu.N. Kulchin), 2009. Vladivostok, IAPU DVO RAN. P.157.
35. Kulchin Yu.N., et al., Vestnik RAN. 2013. V. 83, No. 2. p. 99–111.
36. Bratskaya S.Yu., Vestnik DVO RAN. 2009. No. 2. P. 84.
37. Corres J.M., et al., Sensors and Actuators B. 2007. V. 122. P. 442.
38. Wang Y., et al., Journal of the American Chemical Society. 2008. V. 130 (49). P. 16510.
39. Avramenko V.A., etc., DAN. 2010. V. 435, No. 4. P. 487.
40. Avramenko V.A., et al., Zh. Fiz. Khimii. 2004. V. 78, No. 3. P. 493.
41. Papynov E.K., et al., Khim. Tekhnologiya. 2011. No. 6. P. 367.
42. Schröder H., et al., Nat. Prod. Rep. 2008. V. 25. P. 455.
43. Ehrlich H., et al., Nature Chem. 2010. V. 2. P. 1084–1088.
44. Fairhead M., et al., Chem. Commun. (Camb). 2008. V. 15. P. 1765–1768.
45. Müller W.E.G., et al., Marine Biotechnology (NY). 2010. V. 12. P. 403–451.
46. Veremeichik G.N., et al., Marine Biotechnology (NY). 2011. V. 13. P. 810.
47. Shankar S.S., et al., Nature Materials. 2004. V. 3. P. 482.
48. Shkryl' Yu.N., et al., Method for Obtaining Metal Nanoparticles, in: Patent No. 2011145718. B22F1/00, B22F9/24. 2011.

7

Dynamic Holography and Optical Novelty-Filters

Introduction

Novelty-filters are elements of optical signal processing systems that serve to recognize or isolate real-time changes in the structure of images [1, 2]. Analogues of such filters are widely distributed in animals as primary elements of information processing. For example, frogs with Novelty-filters detect flying insects with high accuracy. The humans also use Novelty-filters to remove from the visual field the images of the blood vessels located on the front of the sensitive retina of the eyeball. For this purpose, we continuously scan the image of the object, moving the eyeball and subtract from the image objects that do not change, that is, the image of the grid from the blood vessels [3].

The Novelty-filters used in the optical systems are analogues of high-frequency spectral filters in radio engineering. However, despite the similarity, optical Novelty-filters have their own characteristics, which are caused by the use of quadratic detectors at the output of the processing system (eyes or television camera) detecting the optical intensity field better than the amplitude field. To understand this difference, consider an example of the detection of insects flying over the flowers of water lilies in a stationary pond.

When the image is transmitted via the Novelty-filter, the image of the pond with water lilies will initially be obtained. But after a while the filter adapts and removes the image of the fixed pond from the image at the output of the system. Flying insects in the output image are always constantly new objects, as a result they will be seen.

A characteristic feature of Novelty-filtration is that if an insect flies from one lily to another, then, as a result of the finality of the filter's response time, the insect will be continuously observed as if in two places: where it was originally located, i.e., in the initial position, and, as it quickly moved, then at the same time on another water lily, where it appeared as a new one.

Despite the fact that detecting changes in the observed object or scene is extremely important, it is equally important to know what has remained unchanged in the continuously changing scene? For example, consider the process of monitoring the trajectory of a moving object. In this case, we continuously scan our attention throughout the observed scene to fix a specific object at every moment of its movement in a continuously changing scene as a result of scanning. Thus, we can classify the Novelty-filter also as a low-pass filter. The most successful low-pass filters are used in observing vibrating objects. Therefore, very often Novelty-filters use holograms for this purpose.

Both, high-frequency and low-frequency Novelty filters are the main elements for linear image processing systems. The presence of such functions makes optical systems more attractive in comparison with electronic devices when processing images. The latter is due to the fact that when using digital image processing systems to detect changes in the current scene, it becomes necessary to perform a number of numerous sequential operations: perception and digitization of the current scene, transfer of digital data to memory, subtraction of the previous scene from the subsequent, presentation of the processing results (and so it happens scene after scene), which significantly increases the processing time and does not allow one to monitor fast-paced processes in real time. Novelty-filters have been specially developed with the use of dynamic holograms recorded in photorefractive crystals to create them. In this case, unlike electronic devices, optical Novelty-filters use dynamic real time holograms to memorize the current scene. On the dynamic hologram, an exponentially damped time-averaged scene, recorded at the input of the processing system, is stored. As a result of further interference of the instantly memorized scene with the current scene, parallel processing of the image takes place. As a result, the Novelty-filter is an extremely simple element for processing fast-changing scenes and works well and quickly. In this case, the output signal consists of strongly contrasting white-to-black elements of the image of those places where the changes occurred, and where such changes were not correspondingly. The depth of the observed contrast is approximately 30:1. The speed and spatial resolution of processing systems with Novelty-filters are mainly determined by the capabilities of space-time light modulators and television monitors used for input and output of information.

7.1. The interaction of two plane waves on dynamic holograms in photorefractive crystals

All Novelty-filer based image processing units discussed in this chapter use photorefractive crystals as active elements. In connection with this, we briefly consider the features of the process of coupling of two interacting waves on dynamic holograms recorded in photorefractive crystals, which will allow us to understand the physical foundations of linear processing systems based on Novelty-filters.

The phenomenon of coupling of two plane waves in a photorefractive crystal essentially consists of several processes. As a result of interference of the object and reference rays in the volume of the photorefractive crystal, a spatially modulated distribution density of electric charges arises in the latter, due to uneven absorption of radiation in the spatially periodic field of interfering waves. The static electric field produced by the spatially periodic charge distribution deforms the crystal lattice of the crystal, which leads to a spatial modulation of the refractive index and the formation, mainly, of a phase hologram in the volume of the crystal. One of the noteworthy properties of photorefractive crystals is that the recorded phase lattice turns out to be spatially displaced with respect to the interference pattern that creates it. In this connection, the created holographic grating connects the amplitudes of the plane waves recording it in such a way that one of them is amplified and the other is attenuated. The direction of energy transfer during the interaction of waves depends on their orientation with respect to the c-axis in the crystal.

To describe the interaction of two plane waves (Fig. 1) on a dynamic hologram, we assume that they have the same cyclic frequency ω and their amplitude varies little as they propagate in the crystal. In the scalar approximation, the complex amplitudes of these waves can be described by the following expressions:

$$E_1(\mathbf{r},t) = \frac{1}{2} E_{10}(z) \exp\left[i(\mathbf{k}_1\mathbf{r} - \omega t)\right] + \text{c.c.,} \qquad (7.1)$$

$$E_2(\mathbf{r},t) = \frac{1}{2} E_{20}(z) \exp\left[i(\mathbf{k}_2\mathbf{r} - \omega t)\right] + \text{c.c.,} \qquad (7.2)$$

where $E_{10}(z)$, $E_{20}(z)$ are the amplitudes of interacting plane waves slowly varying in the direction of the z axis, $\mathbf{k}_1$, $\mathbf{k}_2$ are wave vectors, $\mathbf{r}$ is the radius vector, i is the imaginary unit, c.c. is the notation of complex conjugation.

The waves introduced into the crystal propagate in such a way that the bisector of the angle between them coincides with the z axis, and the plane of their intersection is the plane of (x, z).

The volume phase holographic grating formed in the photo-refractive crystal (Fig. 1) can be described with the aid of the scalar 'grating field' $G(z, t)$, whose amplitude is proportional to the amplitude of the change in the refractive index of the crystal material

$$G(z,t) = \frac{-i\omega\Delta n(z,t)}{(2c_1)}, \qquad (7.3)$$

where c_1 is the speed of light in a vacuum.

According to [5, 6], the slowly varying distribution of the amplitude of the 'grating field' can be described by the following equation:

$$\frac{\partial G(z,t)}{\partial t} = \gamma\left(-G(z,t) + \frac{\Gamma}{2} \frac{E_{10}(z)E_{20}^*(z)}{I(z)}\right), \qquad (7.4)$$

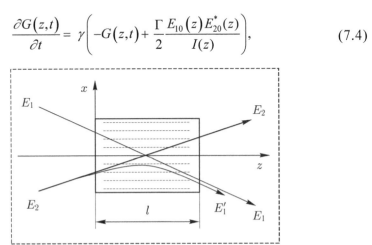

Fig. 1. Scheme of interaction of two plane waves on the dynamical holographic grating created by them in the photorefractive crystal: E_1 and E_2 – waves incident on the crystal; E_1' – the wave diffracted on the grating.

where γ is a structural constant of the 'grating field' proportional to the intensity, Γ is the constant of the coupling of plane waves interacting on the lattice, and * is the sign of complex conjugation. In the general case, both these constants are complex quantities, and their analytic dependences (on the orientation of the photorefractive crystal and the characteristics of its material, and on the radiation intensity and the magnitude of the applied external electric field) can be found in [6].

The intensity of the total radiation field of the interfering waves $I(z)$ weakly changing in the direction of the z axis because of the absorption in the crystal is

$$I(z) = |E_{10}(z)|^2 + |E_{20}(z)|^2. \tag{7.5}$$

In a medium having a large coupling constant Γ, the recorded phase diffraction grating changes the distribution of the optical field, which in turn also affects the lattice. Such interference of the grating and the optical field occurs both in time and in space, which complicates the analytical description of the process. However, for the steady-state process of formation of the phase dynamic hologram $\partial G(z, t)/\partial t = 0$. Then, according to (7.4), the amplitude of the 'grating field' can be expressed in the following form

$$G(z) = \frac{\Gamma}{2} \frac{E_{10}(z) E_{20}^*(z)}{I(z)}. \tag{7.6}$$

The spatial variation of the amplitudes of plane light waves interacting on a phase dynamic hologram can be described using the equations of the theory of coupled waves [7]:

$$\frac{dE_{10}(z)}{dz} = G(z) E_{20}(z) - \frac{\alpha}{2} E_{10}(z), \tag{7.7}$$

$$\frac{dE_{20}(z)}{dz} = -G^*(z) E_{10}(z) - \frac{\alpha}{2} E_{20}(z), \tag{7.8}$$

where α is the absorption coefficient.

Equations (7.7) and (7.8) can be rewritten in terms of intensity and phase taking into account that $E_0(z) = I^{1/2}(z) e^{i\phi}$ and. $dI(z)/dz = E_0^*(z) dE_0(z)/dz + \text{c.c.}$ (7.6), we obtain expressions for the changes in the intensities I_1 and I_2, and also for the phases ϕ_1 and ϕ_2 of the interacting plane light waves as they propagate in the crystal:

$$\frac{dI_1}{dz} = \text{Re}\{\Gamma\}\frac{I_1 I_2}{I_1 + I_2} - \alpha I_1, \tag{7.9}$$

$$\frac{dI_2}{dz} = -\text{Re}\{\Gamma\}\frac{I_1 I_2}{I_1 + I_2} - \alpha I_2, \tag{7.10}$$

$$\frac{d\phi_1}{dz} = \frac{1}{2}\,\text{Im}\{\Gamma\}\frac{I_2}{I_1 + I_2}, \tag{7.11}$$

$$\frac{d\phi_2}{dz} = \frac{1}{2}\text{Im}\{\Gamma\}\frac{I_1}{I_1 + I_2}. \tag{7.12}$$

The coupling constant in a photorefractive crystal directly depends on the Pockels coefficient and the refractive index of the crystal material and can take a real value if the grating of the refractive index in the crystal is phase shifted by $\pi/2$ radians with respect to the interference pattern. In this case, note that for $\text{Re}\{\Gamma\} > 0$, the intensity of the ray I_1 will increase with the distance, and if at the entrance to the crystal $I_1 \ll I_2$, its intensity will grow exponentially and not depend on the intensity of the second ray I_2. In this case, even if $I_1 > I_2$, the more intense first ray will take energy from the second, less intense beam.

The equations (7.9)–(7.12) obtained above are valid only for plane waves and will not be satisfied for waves carrying images. Nevertheless, using these relationships, at this stage it is possible to get some ideas about the properties of Novelty-filters. To this end, according to the coupling constant (7.11), (7.12), the phases of the interacting waves remain constant, and their intensities, when propagating over a distance $z = l$, are, respectively, equal to:

$$I_1(l) = I_1(0)\frac{1+r}{1+r/g}, \tag{7.13}$$

$$I_2(l) = I_2(0)\frac{1+1/r}{1+g/r}, \tag{7.14}$$

where $g = \exp[\text{Re}\{\Gamma\}\,l]$ is the gain factor for the coupled waves, and r is the ratio of the intensities of the light waves at the entrance to the crystal:

$$r = \frac{I_2(0)}{I_1(0)}. \tag{7.15}$$

Suppose that a wave with index 1 carries an image, that is, it is an object image. Since the magnitude of the wave coupling amplification factor can be significant, wave 2, when passing through the crystal, will lose energy and, according to (7.14), its intensity (I_2) will be close to zero.

What happens if the image changes? For simplicity of reasoning, suppose that both waves at the entrance to the crystal have equal intensities: $r = 1$. Also assume that the phase of the object wave 1 suddenly changes to π radian, in this case the intensity distribution in the interference pattern shifts by $\lambda/2$, where λ is the wavelength of the radiation used. But this is equivalent to the fact that the coupling gain changed the sign to opposite. Taking into account (7.13), (7.14) and the assumptions made, we obtain the value for the contrast (C), defined as the ratio of the intensities of the object (I_{novel}) and the reference (I_{old}) waves after their passage through the dynamic hologram:

$$C = \frac{I_{novel}}{I_{old}} = \frac{I_1(l)}{I_2(l)} = \frac{1+g}{1+1/g} = g. \tag{7.16}$$

Thus, the limiting value of the achieved contrast using the Novelty-filter can be $C = g$. In some crystals, the amplification value for a two-wave interaction can reach 100, hence the contrast value in the processing system can also reach this value.

It should also be noted that after a certain time, equal to the response time of the crystal, the phase hologram will be overwritten and the contrast will decrease to values determined from the relations (7.13) and (7.14).

7.2. The transfer characteristic of an optical Novelty-filter

As already noted above, the optical Novelty-filter can operate as a low-pass filter as well as a high-pass filter. Our task will be to find a general approach to describing the principles of the filter operation and calculating its characteristics, as well as to determine the limits of their applicability. During the development of Novelty-filters, many different methods, designs and models for describing their work [8–10] were suggested, which allowed to perform operations primarily with amplitude or phase objects. The most universal approach for creating a Novelty-filter, capable of operating both with amplitude and phase objects, was first proposed in [11] and considered in detail in the review [2]. The principle of such a Novelty-filter is based on the phenomenon of strong coupling of a weak informative wave with a strong pump wave in a photorefractive crystal. The filtering scheme is shown in Fig. 1. A wave with amplitude $E_{10}(x, z, t)$ is an object wave, and a wave with amplitude $E_{20}(x, z, t)$ is a pump wave, and the condition $I_2 \gg I_1$ is fulfilled in this case. Here we assume that the amplitude of the interacting waves depends only on the coordinates (x, z). The photorefractive crystal is oriented in such a way that the dynamic hologram created by the interfering waves ensures the transfer of energy from the pump wave to the signal wave. In this case, the [object wave to be reconstructed by the pump wave will be shifted in phase by a (radian with respect to a directly propagating object wave and both these waves will quench each other. If the objective wave is changed at some time, the object wave reconstructed on the hologram by the pump wave the course of the characteristic, recording-dependent and material of the crystal, the time interval [12] will remain unchanged.

The result of the addition of these two waves will be the destructive interference of the transmitted signal wave and the diffracted pump wave (the reconstructed wave), as a result of which the complex amplitude of the total object wave at the output can be represented in the form

$$E_{10}\left(x,l,t\right)_{\text{out}} \approx E_{10}\left(x,l,t\right)_{\text{in}} - E_{10}\left(x,l,t-\tau\right) \approx \tau\frac{dE_{10}\left(x,l,t\right)_{\text{in}}}{dt}, \quad (7.17)$$

where l is the length of the crystal in the z-direction; x is the coordinate normal to the z axis; $E_{10}(x,l,t)_{\text{in}}$ is the amplitude of the object wave after the passage of the crystal at time t; $E_{10}(x,l,t-\tau)$ is the amplitude of the wave reconstructed on the dynamic hologram stored at time $(t-\tau)$; $E_{10}(x,l,t)_{\text{out}}$ is the resulting amplitude of the object wave at the output of the system at time t.

As can be seen, the proposed filtering scheme simply allows the temporal differentiation of the complex amplitude of the optical signal entering the crystal. In this case, for stationary scenes when there are no changes in the structure of the object wave, the output signal will be the result of destructive interference of waves and its contrast will directly depend on the ratio of the intensities of the object and reconstructed waves. At the same time, for the case of a time-varying structure of the object wave, this filter will remove static fragments from the image and select the changing objects in the observed scene.

Expression (7.17) allows one to understand the principle of the Novelty-filter, but does not provide an opportunity to describe the characteristics of its operation. Since, as noted above, image processing systems based on Novelty-filters are analogous to frequency filters for electrical signals, it is convenient to use the concept of the transfer function and the Laplace transform system to describe their operation [13]. We use the definition of the direct and inverse time Laplace transform for the amplitudes of the light waves entering the crystal $E_0(t)$:

$$E_0(t) = \frac{1}{2\pi i} \int_{\sigma-i\infty}^{\sigma+i\infty} e(s)\exp(st)ds, \qquad (7.18)$$

$$e(s) = \int_0^\infty E_0(t)\exp(-st)dt, \qquad (7.19)$$

where e is the complex frequency of the amplitude of the light wave: $s = \sigma + i\Omega$, wherein σ characterizes the attenuation and Ω is the frequency of the change (modulation) of the amplitude of the light wave; the beginning of the process corresponds to the instant of time $t = 0$.

The advantage of using such a signal representation is that the filter operation can be described using the concept of the transfer function $h(s)$ such that the amplitudes of the spectral components at the input $e(s)$ and at the output $\tilde{e}(s)$ are related by the following relation:

$$\tilde{e}(s) = h(s)e(s). \qquad (7.20)$$

As a result, the amplitude of the signal at the output of the system is

$$E_{out}(t)_0 = \frac{1}{2\pi i} \int_{\sigma-i\infty}^{\sigma+i\infty} h(s)e(s)\exp(st)ds. \qquad (7.21)$$

The characteristics of high- and low-frequency filters are set corresponding to the position of the zeros and poles of the transfer function.

To find the type of transfer function $h(s)$ of the Novelty-filter under conditions of two-wave interaction in a photorefractive crystal, using (7.1), (7.2), we write the expressions for the amplitudes of the object wave and pump wave, respectively, in the form

$$E_{10}(\mathbf{r},t) = \mathrm{Re}\{E_{10}(z)\exp[i(\mathbf{k}_1 \cdot \mathbf{r} - \omega t)]\}, \tag{7.22}$$

$$E_{20}(\mathbf{r},t) = \mathrm{Re}\{E_{20}(z)\exp[i(\mathbf{k}_2 \cdot \mathbf{r} - \omega t)]\}. \tag{7.23}$$

Taking into account the smallness of the amplitude of the object wave and the considerable intensity of the pump wave, it can be approximately assumed that the amplitude of the pump wave remains constant during propagation in the photorefractive crystal ($E_{20}(z) \approx E_{20}$ = const). In this case, neglecting the losses for the object wave: $\alpha = 0$, according to (7.7), we obtain an expression for describing the dynamics of the change in the amplitude of the object wave:

$$\frac{\partial E_{10}(z,t)}{\partial z} = G(z,t)E_{20}. \tag{7.24}$$

Using relation (7.24), we rewrite the equation for the amplitude of the 'grating field' (7.4) in the form:

$$\frac{\partial G(z,t)}{\partial t} 0 = \gamma \left[-G(z,t) + \frac{\Gamma}{2} \frac{E_{10}(z,t)E_{20}^*}{I(z,t)} \right]. \tag{7.25}$$

Considering that $I_2 \gg I_1$, we can assume that $I(z,\ t) \approx |E_{20}|^2 = I_2 =$ const, then, using (7.24), equation (7.25) can be represented in the following form

$$\frac{\partial^2 E_{10}(z,t)}{\partial t\,\partial z} = \gamma \left[-\frac{\partial E_{10}(z,t)}{\partial z} + \frac{\Gamma}{2} E_{10}(z,t) \right]. \tag{7.26}$$

Applying to (7.26) the formalism of the Laplace transforms (7.18), (7.19), we obtain

$$s\frac{\partial e(z,s)}{\partial z} = \gamma \left[-\frac{\partial e(z,s)}{\partial z} + \frac{\Gamma}{2} e_{10}(z,s) \right], \tag{7.27}$$

or

$$\frac{\partial e(z,s)}{\partial z} = \frac{\gamma\Gamma}{2(s+\gamma)} e(z,s). \tag{7.28}$$

This equation has the following solution

$$e(z,s) = e(0,s)\exp\left[\frac{\gamma \Gamma z}{2(s+z)}\right],\qquad(7.29)$$

where

$$e(0,s) = \int_0^\infty E_{10}(0,t)\exp(-st)dt.\qquad(7.30)$$

Performing the inverse Laplace transform for expression (7.29), taking into account that the size of the crystal is $z = 1$, we obtain for the amplitude of the object wave at the output from the Novelty-filter under consideration:

$$E_{10}(l,t) = \frac{1}{2\pi i}\int_{\sigma-i\infty}^{\sigma+i\infty} e(0,s)\exp\left[\frac{\gamma \Gamma l}{2(s+\gamma)}\right]\exp(st)ds.\qquad(7.31)$$

For some particular cases of functional dependences, $e(0, s)$, the expression (7.31) is easily integrable (see, for example, [14]).

Comparing expressions (7.21) and (7.31), we can see that the transfer function of a Novelty-filter based on a photorefractive crystal is described by expression

$$h(s) = \exp\left[\frac{\gamma \Gamma l}{2(s+\gamma)}\right].\qquad(7.32)$$

7.3. Features of optical Novelty-filters

7.3.1. Low-frequency and high-frequency Novelty-filters

To understand how the filter characteristics depend on the parameters γ and Γ, we assume that $s = i\Omega$, and represent (7.32) in the form

$$h(i\Omega) = \exp\left[\frac{\gamma^2 \Gamma l}{2(\gamma^2 - \Omega^2)} - i\frac{\Omega\gamma \Gamma l}{2(\gamma^2 - \Omega^2)}\right].\qquad(7.33)$$

As seen from (7.33), the properties of the Novelty-filter, in the first place, are strongly dependent on the magnitude and sign of the real and imaginary components of the wave coupling coefficient (Γ). If Im $\Gamma = 0$, and Re$\Gamma > 0$, then for $\Omega = 0$ the value of the transfer

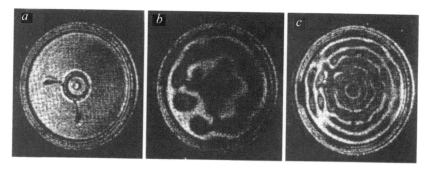

Fig. 2. The image of an oscillating diffuser membrane diffuser transferred through a monotic filer and excited at different frequencies [16].

function will be $h(0) = \exp(\Gamma l/2)$, and for $\Omega \rightarrow \infty$ its value will tend to zero. Apparently, for $\mathrm{Re}\Gamma > 0$ there is an implementation of a low-frequency or monotonic filter. This type of filters has been studied in detail with reference to the use of crystals such as BSO [15], BTO [16] and SBN [17] for their creation. Figure 2 shows experimental photographs demonstrating the operation of a low-frequency (monotonic) Novelty-filter used to study the mode structure of the oscillations of the electrodynamic acoustic diffuser membrane.

If, on the other hand, $\mathrm{Im}\Gamma = 0$, and $\mathrm{Re}\Gamma < 0$, then for $\Omega = 0$ the transfer function value is equal to $h(0) = \exp(-\Gamma l/2)$, and for $\Omega = \infty$ its value will tend to 1. That is, in this case, the high-pass filter function will be implemented. To illustrate the operation of such a Novelty-filter Fig. 3 shows the calculated dependences of the amplitude and phase changes of the filter transfer function, constructed for $\Gamma l/2 = -10$ and $\gamma = 1$ [2].

Using the obtained form of the transfer function and the Laplace transform properties, it is possible to calculate the contrast value for the high-frequency Novelty-filter ($\mathrm{Re}\Gamma < 0$):

$$C = \frac{I_{\mathrm{novel}}}{I_{\mathrm{old}}} = \left|\frac{h(i\infty)}{h(i0)}\right|^2 = e^{\Gamma l} = g. \qquad (7.34)$$

As can be seen, the magnitude of the contrast achieved with high-frequency filtering depends on the practical value of the amplification of the interacting waves. It is also noteworthy that the value obtained for the contrast value agrees very well with the estimates made in section 1 when considering the result of the interaction of two plane waves. The research of high-frequency Novelty filters has been

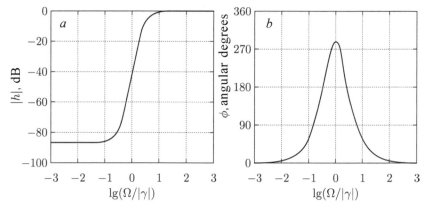

Fig. 3. Amplitude (*a*) and phase (*b*) characteristics of the transfer function of the high-frequency Novelty-filter [2].

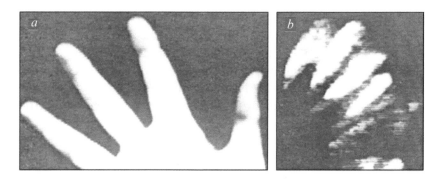

Fig. 4. The image of a human hand on the vidicon of a television camera (*a*) and the result of two-dimensional differentiation of the hand image when it moves in the input plane of the Novelty-filter (*b*) [11].

studied in many studies, a fairly comprehensive survey of which can be found in [2].

Experimental illustration of the operation of a high-frequency Novelty-filter based on a BaTiO$_3$ crystal, $\tau \approx 0.8$ s [11], as shown in Fig. 1, is shown in Fig. 4. The figure shows the result of the two-dimensional differentiation of the image of the human hand moving rapidly before the vidicon of the television camera.

In conclusion of the consideration of low-frequency and high-frequency Novelty-filters, it should be noted that for both filters there is a singular point in the frequency interval determined by the characteristic recording time of the dynamic hologram in which the amplification of the signal can vary greatly.

7.3.2. Bandwidth Novelty filter

In the previous section, we assumed that the coupling coefficient (Γ) and the structural constant of the filter (γ) take only real values, as a result of which, depending on the sign of the coupling coefficient, either a high-frequency or a low-frequency type of the time filter is realized. If, on analyzing the filter operation it is taken into account that both these constants are complex quantities, the result will change radically. In particular, in the steady state, when $s = 0$, if the wave coupling constant has a purely complex value, then, as follows from (7.32), there will be no energy transfer between the waves, and energy exchange will occur only under transient conditions. However, if the amplitudes of the interacting waves are equal, then the process of energy transfer will not be observed even under the conditions of the transient process.

Consider the case when the structure constant of the filter has an imaginary value: $\gamma = \tilde{\gamma} + i\tilde{\Omega}$. In this case, the transfer function of the filter (7.32) can be written in the form

$$h(s) = \exp\left[\frac{\Gamma l}{2} \frac{\tilde{\gamma} + i\tilde{\Omega}}{s + \tilde{\gamma} + i\tilde{\Omega}}\right]. \tag{7.35}$$

If, as in section 7.3.1, the argument of the transfer function is represented in the form $s = i\Omega$ and we analyse the expression (7.35) for the extremum, it can be established that the gain maximum for a two-wave interaction will be observed at a frequency

$$\Omega = \Omega_{max} = |\gamma| = \left(\tilde{\gamma}^2 + \tilde{\Omega}^2\right)^{1/2}. \tag{7.36}$$

For all other frequencies on both sides of the frequency Ω_{max} the amplification of the signal will be weakened.

Having transformed (7.35), taking into account that $s = i\Omega$, to the form:

$$h(i\Omega) = \exp\left[\frac{\Gamma l}{2} \frac{\Omega\tilde{\Omega} + |\gamma|^2 - i\Omega\tilde{\gamma}}{\tilde{\gamma}^2 + \left(\Omega + \tilde{\Omega}\right)^2}\right], \tag{7.37}$$

and assuming that Γ is a purely imaginary quantity, that is, $\mathrm{Re}\{\Gamma\} \approx 0$ (this case occurs in photorefractive crystals of the BSO type in the case of using strong static electric fields applied along the c-axis of the crystal [18]), it can be seen that for limiting cases

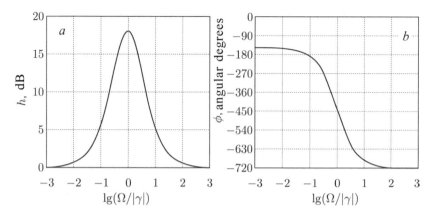

Fig. 5. Amplitude (*a*) and phase (*b*) characteristics of the band Novelty-filter [2].

of deviation from the central frequency of the filter ($\Omega = \Omega_{max}$): $\Omega \to 0$ and $\Omega \to \infty$ the gain in the filter tends to a constant value: $|h^2| \to 1$.

Thus, the Novelty-filter, created on the basis of a photorefractive crystal with imaginary values of the coupling coefficient of waves and the structure constant, is a bandpass filter. Figure 5 shows the calculated dependences of the magnitude and phase of the transfer function for a bandwidth Novelty-filter with the following parameters: $\Gamma l/2 = 10i$ and $\tilde{\gamma} = \tilde{\Omega} = 1/\sqrt{2}$ [2].

The magnitude of the maximum contrast of the bandwidth Novelty-filter can be obtained by calculating the ratio for the maximum and minimum of the gain function obtained for the frequencies $\Omega = \Omega_{max}$ and $\Omega = 0$ or $\Omega \to \infty$ respectively. In this way

$$C = \left| \frac{h(i\Omega_{max})}{h(i0)} \right|^2 = \exp\left[\frac{l \, \text{Im}\{\Gamma\}}{2} \frac{\tilde{\gamma}}{|\gamma| + \tilde{\Omega}} \right]. \tag{7.38}$$

7.4. Novelty-filters based on the use of the phenomenon of fanning in photorefractive crystals

The phenomenon of fanning is observed practically in all photorefractive crystals and in most cases it is a parasitic effect, leading to noise in information processing systems [19]. The process of appearance of fanning is considered in detail in [9, 11]. Initially, a coherent light wave is scattered on defects or impurities that are always present in photorefractive crystals, which creates

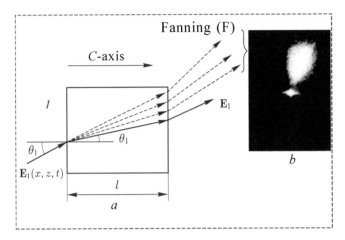

Fig. 6. Novelty-filter based on the phenomenon of fanning in a photorefractive crystal: a – Novelty filter scheme: 1 – photorefractive crystal; E_1 (x, z, t) – an object wave, F – fanning waves; b – photograph of the distribution of the intensity of the radiation waves at the exit from the photorefractive crystal [10].

a temporally coherent noise, which initially has a low intensity. Then, amplification of the scattered waves in the crystal takes place, which leads to the selection of power from the main wave and the generation of coherent noise.

The diagram of the process for the appearance of a fanning wave is shown in Fig. 6. In the process of propagation in a crystal, the coherent noise of the scattering waves is amplified by transferring energy from the fundamental (in our case, object) wave. Analogously to the case considered above, when an intense pump wave interferes with a weak object wave, chaotically oriented in the space of scattered waves, interfering with the object wave, it creates in a photorefractive crystal a set of chaotically oriented phase dynamic gratings. It is the chaotic orientation of the dynamic gratings that leads to an uneven enhancement of the waves scattered in the crystal. In connection with this, as the crystal propagates, the scattered waves propagate in a certain solid angle and in a certain direction. As a result, after a certain characteristic time interval called the recording time, a stationary set of dynamic holographic gratings is formed in the photorefractive crystal, and at its output there will be a stationary scattering wave made up of the well-defined scattered waves amplified in the crystal, which are called the *fanning waves*. The gain, polarization, and shape of the fanning waves are determined by the optical and electro-optical characteristics of photorefractive crystals [20, 21].

7.4.1. Functional Novelty-filters for processing images based on the fanning effect

The time dependence of the intensity of a plane light wave transmitted through a crystal is described by the following equation [9]:

$$I(t) = I(0)\frac{1 + m_0}{1 + m_0 \exp\left[\Gamma_{\text{eff}}(t) l_{\text{eff}}\right]}, \qquad (7.39)$$

where m_0 is the initial scattering efficiency of the wave introduced into the crystal, $l_{\text{eff}} = l/\cos(\theta_1)$ is the interaction length of the waves in the crystal.

The effective coupling coefficient of the incident wave with the fanning wave (coupling coefficient) Γ_{eff} is defined as

$$\Gamma_{\text{eff}}(t) = \Gamma_0\left[1 - \exp\left(\frac{t}{\tau}\right)\right], \qquad (7.40)$$

where t is the time interval chosen from the moment of entering the wave into the crystal; Γ_0 is the amplitude value of the coupling coefficient in the medium, and τ is the recording time of the fanning gratings in the medium (inversely proportional to the intensity of the wave introduced into the crystal).

As follows from (7.39), (7.40), when the steady-state regime of the formation of the fanning gratings is reached $t \geq \tau$, the value of the power transferred from the incident wave to the wave of fading is a function of m_0 and $\Gamma_0 l_{\text{eff}}$. As a result, if the coupling coefficient in the photorefractive crystal is 3 mm^{-1}, and the crystal length is 5 mm, then only about 1% of the power is transferred from the incident plane wave to the fanning wave. In other cases, the amount of transmitted power can be more than 50%.

Let us consider the case of a dynamic change in the input signal. In the stationary case, the process of energy transfer from an object wave to a fanning wave is determined by the interaction length of these waves on a thick dynamic phase hologram. Any change in the object wave on one side will lead to detuning from the Bragg resonance and increasing the intensity of the transmitted object wave, and on the other hand initiates the process of rewriting the dynamic hologram. The new fanning grating begins to adjust to a new type of object wave and to pump energy into a new fanning wave. As a result, the intensity of the transmitted new object wave will decrease exponentially. Both of these processes underlie the Novelty-filter principle, based on the use of the fanning phenomenon.

When a time-varying object wave with amplitude $E_{in}(x, y, t)$ is introduced into the photorefractive amplifying crystal, it begins to form a fanning wave in the crystal intensity of which will increase exponentially and reach its stationary value for a time slightly longer than the characteristic recording time of the dynamic hologram (τ). Due to this, the crystal will react to any changes in the object wave, including amplitude, phase, polarization and wavelength.

An exact analytical description of the dynamics of the process of interaction between the object wave and the fanning waves requires the solution of the coupling equations (7.7), (7.8) for a large number of scattered waves. The solution of these equations is based on the use of the stochastic approach and certain assumptions regarding scattering processes in the medium. Nevertheless, a general idea of the principles of the operation of such filters can be obtained using the approximation result for plane waves (7.39), (7.40). In this case, the expression for the intensity of the object wave at the exit from the crystal can be written in an approximate form [9]

$$I_{out0}(x,y,t) = \left| E_{out0}(x,y,t) \right|^2 = \frac{1+m_0}{1+m_0 \exp[\Gamma_{eff}(x,y,t)l_{eff}]} \left| E_{in0}(x,y,t) \right|^2,$$
(7.41)

where the effective gain $\Gamma_{eff}(x, y, t)$ not only is a function of time but also a function of the difference between the current distribution of the field at the input and the average value of the input field stored on the dynamic holographic grating:

$$\bar{E}_0(x,y,t) = \frac{1}{\tau} \int_0^\infty E_0(x,y,t-T) \exp\left(-\frac{T}{\tau}\right) dT.$$
(7.42)

Note that the value of τ at each filter location depends on the local intensity value of the wave introduced into the crystal at this point. In this connection, the effective coupling coefficient of the object wave and the fanning waves can vary over a wide range: from 0 in those places where there were changes in the objective wave, to Γ_0 in places where there was no change. In this regard, in order to achieve a high contrast in the processed image, for processing systems it is necessary to select such material and the orientation of the photorefractive crystals under which the maximum value of the coupling factor Γ_0, and hence also the enhancement of the fanning waves, is ensured.

As already noted above, using filters based on the phenomenon of fanning one can detect amplitude, phase and polarization changes in the object wave. As an example of the filter operation, consider the scheme of a single-beam interferometer (Fig. 7) intended for processing phase images in real time. The phase-modulated image is introduced into the plane wave *1* by means of a liquid crystal space-time light modulator *2*. Next, the objective *3* performs the Fourier transform of the wave and introduces the radiation into the photorefractive crystal *4*. The objective *5* performs the inverse Fourier transform and forms the processed image in the output plane *7*. A point diaphragm *6* is used to cut off scattered waves.

The amplitude of the wave at the output of the Novelty-filter, up to a constant factor, can approximately be written in the form [9]

$$E_{out0}(x,y,t)\exp\left[i\,\phi_{out}(x,y,t)\right] \approx \frac{1}{2\tau}\,E_{in0}(x,y,t)\times$$

$$\times \int_0^\infty \left\{\exp\left[i\,\phi_{in}(x,y,t)\right] - \exp\left[i\phi_{in}(x,y,t-T)\right]\exp\left(-\frac{T}{\tau}\right)dT\right\}, \qquad (7.43)$$

where $\phi_{in}(x,y,t)$ and $\phi out(x,y,t)$ are the phase distribution functions in the image-carrying object wave at the input and output of the filter, and $E_{in0}(x,y,t)$ and $E_{out0}(x,y,t)$ is the amplitude of the wave at the input and output. When the gain of the fanning waves is sufficiently large for a stationary case, the amplitude of the wave at the output of a single-beam interferometer is close to zero. Any changes in the structure of the object wave that have occurred over a time $t < \tau$, will lead to the appearance of a signal at the output of the processing system. If small, the output signal will be proportional to the derivative of the input image.

The value of the time constant for the filter using the fanning phenomenon is inversely proportional to the total value of the intensity of the wave introduced into the crystal. In this connection, due to the spatial inhomogeneity of the wave incident on the filter input, the values of the time constant turn out to be dependent on the coordinates in the crystal. Therefore, if the crystal is placed near the Fourier plane of the objective lens, high-intensity low spatial frequencies will be weakened in the output image, and the less intense high spatial frequencies will be practically unchanged. As

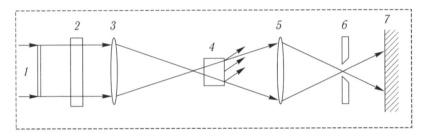

Fig. 7. Diagram of a single-beam interferometer for processing phase images in real time [2]: *1* – front of a plane laser wave; *2* – liquid crystal space-time light modulator; *3, 5* – objectives; *4* – Novelty-filter based on the fanning phenomenon in a photorefractive crystal; *6* – point diaphragm; *7* – output plane.

Fig. 8. The result of differentiating the image of the car model, obtained with its insignificant displacement in the Novelty plane of the filter [9].

a result, in the image transmitted through the filter, the stationary uniform illumination will be predominantly removed, and the changes, on the contrary, will be highlighted.

As an example, Fig. 8 shows a photograph illustrating the result of filtering the image of a car model, resulting from a slight displacement of the photo in the filter plane [9]. The contrast value in the processed image at the output was approximately 100:1.

7.4.2. High-frequency correlation real-time Novelty-filters

In view of the fact that the dynamic gratings created in photorefractive crystals are thick holograms, the effect of the formation of the fanning waves has a high spatial-angular selectivity with respect to the object wave. This feature made it possible, using the phenomenon of amplification of fanning waves in photorefractive crystals, to propose another correlation method, which is also widely used in linear optical systems, to process light signals [22, 23].

Using (7.39) and (7.40) and the filtration method proposed in [21] we can write an approximate expression for the dependence of the intensity of the inhomogeneous fanning wave on the output of the system as a function of the amplitude of the object wave $E_{\mathrm{in0}}(x,y,t)$ at the entry to the system:

$$I_f(x,y,t) \propto m_0 \exp\left(\Gamma_{\mathrm{eff}}(x,y,t)l_{\mathrm{eff}}\right)\left|E_{\mathrm{in0}}(x,y,t)\right|^2. \qquad (7.44)$$

In the case of a small-angle scattering process in a photorefractive crystal (<10 angular degrees), the coupling coefficient for all waves can be assumed constant and independent of the coordinates, as a result of which the time dependence for the total power of all fanning waves at the exit from the system for the steady-state process will be described by the expression:

$$P_f(t) = \iint_S I_f(x,y,t)dx\,dy \propto m_0 \exp\left(\Gamma_0 l_{\mathrm{eff}}\right)\iint_S \left|E_{\mathrm{in0}}(x,y,t)\right|^2 dxdy. \qquad (7.45)$$

If for some time T there is a change in the parameters of the object wave at the input, then the total power of the fanning waves at the output of the system will be proportional:

$$P_f(t) \propto m_0 \exp\left(\Gamma_0 l_{\mathrm{eff}}\right)\iint_S \left|E_{\mathrm{in0}}(x,y,t) - E_{\mathrm{in0}}(x,y,t-T)\right|^2 dx\,dy, \qquad (7.46)$$

where S is the cross-sectional area of the filter.

If the photodetector measures the power of the fanning waves at the output of the filtration system, the amplitude of the electrical signal at its output will be described by the expression:

$$V_S \propto K\left[\overline{P}_f - \mathrm{Re}\iint_S E_{\mathrm{in0}}(x,y,t)E_{\mathrm{in0}}^*(x,y,t-T)dx\,dy\right], \qquad (7.47)$$

where K is a constant, $\overline{P}_f$ is the average value of the power generated by the fanning wave.

As seen from (7.47), the magnitude of the signal at the output of the processing system is proportional to the correlation function of the complex amplitudes of the object waves at different instants of time.

Thus, a photorefractive crystal with a set of dynamic holographic fanning gratings formed in it is able to perform the functions of a correlation spatial and temporal filter (CF).

Since each CF is characterized by its own time to establish a stationary mode – the recording time τ, which is determined by the material of the photorefractive crystal, the orientation of its crystallographic axes, the wavelength and the power of the radiation used, and also the magnitude and frequency of the applied electric field, its value can vary in wide range: from fractions of a second to a few seconds. As an example, Fig. 9 shows the oscillogram of the response of the fanning signal to the impulse change in the structure of the object wave for a $Bi_{12}TiO_{20}$ crystal. As can be seen, the characteristic time constant for the steady-state regime is ~2 s [23]. Recently, filters of this type have found wide application as adaptive correlation filters for signal processing of single-fibre multimode interferometers [24, 25]. The principle of operation of these interferometers is based on the detection and quantitative measurement of the magnitude of the change in the intermode interference pattern of waves directed in a multimode fibre waveguide in time as a result of the external action of the physical field on the lightguide. Observed at the output of a single-fibre multimode interferometer, the field of a light wave is highly inhomogeneous and modulated by a set of speckles, the distribution of which in turn is highly prone to low-frequency random and uncontrolled

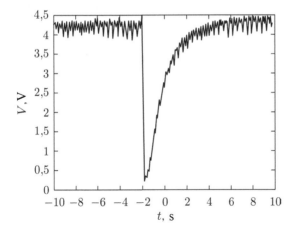

Fig. 9. Oscillogram of the response of the intensity of the fanning wave to the pulsed change in the structure of a speckled object wave [23].

temperature and technological influences. Therefore, the use of standard methods for processing such light signals is ineffective. Correlation filters, based on the use of the phenomenon of fanning, due to the extremely high dynamic range and resolution of the material of photorefractive crystals allow us to easily process optical fields of any complexity. An important feature of the use of photorefractive crystals is that after each stage of the amplitude distribution of the light wave processed by the filter, during the time $t = \tau$ (the recording time of the filter), the previous filter will be erased and a new one coordinated with the changed structure of the optical field will be recorded. Thus, only those elements of the radiation wave field from the interferometer will participate in the formation of the output signal of the processing system, in which changes occurred in the time $t < \tau$. All other changes in the wave field that occurred during time $t > \tau$ will be blocked and will not contribute to the output signal. Thus, selecting the filter material, its geometry and operating mode, it is possible to exclude uncontrolled external influences from the observation. As a result of any other, even very strong impulse, due to the use of dynamic holograms, the system will quickly return to the working state. In this sense, the correlation filter under consideration is adaptive.

As an illustration, Fig. 10 shows the experimentally measured amplitude-frequency characteristic of a sensor with a sensitive element based on a single-fibre multimode interferometer, the $Bi_{12}TiO_{20}$ crystal was used as the material of the correlation filter. The present dependence illustrates the high efficiency of the filter when the low frequencies are cut off in the spectrum of the detected signals.

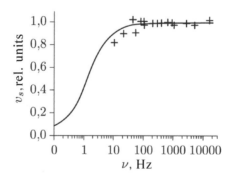

Fig. 10. Amplitude–frequency response of a sensor with a sensing element based on a single-fibre multimode interferometer using a Novelty-filter for signal processing [24].

It should also be noted that the considered method of stabilizing the performance of a single-fibre multimode interferometer is quite universal and can be used in other circuits of both fibre and traditional optical measuring interferometers.

References

1. Kohonen T., Self–Organization and Associative Memory. New York, Springer, 1983.
2. Anderson D.Z., Feinberg J., IEEE J. Quantum Electronics, 1989. V. 25, No. 3. P. 635–647.
3. Falk S.D., et al., Seeing the Light. New York: Harper and Row, 1986. P. 192–193.
4. Anderson D.Z., Saxea R., J. Opt. Soc. Amer. B. 1987. V. 4. P. 164–176.
5. Kukhtarev N.V., et al., Ferroelectrics, 1979. V. 22. P. 949–960.
6. Photorefractive Materials and Their Applications I: Fundamental Phenomena, ed. P. Gunter and J.P. Huignard. New York, Springer, 1988.
7. Kogelnic H., Bell Syst. Tech. J., 1969. V. 48. P. 2909.
8. Soliman L., Heaton J.M., Opt. Commun. 1984. V. 51. P. 76–78.
9. Ford J.E., et al., Opt. Lett. 1988. V. 13, No. 10. P. 856–858.
10. Sedlatschec M., et al., Appl. Phys. B. 1999. V. 68. P. 1047–1054.
11. Cronin–Golomb M., et al., Optics Letters. 1987. V. 12, No. 12. P. 1029–1031.
12. Stepanov S.I., Petrov M.P., Opt. Communs. 1985. V. 53 (5). P. 292–295.
13. Korn G., Korn T., Handbook of Mathematics. Moscow, Nauka, 1978.
14. Varrakchi A., et al., Opt. Communs. 1980. V. 34. P. 15–18.
15. Kamshilin A.A., Mokrushina E., Proc. Symp. Optika'84. Budapest, Hungary, SPIE, 1984. V. 433. P. 83–86.
16. Kukhtarev N.V., Processes in Photorefractive Materials and Their Application, ed. P. Gunter and J.P. Huignard, New York, Springer, 1988.
17. Kwong N.S.K., et al., J. Opt. Soc. Amer. 1988. V. 5. P. 1788–1791.
18. Cronin–Golomb M., Yariv A., J. Appl. Phys. 1985. V. 57 (11). P. 4906–4910.
19. Fainman Y., et al., Optical Engineering, 1986. V. 25, No. 2. P. 228–234.
20. Feinberg, J., J. Opt. Soc. Amer. 1982. V. 72, No. 1. P. 46–51.
21. Kamshilin A.A., et al., Appl. Phys. Lett. 1988. V. 73. P. 705–707.
22. Kulchin Yu.N., Distributed fibre-optic measuring systems. Moscow, Fizmatlit, 2001.
23. Kulchin Yu.N., Usp. Fiz. Nauk. 2003. V. 173, No. 8. P. 894–899.
24. Kulchin Yu.N., et al., Adaptive methods for processing speckle-modulated optical fields, Moscow, Fizmatlit, 2009.

8

Adaptive Optoelectronic Smart Grid Systems for Monitoring Physical Fields and Objects

Introduction

The rapid development of modern technology in the late twentieth and early 21st centuries is characterized by a variety of controlled processes and measured physical quantities. Today, monitoring and studying phenomena occurring at high rates of their flow, a wide range of temperature and pressure changes, and other environmental parameters acquire paramount importance. The monitoring objects are shown schematically in Fig. 1. At this stage, the requirements to the accuracy and reliability of executive machines and mechanisms, especially those used in aviation, fleet and space vehicles, have increased to a great extent. The creation of industrial and power structures, vehicles, sports complexes and high-rise buildings required the development of extensive systems for controlling the strength and stability of various building structures and equipment in them in real time. Modern technology, with its extreme high-intensity operating modes, presents very special requirements, both to structural materials and to operating modes. This requires controlling the strength and stability of various building structures, including shells, foundations, dams and retaining walls, hulls of ships, aircraft and missiles, high-rise buildings, structures, etc. It is often quite impossible to calculate or predict the behaviour of one or another structure or structure. Based on this, the problem of monitoring the process of destruction of materials, structures and mechanisms, aimed at preventing sudden catastrophic events, remains the focus

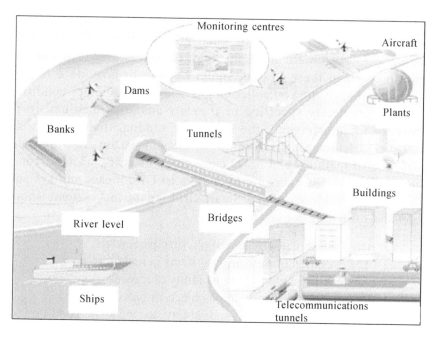

Fig. 1. Objects of Smart Grid monitoring systems.

of attention. The operational problems caused by the vibrations of structures arising from periodic power loads, earthquakes, strong winds or explosions have a special place.

Emergency situations in the operation of pressure vessels, pipelines, turbines of hydroelectric power stations, power plants of nuclear power plants, ships, aircraft, helicopters and missiles are caused by a number of reasons: the occurrence and development of cracks; joint action of loads and corrosive-active media; radiation; manifestation of the viscous and plastic properties of the material. Accidents of aircraft and helicopters, and often of ships and breakdowns of machines and their structural components, most often occur as a result of fatigue failure, which is a consequence of the gradual development of cracks inside the material due to stress variables occurring both during normal operation of the facility and due to accompanying vibration.

Therefore, the prediction and prevention of catastrophic destruction phenomena, as well as ensuring the safe operation of structures and structures, requires the development of appropriate measuring equipment that can monitor and measure the above characteristics, and also create distributed systems for their monitoring in real time on their basis.

As the results of research by Russian and foreign scientists have shown, the creation of such measurement systems faces the need to search for new physical, technical and technological solutions, since these systems must be able to provide a reliable representation of the controlled object and its state under conditions of an incompletely specified or *a priori* unknown medium, in the presence of interference, the impact of random disturbances, the fuzziness of incoming information, etc., based on the integrated attraction of methods and those chronology of artificial intelligence. Such monitoring systems are called Smart Grid measuring systems.

The main qualities of the Smart Grid measuring systems should meet the following requirements:

1. High speed measurement of physical parameters;
2. The operation of measuring systems in real time;
3. Adaptability and/or learning ability of measuring systems;
4. Remote access to monitored objects or processes;
5. High noise immunity and ability to work in conditions of insufficient data.

The process of intensive development and introduction of fibre-optic telecommunication systems has led to the emergence of one of the most dynamically developing areas of optoelectronics-fibre-optic sensors (FOS) of physical quantities [1]. The organic combination of a communication fibre system and a physical quantity monitoring system in a single fibre optical fibre path opens up broad prospects for the creation of extended and distributed information–measurement systems (IMS), the functional purpose and configuration of which can be continuously improved without involving additional communication lines. At the same time, an important advantage of FOS is the introduction in IMS of qualities such as high sensitivity, small size, immunity to electromagnetic interference and aggressive environmental influences, the possibility of multiplexing individual sensors into complex measurement systems and potentially low cost [2]. The above advantages and the transition in recent years from discrete FPSs of physical quantities to extended distributed FOS have recently allowed the development of distributed fibre-optic measurement networks (DFOMS) and 'sensitive surfaces' capable of reconstructing the spatial distributions of the parameters of the investigated physical fields (PF) [3].

The transition to DFOMS with a higher degree of numerical mobility, the development and improvement of methods for processing information arrays at the output of the DFOMS based on the use of new physical phenomena and the application of modern neural network signal processing technologies open up wide opportunities for giving IMS such new qualities as the ability to learn and adaptability [5], which is an important step towards the development of modern Smart Grid measuring systems designed for PF research and monitoring of the state of technical and technological objects.

This chapter is devoted to the presentation of new approaches to the development of Smart Grid measuring systems based on the latest achievements in laser physics and optoelectronics.

8.1. Tomographic DFOMS for reconstructing the distributions of scalar and vector physical fields

The traditional approach to solving the problem of restoring multidimensional distribution functions of physical fields through the use of DFOMS consisting of a set of 'point' FOSs in which measurements are related to a completely defined discrete set of points in space is not always successful due to the emerging technical difficulties of multiplexing/demultiplexing of signals obtained from a large number of 'point' FOSs. This does not allow achieving high spatial resolution and speed [1, 3]. Sequential integration of 'point' FOSs into an extended fibre-optic measuring line (FOML) or the use of distributed FOSs makes it possible to obtain an integral phase or amplitude signal of the action of an external PF on the FOS along the fibre lightguide (FL) [3].

In the general case, the integral signal (the Radon transform) about the effect of a physical field on an FOML can be represented in the form [6–12]

$$g(p,\phi) = \int_{L(p,\phi)} h(x,y,\phi)dL, \qquad (8.1)$$

where $h(x, y, \phi)$ is the response function of the FOML to the action of the physical field:

$$h(x,y,\phi) = \begin{cases} qf(x,y), & \text{for scalar physical field,} \\ q\hat{F}\left[\mathbf{A}(x,y), \mathbf{m}(x,y,\phi)\right] & \text{for vector physical field,} \end{cases}$$

where $f(x, y)$ is the function of the spatial distribution of the detected parameter of the physical field; $A(x, y)$ is the intensity vector of the field under investigation; $m(x, y, \phi)$ is the unit vector of the tangent (**e**) or normal (**n**) to the contour of the FOML stacking; $\hat{F}$ [...] is the operator; (x, y) are the Cartesian coordinates in the registration plane (S); q is a constant coefficient, which determines the linear sensitivity of the FOML to the measured parameter of the physical field; L is the coordinate along the contour along which the FOML is laid; (p, ϕ) are the polar coordinates that define the position of the FOML contour in the registration plane (Fig. 2 *a*).

The signals described by expression (8.1) carry indirect information about the distribution function of the parameters of the physical field. Therefore, its restoration requires special mathematical methods. In the case of a scalar physical field, such methods are based on the theory of tomography [13]. According to this theory, to solve the problem of recovering the function $f(x, y)$, it is necessary

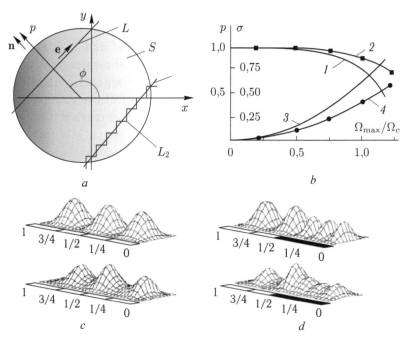

Fig. 2. Recovering the distributions of physical fields using DFOMS: *a* – location of FOML in the physical field; *b* – calculated (*1, 3*) and experimental (*2, 4*) dependences of the correlation coefficient ρ (1, 2) and the mean-square error σ (*3, 4*) for the initial and reconstructed distribution functions of the physical field; *c* and *d* – calculated (top) and experimental (bottom) distributions of the intensity of transverse vibrations of the walls of an empty and half-filled metal reservoir, respectively [1].

to obtain a set of integral images $g(p, \phi)$ for different values of the pairs (p, ϕ). This set of images will form a direct Radon transform from the desired function $f(x, y)$ which can be restored by adding to the resulting the Radon inverse transformation procedure [17, 18].

In the case of a vector phase field, the problem of reconstructing the distribution of the parameters $A(x, y)$ does not have a direct solution by tomography methods. Therefore, for each type of dependence $h(x, y, \phi)$ it is necessary to apply a special method for processing the measurement results [1].

The existing algorithms for computational tomography allow us to reconstruct the original of the scalar functions of the physical field reasonably accurately only by providing a sufficiently high sampling rate for each of the parameters p and ϕ of the Radon transform (8.1). In our case, the required sample can be obtained by covering the plane S with a distributed fibre optic measuring network made up of FOML stacked with a certain step along the radial and angular coordinates [7–10]. One of the advantages of this approach is the ability to obtain a real-time scan result. The required periodicity of readings in the RDHIOC is determined by the polar angle and the polar radius from the following expressions [13]:

$$\delta\phi = \frac{\pi}{D\Omega_{max}}, \quad \Delta p = \frac{\pi}{\Omega_{max}}, \quad (8.2)$$

where D is the characteristic transverse dimension of the region S; Ω_{max} is the maximum spatial frequency of variations of the physical field.

Since the construction of IMS on the basis of the FOML network leads to the limitation of the number of integral samples in (8.1), this entails the incorrectness of solving the inverse problem of physical field reconstruction [8]. Therefore, for the most reliable recovery of the original from an incomplete set of integral data, the computational recovery algorithm must be supplemented with a rule for estimating the originals of the recovered functions using dependences having a minimum deviation norm from its mean value in the region S [7]. And to the DFOMS itself, it is necessary to present a requirement that its characteristic spatial frequency, determined from the expression $\Omega_c = \sqrt{6K/S}$ (where K is the number of FOML in the network) exceeds the maximum frequency Ω_{max} [7]. The joint action of these requirements is illustrated by the calculated and experimental dependences of the quality parameters of the reconstruction of the physical field from the ratio between the frequencies Ω_c and Ω_{max} (Fig. 2 *b*) for $f(x,y) = \sin(\Omega_{max}x)$ $\sin(\Omega_{max}y) / (xy)$[8].

In Figs. 2 *c* and *d*, experimental and calculated results obtained by reconstructing the distribution of the intensity of transverse vibrations of the walls of metal tanks for various levels of their filling with a liquid are shown as an example of the use of the DFOMS for reconstructing the distributions of scalar phase transitions [7, 8].

If the effect of the vector physical field on the output signal of the FOML is proportional to the projection $\mathbf{A}(x, y)$ on the direction of the unit vector tangential to the axis of the FOML, then the expression (8.1) can be represented in the form [10, 12]:

$$g(p,\phi) = \int_{(L)} \mathbf{A}(L)\mathbf{e}\,dL = C(p,\phi)\sin\phi - B\cos\phi, \qquad (8.3)$$

where $C(p,\phi) = \int_{(L)} A_x(x,y)d_L$ and $B(p, \phi) = \int_{(L)} B_x(x,y)d_L$ are the coefficients containing information about the Radon transform of the field strength projections $\mathbf{A}(x, y)$ on the Cartesian coordinates x and y, respectively. To separate the contribution of the C and B values to the common integral signal, it is advisable to use the combined DFOMS with additional FOML stacked in the same way as the initial ones, but sensitive to the projection of the field strength vector $\mathbf{A}(x, y)$ onto the normal vector $\mathbf{n}$ to the axis of the FOML. The integrated signal at the output of such additional FOML can be described by expression

$$g_l(p,\phi) = \int_{(L)} \mathbf{A}(L)\mathbf{n}\,dL = C(p,\phi)\cos(\phi) + B(p,\phi)\sin\phi. \qquad (8.4)$$

The joint solution of (8.3) and (8.4) yields: $C(p,\phi) = g\,\sin\phi + g_l\cos\phi$ and $B(p,\phi) = -g\,\cos\phi + g_l\sin\phi$ (Note that the contribution of the C and B values is also separated by using an additional non-rectilinear FOML with a nonuniform sensitivity [10, 12].) Thus, the problem of reconstructing the Cartesian component distribution vector of the intensity of the vector physical field reduces to the already considered tomographic problem of reconstructing the distribution function of scalar quantities from known integral images.

According to (8.3) and (8.4), if only one type of FOML is used in DFOMS, for example, sensitive to the projection of the vector onto the physical field axis, this allows us to reconstruct only the vortex component of the vector physical field [11, 12]. And the use of DFOMS based on FOML, sensitive only to the projection of the vector on the normal to the axis of the FL, allows us to reconstruct only the potential component of the vector physical field [11, 12]. Experimental confirmation of the operability of such DFOMS using the example of reconstructing the distribution of the electrostatic field is given in [12].

In each specific case of the study of vector physical fields, the choice of the trajectory of the stacking of the measuring line in the DFOMS depends on the type of the used FOML. For example, in the problems of reconstructing the distributions of longitudinal displacement fields, it is necessary to use FOML, in which the effect of the vector field on FL is proportional to the derivative of the longitudinal component of the vector **A** in the direction of the tangent to the lightguide: $\partial A_L / \partial L$ (such FOML can be created, for example, on the basis of extended fibre interferometers [11, 12].) In this case, it is expedient to lay FOML along a broken trajectory composed of identical rectangular steps (L_2 in Fig. 2 *a*). For such FOML, the magnitude of the integral signal (8.1) is described by expression

$$g(p,\phi) = \int\limits_{(L_2)} dL_2 \left(\frac{\partial(\mathbf{A} \cdot \mathbf{e})}{\partial L_2} \right) = \sqrt{2} \int\limits_{(L_2 0)} \left(\frac{\partial A_X}{\partial x} + \frac{\partial A_Y}{\partial y} \right) dL_2 =$$

$$= \sqrt{2} \int\limits_{(L_2)} \operatorname{div}\mathbf{A}(x,y) dL_2.$$

(8.5)

In (8.5), a new contour of integration is introduced – a straight line along L_2 oriented along the placement of the stepped contour (Fig. 2 *a*). As can be seen, (8.5) represents the element of the Radon transform of the function div **A**(*x*, *y*) for given values of the coordinates (*p*, ϕ). Thus, using the DFOMS consisting of stepwise FOML, we can solve the problem of tomographic reconstruction of the distribution div **A**(*x*, *y*), that is, to reconstruct the potential component of the vector field.

At the same time, in order to restore the distribution of the vortex component of the vector phase transition, it is necessary to use stepwise FOML whose signal is generated by the derivative of the transverse component of the phase vector in the direction of the tangent to the sun: $\partial A_n / \partial L$. In this case, after performing the inverse Radon transform operation, (**k** · rot **A** (*x*, *y*)), which describes the

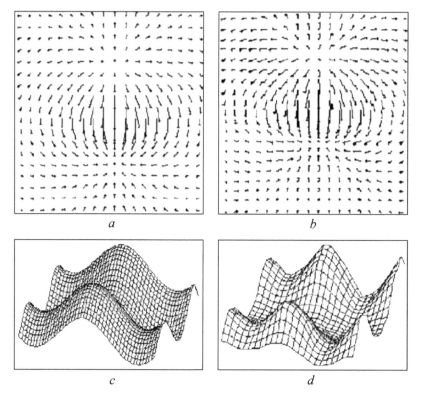

Fig. 3. Tomographic reconstruction of the distributions of the parameters of vector physical fields with the use of DFOMS: *a* – calculated distribution of the potential component of the field of longitudinal deformations of a flat object; *b* – distribution of the potential component of the field of longitudinal deformations of a flat object, reconstructed according to the signals of the DFOMS; *c* – the calculated distribution of the square of the modulus of the gradient vector of the transverse displacement field of the elastic plate; *d* – distribution of the square of the modulus of the gradient field of the transverse displacement of the elastic plate, reconstructed from the signals of the DFOMS.

vortex nature of the physical fied [11], and **k** is the normal to the (*x*, *y*) plane.

In studies [11, 12], when studying the longitudinal deformations of the elastic surfaces the DFOMS made of the stepwise interferometric FOML was rigidly attached to the surface under investigation. Figure 3 *a* shows the pattern of the displacement field under study, and Fig. 3 *b* shows the result of its recovery from the signals of the measuring network (the correlation coefficient between the given distributions is 0.95).

The effect of a vector physical field on a FOML can also be proportional to the square of the projection of the longitudinal component of the vector $\mathbf{A}(x, y)$ on the direction of the tangent to FL (A_{L_2}). A problem of this type arises, for example, under the influence of the gradient of the field of transverse displacements of elastic surfaces on interferential-type FOML [10, 12]. When laying FOML along the trajectory L_2 (Fig. 2 *a*), at its output we get a signal of the form

$$g(p,\phi) = \frac{1}{2} \int_{(L_2)} \left(\mathbf{A}(x,y)\cdot\mathbf{e}\right)^2 L_2 = \int_{(L_2)} |\mathbf{A}(x,y)|^2 \, dL_2. \tag{8.6}$$

Thus, by carrying out the inverse Radon transform for the data array obtained by the DFOMS, it is possible to restore the distribution function of the square of the modulus of the intensity vector of the physical field. This conclusion was verified experimentally in [10, 12]. Figure 3 *c* and *d* show the calculated and experimentally reconstructed distributions of the square of the modulus of the gradient of the transverse deformation of a thin elastic plate.

8.2. Extended FOML based on single-fibre multimode interferometers and adaptive spatial filtering methods

The development of the DFOMS for tomographic IMS for monitoring the distribution of physical fields is associated with the solution of one of the key problems: the creation of extended FOML. Such FOML can be performed on the basis of practically any distributed or quasidistributed FOS of physical quantities, the description and classification of which are given in [3]. At the same time, studies in [6–8, 10–12] have shown that FOML can be created efficiently in tomographic FOMS using FOS on the basis of single-fibre multimode interferometers (SMI). In the SMI, the result of intermode interference of the directional modes of the same multimode fibre lightguide is used to measure the magnitude of the physical field effect. FOML based on the SMI are highly sensitive, capable of directing high-power radiation and do not require the use of a support arm, which distinguishes them from other types of interference measurement devices.

The radiation at the output of the SMI is a complex (speckle) optical signal, which is the result of interference of a large number of guided modes having different phase velocities. The effect on SMI leads to additional phase shifts between the fibre lightguide modes, which is reflected in the correlated spatial rearrangement of the speckle pattern [15]. This made it possible to propose for correlation processing of radiation from SMI correlation methods of spatial filtration based on the use of amplitude and holographic spatial filters (see Chapter 7), the output power of which can be represented in the form [16–18]:

$$P_{out}(t) = \int\int_S I(x,y,t) T(I_0(x,y,0)) dx dy, \qquad (8.7)$$

where $I_0(x, y, 0)$ and $I(x, y, t)$ is the distribution of the radiation intensity in the plane of the spatial filter at the stages of recording the filter and processing the signals of the SMI, respectively; $T(I_0(x, y, 0)$ is the spatial filter transmission function, S is the filter area, t is the time.

The study of the modulation of phase phases in multimode fibre lightguides and the correlation processing of radiation from SMI showed that for both types of spatial filters the power of the correlation signal is a function of the maximum additional phase difference $\Delta\Phi_{max}(t)$ between the guided modes in the SMI [17, 18]:

$$P_{out}(t) \propto \text{sinc}^2\left(\frac{\Delta\Phi_{max}(t)}{2}\right). \qquad (8.8)$$

Therefore, the metrological characteristics of the FOML based on the SMI are determined by the numerical aperture, length and material of the fibre lightguide used. As a result, the sensitivity threshold of the SMI, for example, to the relative deformation of the fibre lightguide, can vary from 10^{-10} to $10^{-0.5}$, and the dynamic range of measurements in the static regime can reach 30–40 dB [3].

As for all types of fibre-optic interferometers, the problem of low-frequency fading of an interference signal, caused by random uncontrolled influences, such as temperature drift, technological vibrations, random mechanical influences, frequency drift of a laser, etc., is of current interest for FOML on the basis of SMI [3]. Therefore, to create practical FOMS systems, intended for the long-term monitoring of the physical field, original adaptive ways of processing and stabilizing the performance characteristics of extended FOML were developed, based on the use of adaptive spatial filtration (ASF) methods for the radiation field from SMI [19, 20].

An effective approach to the realization of the ASF method is the use of phenomena of nonlinear-optical interaction of light waves in photorefractive crystals (PRC), for example $Bi_{12}TiO_{20}$ (see chapter 7). When such crystals are illuminated by radiation with an inhomogeneous intensity distribution, a spatially inhomogeneous electric charge distribution arises in them, causing a change in the refractive index of the PRC material proportional to the radiation intensity gradient [21–23]. A consequence of this are the processes of amplification of the diffraction of the light wave itself (the fanning effect) [24, 25] (chapter 7, section 4) and polarization self-modulation of radiation [26, 27].

The principle of the operation of ASF on the basis of the effect of fanning is as follows. When a radiation is introduced from the SMI into the PRC, it is scattered by the inhomogeneities of the crystal. Fields of interference of scattered waves with the main wave create chaotically oriented volume dynamic diffraction gratings in the PRC. When selecting certain crystal dimensions, its orientation, the angle of the radiation input in the PRC, and the state of the radiation polarization, these scattered diffraction gratings amplify the scattered waves (the fanning effect). The efficiency of converting the main wave into a fanning wave can reach 90% [28]. The spatial–angular selectivity of such a set of dynamic gratings (at a crystal length of 10 mm) reaches $\sim 10^{-4}$ rad, which makes them extremely sensitive to a change in the intensity distribution in the radiation field from SMI.

As shown in [19], the fanning wave formed by the radiation from the SMI at the output from the PRC is a correlation signal whose power is proportional to the correlation function of the distributions of the complex amplitudes of the radiation fields from the SMI to $(U_1(x, y, 0))$ and after $(U_2(x, y, t)$ effects of the physical fie4ld on the SMI:

$$P_{out}(t) \propto \mathrm{Re} \int \int_S U_2^*(x,y,t) U_1(x,y,0) \, dx \, dy. \qquad (8.9)$$

This opened up the possibility of creating effective spatial filters for the implementation of the ASF method in the processing of SMI signals. The process of competition of scattering waves observed in the PRC leads to the formation of a stationary set of volumetric dynamic diffraction gratings with a recording time constant of the correlation filter τ_R. Therefore, any rapid changes ($\tau < \tau_R$) of the radiation field from the SMI at the entrance to the PRC, where τ is the exposure duration, will not affect the state of the spatial filter, which will lead to modulation of the output signal $P_{out}(t)$ in accordance with (8.9). If the changes are slow $\tau > \tau_R$ or rare, then the old correlation filter is erased and a new one is formed, corresponding to the changed radiation intensity distribution at the input to the PRC. In this case, the output signal $P_{out}(t)$ remains unchanged.

The results of an experimental study of the described ASF signal processing algorithm for SMI [19, 20] demonstrated not only its high efficiency, but also its long-term stability (more than 8 hours) and the possibility of processing signals with frequencies above 0.1 MHz.

Another possibility of creating an ASF is based on the use of the phenomenon of polarization self-modulation in PRC. The phenomenon of polarization self-modulation consists in a change in the polarization state of the radiation propagating along the PRC with a nonuniform intensity distribution, which is a consequence of the induced dynamic appearance of the inhomogeneous distribution of the refractive index in a crystal dynamic in time [21, 26, 27]. The appearance of anisotropy of the optical properties of the PRC is manifested in a change in the intensity of the radiation passing through the crystal at the output from the analyzer located behind it. The transparency function of the 'PRC–analyzer' pair can be represented as [21]:

$$T_{PSM}(x, y) = \cos^2(a[E_A + E_{SC}(x, y)], \tag{8.10}$$

where a is a constant, taking into account the properties of the crystal and radiation, E_A is the external electric field; $E_{SC}(x, y)$ is the strength of the internal electric field, which depends on the intensity gradient of the radiation field from the SMI to the PRC.

The complex of studies carried out in [20] showed that the 'polarizer–PRC–analyzer' placed at the output from the SMI is capable of performing the function of an amplitude spatial filter in accordance with (8.7), ensuring its adaptability to uncontrolled influences and thereby implementing the ASF signal processing SMI.

In [29], investigations were made of the mutual influence of the radiation fields of different SMIs upon their overlapping in the bulk of the PRC. It was found that overlapping of optical fields leads to a weakening intensity of correlation signals and does not affect their information component, which opens the possibility for creating multichannel systems for processing signals by the ASF method.

In [30], a study was reported of the feasibility of implementing an ASF method based on an analog electronic device consisting of a photodiode array spatially matched to the specular structure of the radiation field from the SMI and a multichannel electrical signal processing unit effective in processing FOML signals based on low-mode fibre lightguides..

To create ASF-filters, CCD-matrixes are very promising [31]. Charge-coupled devices (CCDs) have high quantum efficiency and a wide dynamic range of the linear portion of the light-electric charge conversion characteristic. In the CCD standard execution the above parameters, respectively, are 50% and 60 dB. Therefore, the photosensitivity and linearity of the CCDs are much better than those of silver-halide photographic materials in the best of which these parameters do not exceed 3% and 40 dB, respectively [32–38].

The systems of correlation processing of speckle signals of SMI using the CCD matrices for registration of optical field distributions allow the data on the distribution of speckle patterns in the computer's memory to be entered and rewritten at high speed. As a consequence, they ensure the ability of measuring systems based on them to adapt quickly to uncontrolled changes in the spatial distribution of such patterns. For example, this makes it possible to register the parameters of significant deformations of objects. As it was shown, the characteristics of the device based on the CCD matrix are: the error of measurements of the elongation is ±3 μm, the dynamic range of measurements is 80 dB, the operating frequency is up to 25 Hz.

8.3. Methods for multiplexing fibre-optic measuring lines in Smart Grid monitoring systems

Any measuring system generally includes a sensor that converts the measured physical quantity (for example, vibrations, deformations, etc.) to a primary signal (phase modulation/light wave amplitude, rearrangement of the speckle field pattern, etc.), and a demodulator, transforming the primary signal from the sensor into an output signal.

In Smart Grid monitoring systems, simultaneous measurement of one or several physical quantities in a variety of different areas of the object under study is required. In this case, the measuring system must be multichannel and have a plurality of point sensors, quasi-distributed or distributed measuring lines, providing measurement of the required number of physical quantities [3].

Multichannel measuring systems (MMS) in the general case are built on extensive or intensive principles. In an extensive MMS, a signal from each sensor enters its own individual demodulator. The main advantage of this approach is the fundamental absence of crosstalk between the measuring channels. However, the extensive approach justifies itself only with a relatively small number of measuring channels, since an increase in the number of demodulators, equal to the number of sensors in an extensive MMS, leads to a disproportionate increase in the cost of the system. In addition, such a measuring system becomes generally quite complex and, as a consequence, its reliability decreases.

In a multichannel measuring system built on an intensive principle, all signals from sensors multiplexed in it are fed to a multichannel demodulator, where, in addition to demodulating signals, demultiplexing occurs (Fig. 4).

There are a number of techniques for multiplexing sensors in optical multichannel measuring systems. The most common of these are [39]:

- (spatial-division multiplexing – SDM);
- (time-division multiplexing – TDM);
- (wavelength-division or spectral multiplexing – WDM);
- (frequency-division multiplexing – FDM);
- hybrid methods.

The spatial-divison multiplexing method (SDM) is probably the simplest way to combine individual sensors into a measuring system [40]. In this case, each sensor is connected to a common recording system via a physically dedicated channel, which allows independent

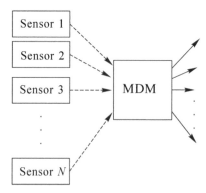

Fig. 4. The general principle of constructing a multichannel measuring system, built on an extensive basis. MDM – multichannel demodulator of signals.

unambiguous addressing of the sensors. Measuring systems based on SDM-principle have high reliability: failure of one or several sensors does not entail complete destruction of the entire system. In addition, in such a system there is practically no mutual influence of the sensors on each other (cross noise). However, the implementation of fibre-optic measuring systems based on the SDM-principle increases the requirements for the power of radiation sources, which, together with the losses in the used splitters, is the determining factor limiting the maximum number of sensors in a measuring system of this type. To date, work has been performed in which spatial multiplexing of up to 60 optical fibre sensors [41] is performed, which is the best result.

The technology of time-division multiplexing (TDM) is based on the time-sharing of the polling processes of all sensors integrated into the measuring system [49, 40]. A short optical pulse is sent to a measuring line (ML) consisting of a series or parallel connected point sensors whose state determines their reflectance or transmittance. The reflected or transmitted signal is a series of pulses, the order of which corresponds to the position of the sensors in the ML, and the amplitude to their state. The advantages of TDM-type measuring systems include the use of a single pulsed light source in general for all sensors. The main limitation of such a measuring system is due to the fact that as the number of sensors increases, the time between the impulses being interrogated and, accordingly, the overall speed decreases. Thus, there is a competition between the resolving power of the measuring system and its ability to record rapid changes in the parameters of the physical field under study. As in the previous case, the maximum possible number of combinable sensors is limited by power losses, which increase with increasing number of sensors. It is noted in [42] that the practical limit of integration is 10 sensors, but by adding fibre-optic amplifiers to the measuring network, an increase in the number of sensors can be achieved [43].

The technique of multiplexing signals of fibre-optic sensors based on their wavelength division (WDM) [3, 40] has become widespread. In a measuring network built on this principle, each individual sensor operates in its spectral range independently of other sensors. Light sources in such a network include broadband light-emitting or laser diodes [44] or systems of several narrow-band light emitters [45]. The sensors in WDM type measuring systems usually include fibre-optic Bragg gratings [46] which have the ability to reflect the light of a certain wavelength. The reflected signals from different sensors are collected and sent along a common fibre to the demultiplexer, where they are separated. Demultiplexers include diffraction gratings [47], scanning interferometers [48], volumetric holograms [49], selective fibre splitters [50], biconical filters [51] and fibre-optic gratings with a long period [52]. Among the merits of fibre-optic measuring systems built on the WDM-principle the most significant are the following: simultaneous interrogation of all sensors; simple methods of electronic processing; compatibility with sensors based on FBG (fibre Bragg gratings) which allows the creation of systems with low losses. In the literature there are works in which the integration of up to 31 sensors into a single network is performed [53]. The maximum possible number of sensors in the network is limited by the finiteness of the width of the emission spectrum of the light source used [54]. When using a topology with efficient power sharing, the theoretical possibility of combining up to 40 sensors is considered [55]. The disadvantages of WDM-type measuring systems include the potential possibility of the appearance of cross-talk noise between channels when overlapping signal spectra from the corresponding sensors.

Frequency-division multiplexing (FDM) is performed by allocating an individual frequency band for each sensor in the measuring system [56]. Optical radiation in the channels through which it is applied to the sensors is modulated in amplitude by means of modulators each of which operates at its own frequency. In the sensors, the radiation receives an additional modulation proportional to the effect of the measured physical quantity. Then the signals from different sensors are assembled together and sent to the photodetector via a common fibre. The total electrical signal from the photodetector arrives at narrowband filters tuned to the corresponding frequencies, where the signals of the individual sensors are isolated. The shortcomings of the measuring system, built on the FDM-principle, include the need to provide a sensor signal to each sensor and the availability of an appropriate number of generators. The results on combining up to 16 sensors into a single system are known [57].

Despite the rather wide range of existing methods for constructing multichannel measuring systems, in some cases the number of multiplexed sensors is insufficient. Attempts to increase this number led to the development of hybrid multiplexing techniques. For example, the use of TDM-multiplexed sensors in several channels, separated by the WDM-principle, allows to unite up to 128 sensors [58]. However, the majority of works in this area [59] remain at the stage of demonstrating the concept, revealing the potentialities of hybrid methods. The integration of 60 sensors is realistic [60].

It is obvious that when building a multichannel measuring system, it is necessary to take into account both the principle of sensor multiplexing into the system and the basic physical principle of demodulation in the channel. Thus, in holographic measuring systems (including adaptive ones, based on dynamic holograms), the function of the demodulator of the primary signal is performed by the hologram itself. Thus, the natural direction of creating a multichannel holographic system is the recording of multiple holograms in the volume of a single photorefractive crystal.

In the approaches to multiple recording of holograms, the methods of spatial, angular, and spectral multiplexing are distinguished.

In the first case, holograms are recorded in different regions of the same crystal. This approach allows one to maximize the independence of the channels, The cross noise between them is practically excluded. However, the number of multiplexed holograms (and hence channels) is limited by the size of the crystal and in most cases does not exceed several units [29]. In addition, with a relatively large separation of holograms in a crystal, in general, an individual reference light beam may be required to form each hologram, which in turn leads to a complication of the multichannel system as a whole and, consequently, to a decrease in its efficiency.

Angular multiplexing of holograms can significantly increase the number of channels realized on a single crystal, which can be formed in the same region of the crystal, partially or even completely overlapping each other, using a common reference beam. However, in the general case, there may be crosstalk between multiplexed holograms (and hence between channels). This defines one of the main tasks that should be solved when developing a multichannel holographic system, namely, the search for conditions that ensure the elimination of crosstalk noise or the reduction of their level to an acceptable limit.

The basis for spectral multiplexing of holograms is the use for their recording of radiation with different wavelengths. This approach is in good agreement with the WDM-principle of integrating sensors, allowing the creation of highly efficient multichannel measuring systems. When implementing this approach, it is necessary to take into account a number of features. In particular, the wavelength range used should correspond to the spectral sensitivity of the photographic recording medium (crystal). In addition, it is necessary to take into account the possible appearance of cross-effect between holograms due to overlapping spectra of different channels, which, together

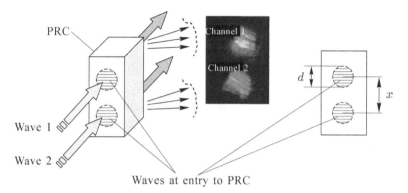

Waves at entry to PRC

Fig. 5. Formation of a multichannel adaptive demodulator based on spatial multiplexing of dynamic holograms in a photorefractive channel and a photograph of the corresponding space-separated measuring channels of the fanning waves; x is the distance between the centres of the channels; d is the efficient channel diameter, determined by the size of the light beam introduced into the crystal [29].

with the first requirement, imposes a restriction on the number of channels.

In the following sections we will consider the implementation of multichannel adaptive demodulators and measuring systems based on the above principles of multiplexing dynamic holograms in a photorefractive crystal.

8.3.1. Spatial multiplexing

In work [29], the multichannel adaptive demodulator is constructed on the basis of spatial multiplexing of holograms, which are formed in various regions of a $Bi_{12}TiO_{20}$ crystal (Fig. 5). At the same time, the formation of channels is based on the fanning effect [23, 24], which is a self-diffraction process of a light wave in a photorefractive medium.

The nature of this effect is described in chapter 7 and consists in transferring energy between the beam injected into the crystal and the rays scattered by the defects and inhomogeneities of the crystal [24]. Due to the holographic nature of the fanning effect, any rapid changes in the parameters of the input optical field will lead to a change in the intensity of the scattered waves (the fanning waves), a demodulation signal will appear. Figure 5 shows the radiation field at the crystal output for the case of the demodulation in the last two channels. The channels are located so close to each other that in the near zone their fields overlap, and only in the far zone it becomes possible to separate them in space.

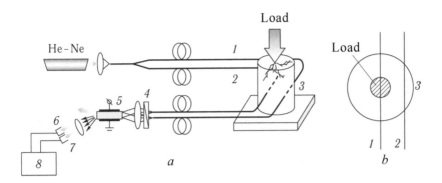

Fig. 6. Two-channel adaptive fibre-optic measuring system (*a*) and position of optical fibres inside the tested object – top view (*b*): *1, 2* – single-fibre multimode interferometers (measuring channels); *3*– tested object; *4* – polarizer; *5* – photorefractive crystal $Bi_{12}TiO_{20}$, under the influence of an external alternating electric field (amplitude 9 kV/cm, frequency 30 Hz); *6, 7* – photodetectors detecting the intensity of the fanning waves in each of the two channels; *8* – registration system [61].

It was shown in [29] that when the channels in the crystal approach each other, starting from a distance equal to the effective channel size *d* (see Fig. 5), the level of the demodulation signal in the channel begins to decrease. Such a decrease is due to a decrease in the contrast of the interference field in the channel when radiation enters it from another channel. However, even with full overlapping of channels, the signal level in the channel decreases only by 2.5 dB.

The peculiarity and advantage of forming a demodulation channel on the basis of the fanning effect is the fact that for holograms an external reference light beam is not required; its role is played by waves scattered on the crystal defects. Thus, the radiation in an individual channel can be mutually incoherent with radiation in all other channels. Due to this, with close packing of channels in a crystal, the radiation scattered in one channel, overlapping with radiation from another channel, does not form a cross hologram, which ensures that there is no crosstalk between channels whose level does not exceed the noise level of the detecting electronics (−11 dB).

The two-channel adaptive fibre-optic measuring system (Fig. 6) developed in [29], based on the multichannel demodulator considered, was used to monitor the process of crack formation in an object under the influence of increasing mechanical stress [61]. Two fibre-optic sensors realized on the basis of SMI were built into the object under study at the stage of its manufacturing. Mechanical deformation of the object, caused by the applied external load,

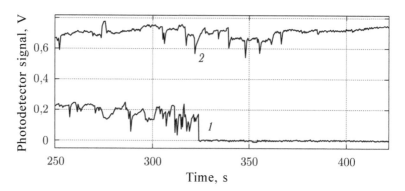

Fig. 7. Oscillograms of signals received from photodetectors on two channels of an adaptive fibre-optic measuring two-channel measuring system that monitors the processes of cracking in a solid [61].

leads to the appearance of cracks, which, in turn, are sources of sound pulses propagating through the object. Having reached fibre fibres and acting on them, acoustic pulses cause modulation of the parameters of the radiation channeled into the SMI. The radiation of the fanning corresponding to the channels modulated in amplitude carries information on the formation of cracks.

Figure 7 shows oscillograms of signals obtained by two channels. The adaptive properties of the holographic demodulator formed in the PRC allow not only to exclude the influence of low-frequency external factors (for example, temperature drift) on the measuring system, but also to make it insensitive to slow deformations of the object under the action of a constantly growing load, ensuring detection of only the appearance moments cracks.

8.3.2. Angle multiplexing

In Ref. [62], a multichannel adaptive measuring system is implemented by the principle of angular multiplexing of dynamic holograms in a photorefractive crystal. The scheme of the multichannel demodulator of the measuring system is shown in Fig. 8.

Signal light beams emerging from the fibre optic sensor channels are sent to the photorefractive crystal, where they interact with a common reference beam. The interference of each signal light beam separately with the reference beam leads to the formation of a set of basic dynamic holograms on which the phase of the corresponding

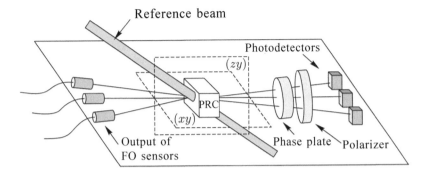

Fig. 8. Optical scheme of a three-channel holographic adaptive demodulator based on the angular multiplexing of transmissive dynamic holograms.

signal waves is demodulated. The use of a common reference beam for all multiplexed holograms assumes that all signal beams will be mutually coherent. In this case, overlapping of the signal light beams in the thickness of the crystal with each other can cause the formation of cross holograms and, as a consequence, the appearance of a cross signal between the multiplexed channels.

To minimize the effect of cross holograms and the associated crosstalk noise, the approach proposed in [75] is used. As can be seen from Fig. 9, the holograms are formed in the transmissive geometry – the signal and reference light beams propagate in the

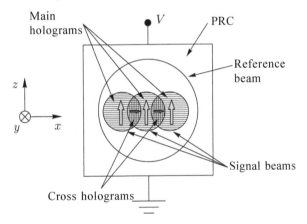

Fig. 9. Configuration of the multiwave interaction in the PRC: all signal light beams lie in the horizontal plane (xy) and propagate in a direction close to the y axis; the reference light beam makes an angle of $5°$ with the plane (xy); as a result, the wave vectors of the main holographic gratings are oriented vertically and coincide with the direction of the external electric field applied to the crystal, while the wave vectors of the cross holograms are oriented horizontally.

crystal in one direction under some relatively small angle to each other. Is this a relatively large spatial period of the holographic grid Λ and, as a consequence, its low diffraction efficiency. To increase the efficiency of the hologram, an external alternating electric field with an amplitude of 6 kV/cm and a frequency of 3000 Hz is applied to the crystal. The wave vectors of the main and cross holograms are oriented orthogonally to each other in the crystal (Fig. 9) due to the fact that the falling plane (*xy*) of the signal light beams is orthogonal to the plane (*yz*) of the falling of the reference beam (see Fig. 8). Thus, an external electric field, applied along the vectors of the basic holograms, will provide an increase in their diffraction efficiency, while the cross holograms remain unresponsive.

The disadvantages of the system proposed in Ref. [62] include the need to use a strong electric field applied to the crystal to selectively enhance the basic holograms. At the same time, as shown in [63], a sufficiently effective adaptive interferometer, comparable in parameters to the classical one, can be constructed on the basis of the use of diffusion reflective geometry that allows recording dynamic holograms in photorefractive crystals without an external electric field [64]. This fact can also be used for the effective multiplexing of diffusion reflective holograms in a crystal of cubic symmetry and the construction of a multichannel adaptive measuring system [64].

A scheme for multiplexing the reflecting holograms in a crystal of cubic symmetry (point groups 23 and 43m) is shown in Fig. 10. Two signal waves S_1 and S_2 propagate along the main crystallographic axis [001] and mix with the common reference wave R propagating

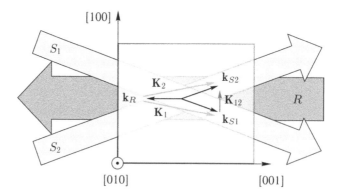

Fig. 10. The geometry of the angular multiplexing of two reflective holograms in a photorefractive crystal: k_R, k_{S1}, k_{S2} – wave vectors of the reference (R) and signal (S_1, S_2) light beams, respectively; K_1, K_2 – wave vectors of the basic holograms; K_{12} is the wave vector of the cross hologram [64].

towards them. As in the case considered above, all the waves in the channels are mutually coherent, and their pairwise interference leads to the formation of dynamic holograms of two types. The first type is the basic hologram formed by a pair of two beams, the signal and the reference. The second type of hologram is a cross hologram formed by two signal beams. The basic holograms form demodulation channels, while the appearance of cross holograms leads to cross-talk noise between the channels. In addition, multiple recording of holograms in the same crystal can generally lead to a decrease in the diffraction efficiency of a single hologram and, as a consequence, to a decrease in the sensitivity in a separate channel. The degree of reduction in sensitivity, together with the level of crosstalk noise, determines the efficiency of the multi-channel holographic modulator. As was shown in [63], the level of crosstalk noise (C) can be estimated with sufficient accuracy

$$C \approx 20 \log\left(h \frac{|\sin\Theta|}{2} \right), \tag{8.11}$$

where Θ is the angle of the deviation of the vector $\mathbf{K}_{12}$ from the normal to the axis [001]; $h < 1$ is a coefficient, taking into account the degree of overlap of the signal light beams in the crystal.

For example, for a CdTe crystal with a refractive index $n = 2.8$ the range of angles of incidence of signal light beams at which the crosstalk remains below the level of the intrinsic noise in the channel (-30 dB), can reach 60°. In this case, the signal beams can be in different planes (located, for example, in a fan), in contrast to the angular multiplexing system based on the transmission geometry described above, where all the beams must lie in one plane. The wide solid angle of the signal beams entering the crystal greatly simplifies the process of forming holographic demodulation channels and increases the potential number of them.

The contrast value (m) of the interference field in one of the N multiplexed channels with the equality of signal intensities in the remaining channels can be estimated as follows [63]:

$$m(N) \approx \frac{2\sqrt{I_S I_R}}{I_R + hNI_S},$$

where I_R is the intensity of the reference light beam; I_S is the intensity of the signal light beam in an arbitrarily selected main channel; N is the total number of channels.

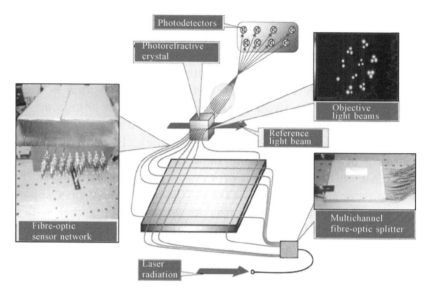

Fig. 11. 32-channel fibre-optic measuring system Smart Grid [65].

Figure 11 shows the scheme of the 32-channel fibre-optic Smart Grid measuring system, designed to monitor in real time the spatial distribution of weak mechanical oscillations of the surface of objects with an amplitude of up to 100 μm [65]. A single CdTe crystal was used to demultiplex and demodulate signals coming from fibre optic measuring lines.

8.3.3. Spectral multiplexing

As already noted at the beginning of this chapter, spectral multiplexing (or wavelength multiplexing) is one of the effective methods for creating multichannel optical and fibre-optic measurement systems. This method plays a special role when using fibre optic Bragg gratings (FBG) [3, 40] as sensors [3, 40], since the information retrieval in the FBG sensors is mainly based on the spectral coding of the external effect result, consisting in the peak shift in the transmission spectrum (or reflection) of the sensor. FBG-sensors and measuring systems based on them are widely used, for example, in problems of monitoring the state of engineering structures and technical structures [40, 66]. A large number of FBG-sensors, being formed in a single fibre, makes it possible to create distributed and quasi-distributed measuring lines. The spectral shift is detected using appropriate demodulators, which, according to the principle of operation, can be divided into three categories: scanning, spectrometric and interferometric. Scanning demodulators include Fabry–Perot scanning filters [67], acousto-optical tunable filters [64], and tunable laser radiation sources [68]. The main disadvantage of the scanning methods is the fact that only one sensor is interrogated at each individual time. This excludes the possibility of using such systems in problems where simultaneous interrogation of all sensors is required (for example, in monitoring the effect of pulsed signals, studying acoustic emission, etc.) [69]. Spectrometric methods [70] have low sensitivity and are unsuitable for dynamic measurements if several sensors are to be active all the time. The methods for demodulating FBG-sensor signals based on the use of interferometers (for example, the Mach–Zehnder interferometer [71]) are ideal for monitoring dynamic effects. However, when implementing such systems to compensate for the quasistatic drift of the operating point of the interferometer caused by uncontrolled changes in external conditions and the provision of quadrature detection conditions, the use of electronic feedback systems is required. As a result, the cost of a multichannel system increases dramatically, and efficiency decreases, since for each multiplexed sensor an individual feedback system is required.

Thus, in order to create an effective multichannel measuring system based on the FBG-sensors, it is necessary to have a multichannel demodulator capable of both providing simultaneous interrogation of all sensors and realizing the stabilization of their operating parameters. It was shown in [72] that such a demodulator can be created on the basis of a dynamic hologram formed in a photorefractive crystal. Due to its adaptive properties, a demodulator based on the PRC makes it possible to compensate for all slow, including quasistatic, effects on the sensing elements of the measuring system without using any external active stabilization systems, while ensuring stable detection of high-frequency or pulsed effects such as vibrations, shocks, ultrasonic signals, acoustic emission, etc. At the same time, a system based on a two-wave interaction in a PRC makes it possible to provide an effective spectral demultiplexing of the signals of the FBG-sensors.

The shift $\Delta\lambda_B$ in the spectrum of reflection of the radiation from a FBG sensor, caused by deformation (ε_F) and a change in temperature (ΔT), can be written as [73]

$$\Delta\lambda_B = \lambda_B \left\{ 1 - \frac{n_{eff}^2}{2} \left[p_{12} - u(p_{11} + p_{12}) \right] \varepsilon_F + (\alpha_V + \alpha_n)\Delta T \right\}, \quad (8.12)$$

where λ_B is the central wavelength in the reflection spectrum of the FBG-sensor; n_{eff} is the effective refractive index of a fibre lightguide; $p_{ij}(i, j = 1, 2)$ are the components of the photoelastic tensor; u is Poisson's ratio; α_V, α_n is the coefficient of thermal expansion and the thermo-optical coefficient of the material of the optical fibre, respectively.

Estimates made using expression (8.5) show that the mechanical deformation of the length of the fibre of one strain ($\Delta L/L$ extension, measured in strain (1 strain = 10^{-6})), measured using a standard FBG-sensor operating on the length wave $\lambda_B = 1550$ nm, leads to a shift of the central wavelength by ~1.2 pm. At the same time, with a temperature change of only 1°, this shift is already 13 pm, which once again underscores the urgency of the task of compensating for the drift of the operating characteristics of the measuring system caused by uncontrolled fluctuations in the environmental parameters.

A demodulator based on a dynamic hologram formed in a PRC is an adaptive interferometer that, like any other interferometer, ensures the transformation of the phase of the light wave into a change in intensity. At the same time, the signal of the FBG sensor is the displacement of the central wavelength in the spectrum of the reflected wave. In order for the adaptive interferometer based on the PRC to function as a *spectral* demodulator, it is necessary to divide the radiation reflected from the FBG sensor into two beams, the reference and the signal, which must pass unequal paths before entering into the PRC. Then any change in the *wavelength* of the light reflected from the FBG will result in an equivalent phase shift between the reference and signal beams:

$$\varphi = \frac{2\pi b}{\lambda_B^2} \Delta\lambda_B, \quad (8.13)$$

where b is the optical path difference (OPD) of the reference and signal beams. We note that a similar principle of transforming spectral changes into phase ones is used in Mach–Zehnder interferometers and other unbalanced interferometers [71].

It follows from (8.13) that the larger the OPD, the greater the equivalent phase shift and, consequently, the stronger the demodulation signal in the interferometer. However, FBG-based systems use broadband light sources, and the line width in the reflection spectrum from the FBG is typically 0.1–0.4 nm. This imposes a limitation on the coherence length, which must be taken into account both during the formation of the photorefractive grating, and in subsequent interference between the transmitted signal and diffracted reference beams. In particular, the contrast of the interference field formed by beams of a finite spectral range with a width in the wave vector of Δk:

$$\tilde{m} = \frac{2\sqrt{\beta}}{\beta+1} \exp\left(-\frac{\Delta k^2 b^2}{16\ln 2}\right), \tag{8.14}$$

where $\beta = I_R/I_S$ is the intensity ratio of the reference (I_R) and signal (I_S) beams.

The factor in front of the exponent in expression (8.14) is none other than the contrast of the interference field formed entirely by coherent light beams. The allowance for the attenuation of the contrast caused by the decrease in the coherence of the radiation makes it possible to obtain the following expression for the change in the intensity of the signal-modulated beam (ΔI_S) after its interaction in the PRC with the reference beam [74]:

$$\Delta I_S \sim \exp\left(-\frac{\Delta k^2 b^2}{16\ln 2}\right)\frac{b}{\lambda_B^2}\Delta\lambda_B. \tag{8.15}$$

We note that the weakening of the contrast of the interference field due to the finiteness of the emission spectrum does not have a significant effect on the efficiency of the holographic grating. Since in most practical cases the contrast is initially low, since the intensity of the reference beam is many times greater than the signal intensity, it avoids the formation of higher spatial orders in the crystal lattice, which in turn could lead to undesirable cross-noise in the multiplex system.

As follows from expression (8.15), the level of the demodulation signal decreases exponentially with increasing optical path difference b due to a decrease in the mutual coherence of the interacting beams and increases linearly as a result of the increase in the phase shift caused by the spectral shift of the reflected wave.

It was shown in Ref. [75] that the minimum detectable value of the dynamic deformation of the recorded FBG is ~0.25 μstrain,

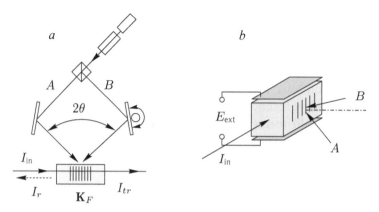

Fig. 12. *a* –scheme for the formation of a dynamic band-pass Bragg filter in PRC; *b* –orientation of the external electric field applied to the crystal [76].

which corresponds to a shift in the central wavelength in the reflection spectrum of the FBG of ~0.3 pm. And the main sources of noise limiting the minimum detection threshold are fluctuations in the intensity of the output power of the radiation source and the amplitude noise of the optical amplifier, the effect of which can be minimized by using balanced detection.

Another promising scheme for demultiplexing signals of FBG-sensors of fibre-optic measuring lines, based on dynamic gratings formed in photorefractive crystals, was proposed and investigated in [76, 77]. The scheme of dynamic band-pass filtering is shown in Fig. 12 [76].

As can be seen from Fig. 12, two coherent light beams *A* and *B* (λ_w = 532 nm) forming an angle of 2θ between them are directed to the side face of the $BaTiO_3$:Co crystal, where a dynamic lattice with a spatial period $\Lambda_F = \lambda_w/(2\sin\theta)$ is formed. The radiation subject to spectral filtration is introduced into the crystal along the wave vector of the grating $\mathbf{K}_F$ (see Fig. 12 *a*), which plays the role of a Bragg mirror with a central wavelength in the reflection spectrum

$$\lambda_B = \frac{n_C \lambda_w}{\sin\theta}, \qquad (8.16)$$

where n_C is the refractive index of the crystal.

Figure 13 shows the characteristic transmission spectrum of a band-pass spectral filter based on the dynamic grating formed in

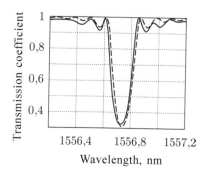

Fig. 13. Spectrum of transmission of a band-pass spectral filter based on a dynamic grating formed in a PRC [76].

the PRC. As one can see, the filter has a high quality factor due to a rather narrow line width (~0.1 nm).

The adaptive properties of the dynamic grating formed in the PRC allow the filter parameters to be reconstructed and realized thus the detection in real time of the signals coming from different FBG-sensors. So, if the angle θ between the recording beams is changed, the old dynamic grating will disappear, and the new (with a new period and, accordingly, adjusted to the new value of λ_B) will be formed, which allows smooth adjustment of the filter to the desired wavelength in a fairly wide range. As was experimentally established, with a change in the angle between the recording gratings by 5.5°, the λ_B value will change by 80 nm (from 1480 to 1560 nm). It was noted in [76] that the displacement of the Bragg wavelength does not entail any noticeable changes in the shape of the reflection/transmission line, as well as the diffraction efficiency of the grating. The time for the adjustment of the filter is determined by the rate of recording of the dynamic grating in the crystal. In work [76] this time was ~0.5 s.

Additional control of the central wavelength of the bandpass filter can be achieved due to the electro-optic effect: application of an electric field to the crystal (see Fig. 12 *b*) leads to a change in the refractive index of the crystal and, accordingly, to a change in λ_B.

Figure 14 shows the transmission spectra of the bandpass filter obtained for different values of the external electric field. As can be seen from the figure, the change in the field strength from −614 to +655 V/cm provides for the tuning of the central wavelength in the reflection spectrum in the range of 0.55 nm. At the same time, the speed of adjustment is limited only by the speed of the control electronics and can be quite high. In [76] this speed was 2.2 nm /μs.

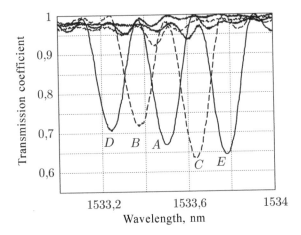

Fig. 14. A change in the transmission spectrum of the dynamic grating formed in a photorefractive crystal in dependence on the strength of the electric field applied to the crystal: A, $E_{ext} = 0$; B, $E_{ext} = -370$ V / cm; C, $E_{ext} = +389$ V/cm; D, $E_{ext} = -614$ V/cm; E, $E_{ext} = +653$ V/cm [76].

References

1. Kersey A.D., Optical fibre technology (ed. A.D. Kersey), 1996. V. 2, No. 3. P. 291-317.
2. Butusov M.M., Galkin S.L., Fibre optics and instrument making. Leningrad, Mashinostroenie, 1987.
3. Kulchin Yu.N., Distributed fibre-optic measuring systems stems. Moscow: Fizmatlit, 2001.
4. Pinchevsky A.D., Izmerit. tekhnika. 1991. No. 8. P. 44–50.
5. Ivanov V.N., Kavalerov G.I., Izmerit. tekhnika. 1991. No. 10. P. 8–10.
6. Kulchin Yu.N., et al., Kvant. Elektronika. 1993. Vol. 20, No. 5. P. 711–714.
7. Kulchin Yu.N., Vitrik O.B., Izmerit. tekhnika. 1999. No. 3. P. 24–30.
8. Kulchin Yu.N., Izmerit. tekhnika. 1995. No. 3. P. 32–35.
9. Kotov O.I., et al., Pis'ma Zh. Tekh. Fiz. 1990. Vol. 16, No. 2. P. 90–93.
10. Kulchin Yu.N., et al., Kvant. Elektronika. 1995. P. 22, No. 10. P. 1009–101.
11. Kulchin Yu.N., et al., Kvant. Elektronika. 1997. Vol. 24, No. 5. P. 467–470.
12. Kulchin Yu.N., et al., Izmerit. tekhnika. 1999. No. 6. P. 21–28.
13. Natterer F., Mathematical aspects of computed tomography. Moscow, Izd–vo Mir, 1990..
14. Tikhonov A.N., Arsenin V.Ya., Methods for solving ill-posed problems. Moscow Nauka, 1979.
15. Bykovsky Yu.A., Kulchin Yu.N., Kvant. Elektronika. 1990. V. 17, No. 8. P. 1080–1083.
16. Kulchin Yu.N., Obukh V.F., Kvant. Elektronika. 1986. Vol. 13. P. 650–658.
17. Bykovsky Yu.A., et al., Kvant. Elektronika. 1990. V. 17, No. 10. Pp. 1377–1378.
18. Bykovskii Yu.A., et al., Kvant. Elektronika. 1990. V. 17, No. 1. P. 95–98.
19. Kamshilin A.A., et al., Appl. Phys. Lett. 1998. V. 73, No. 6. P. 705–707.

20. Kulchin Yu.N., in: Distributed Fibre Optical Sensors and Measuring Networks (ed. Yu.N. Kulchin). Bellingham, Washington, SPIE, 2001. V. 4416. P. 58–61.
21. Nazhestkina N., et al., J. Appl. Phys. 2000. V. 88, No. 5. P. 2194–2199.
22. Borodin M.V., et al., Izv. VUZ. Fizika. 2001. V. 44, No. 10. P. 38–42.
23. Feinberg J.J., Opt. Soc. Amer. 1982. V. 72, No. 1. P. 6–51.
24. Voronov V.V., et al., Sov. J. of Quant. Electron. 1980. V. 7, No. 11. P. 2313–2318.
25. Xie P., et al., J. Appl. Phys. 1993. V. 74. P. 813–819.
26. Kamshilin A.A., et al., Opt. Lett. 1999. V. 24, No. 12. P. 832–834.
27. Kamshilin A.A., et al., Appl. Phys. Lett. 1999. V. 74. P. 2575–2577.
28. Cronin-Golomb M., et al., IEEE J. Quant. Electron. 1984. V. 20, No. 1. P. 12–30.
29. Kulchin Yu.N., et al., Pis'ma Zh. Tekh. Fiz. 2000. V. 26, No. 12. P. 23–27.
30. Nakamura J., Image Sensors and Signal Processing for Digital Still Cameras. CRC Press, 2005.
31. Kulchin Yu.N., et al., Kvant. Elektronika. 2001. Vol. 36, No. 4. C. 339–342.
32. Zuo J.M., Microscopy research and technique. 2000. V. 49. P. 245–268.
33. Murphy D.B., Fundamentals of Light Microscopy and Electronic Imaging. Wiley–IEEE, 2002.
34. Shapiro B. Theoretical principles of the photographic process. Minsk: Editorial URSS, 2000.
35. Miz K., James T. Theory of the photographic process (transl. from English). Leningrad, Khimiya, 1973.
36. Balyakin I.Ya., Appliances with transfer series in radio engineering information processing devices. Moscow, Radio i svyaz'. 1987.
37. Kruglikov S.V., Loginov A.V., Multi-element image receivers. Novosibirsk, Nauka, Siberian division, 1991.
38. Charge-coupled devices, ed. D.F. Barb. Moscow, Mir, 1982..
39. Grattan K.T.V., Meggitt B.T., Optical fibre sensor technology. London, Chapman & Hall, 1995.
40. Grattan K.T.V., Sun T., Sensors and Actuators A. 2000. V. 82. P. 40–61.
41. Tudor M.J., et al., IEE Proc. part D. 1988. V. 135. P. 364–368.
42. Kersey A.D., Optical Fibre Technology. 1996. V. 2. P. 291–317.
43. Sæther J. and Bløtekjær K., Optical Review. 1997. V. 4. P. 138.
44. Jarret B., Burn E., Proc. SPIE. 1992. V. 1586. P. 164–173.
45. Senior J.M., Cusworth S.D., Optics and Laser Technology. 1990. V. 22. P. 113–125.
46. Meltz G., Proc. SPIE. 1996. V. 2838. P. 2–22.
47. Kersey A.D.,et al., Journal of Lightwave Technology. 1997. V. 15. P. 1442–1463.
48. Kersey A.D., Berkoff T.A., Electronics Letters. 1992. V. 28. P. 1215–1216.
49. James S.W., et al., Proc. SPIE. 1996. V. 2838.P. 52–57.
50. Ferreira L.A., Santos J.L., Pure and Applied Optics. 1996. V. 5. P. 257.
51. Lobo Ribeiro A.B., et al., Electronics Letters. 1996. V. 32. P. 382.
52. Zhang L., et al., Proc. of ECOC'98. 1998. P. 609.
53. Vohra S.,et al., IEICE Transactions on Electronics. 2000. V. E83-C, No. 3. P. 454–461.
54. Dakin J.P., Volanthen M., IEICE Transactions on Electronics. 2000. V. E83iC, No. 3. P. 391–399.
55. Walker J.C., et al., Electronics Letters. 1992. V. 28. P. 1627–1628.
56. Mlodzianowski J., et al., IEEE Journal of Lightwave Technology. 1987. V. 5. P. 1002–1007.
57. Juskaitis R., and Shatalin S.V., Proc. SPIE. 1994. V. 2360. P. 538–540.
58. Jarret B., Burn E., Proc. SPIE. 1992. V. 1586. P. 164–173.

59. Voet, M.R., et la., Proc. SPIE. 1994. V. 2210. P. 126–135.
60. Davis M.A., et al., Electronics Letters. 1996. V. 32, No. 15. P. 1393–1394.
61. Kulchin Yu.N., Romashko R.V., Piskunov E.N. Multichannel adaptive fibre optical system for monitoring of fast processes in a solid state // Proc. SPIE. 2001. V. 4513. P. 12–17.
62. Fomitchov P., et al., Applied Optics. 2002. V. 41, No. 7. P. 1262–1266.
63. Di Girolamo S., et al., Opt. Express. 2007. V. 15. P. 545–555.
64. Di Girolamo S., et al., Opt. Express. 2008. V. 16. P. 18040–18049.
65. Romashko R.V., Kulchin Yu.N., Laser Physics. 2014. V. 24, No. 11. P. 115604.
66. Sun C.S., Ansari F., Opt. Eng. 2003. V. 42. P. 2987–2993.
67. Kersey A.D., et al., Optics Letters. 1993. V. 18. P. 1370–1372.
68. Fomitchov P., Krishnaswamy S., Opt. Eng. 2003. V. 42. P. 956–963.
69. Perez I., et al., Proc. SPIE. 2001. V. 4328. P. 209–215.
70. Davis M. A., Kersey A.D., J. Lightwave Technol. 1995. V. 13. P. 1289–1295.
71. Kersey A.D., et al., Electron. Lett. 1992. V. 28. P. 236–238.
72. Qiao Y., et al., Applied Optics. 2006. V. 45, No. 21. P. 5132–5142.
73. Murray T.W., et al., Appl. Opt. 2000. V. 39. P. 3276–3284.
74. Kulchin Yu.N., et al., Adaptive optical methods for processing speckle-modulated optical fields. Moscow, Fizmatlit, 2009.
75. Coppola G.,et al., roc. SPIE. 2001. V. 4328. P. 224–232.
76. Petrov V.M., et al. J. Opt. A: Pure Appl. Opt. 2003. V. 5. P. S471–S476.
77. Runde D., et al., in: OSA Trends in Optics and Photonics (TOPS), Photorefractive Effects, Materials, and Devices. 2005. V. 99. P. 772–776.

9

Laser Cooling, Trapping and Control of Atoms

Introduction

The interaction of laser radiation with matter is widely used to cool atomic and molecular gases to extremely low temperatures. In connection with this, a considerable number of works on quantum optics have recently been devoted to laser cooling of matter. If the gas is cooled by traditional cryogenic methods, it will first turn into a liquid, and then become a solid. As a result, the energy levels of individual atoms, due to strong interactions in the lattice, are transformed into energy bands. Therefore, the usual cooling of gases radically changes their properties and does not allow reaching a temperature below several tens of kelvins. At the same time, cooling the rarefied gas to a temperature close to absolute zero makes it possible to study components of its atoms and molecules of a quantum nature and use them in a number of applied problems.

As a rule, in gases, atoms and molecules move at a very high velocity. At room temperature in the air this velocity is ~300 m/s. To estimate the magnitude of the thermal velocity, we note that at the liquid nitrogen temperature (77 K) the N_2 molecules in the air move with velocities ~150 m/s, and at liquid helium temperature (4 K) the condensed helium atoms move at a velocity of ~90 m/s. If, however, the thermal velocity of the atoms is brought to 1 m/s, all gases are in a condensed state, and practically no gas remains in the gas phase. Therefore, all studies of free atoms were previously carried out with rapidly moving atoms. As a result, the Doppler frequency shift and the relativistic time dilation caused the displacement and broadening of the spectral lines of thermal atoms. In addition, the high velocities of the atoms limited the observation time, and, consequently, the spectral resolution. This has always been an important factor stimulating the search for methods that allow cooling of neutral atoms and ions to extremely low temperatures. In addition, when the atoms of matter are sufficiently cold, new quantum phenomena begin to arise. Indeed, if the de Broglie wavelength $\lambda_D = h/p$ determined by the pulse p becomes of the order of the length of the interatomic interaction, or the interatomic distances, or the size of the confinement region of the atoms, then the wave or quantum nature of the particle begins to appear, and hence the wave nature of the particles becomes explicit and begins to manifest new properties of matter. For the first time, the process of cooling atoms by laser radiation was proposed by the Russian scientist V. Letokhov in 1979. Subsequently, the method was developed in the works of S. Chu, K. Cohen-Tannoudji, and W.D. Phillips, which resulted in the awarding of the Nobel Prize in 1997 [1–6].

9.1. Doppler cooling

The principle of Doppler cooling for neutral atoms was proposed by T. Nanch and A. Shavlov [1], and for trapped ions by D. Wineland and H. Dehmelt [5]. The scheme of the principle of Doppler cooling is shown in Fig. 1 [7]. The atomic beam is slowed by transmission of a pulse that occurs at the time when the atom absorbs a quantum of light. A beam of atoms moving at a velocity V is irradiated by a laser beam (Fig. 1 *a*). If the atom in the ground state moves toward the laser beam, then for each act of absorbing the photon it acquires an impulse $p_i = \hbar k$, where k is the wave number of the light quantum ($i = 1, 2, 3, ...$). As a result, each photon absorbed by an atom slows down the velocity of an atom by the amount $V_{rec} = \hbar k/m$, where m is the mass of the atom. To absorb the photon of laser radiation again,

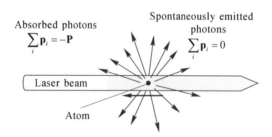

Fig. 1. Scheme of interaction of a cooled atom with a laser beam [7].

the atom must return to the ground state, while emitting a photon. Since photons are emitted by an atom in arbitrary directions, but with a symmetric angular distribution, the average contribution from the emitted photons to the momentum of the atom is zero. Therefore, the atom will lose speed only in the direction of its propagation.

For example, for sodium atoms, interaction with resonant yellow radiation leads to a loss of velocity in one act of absorbing a photon by a value of V_{rec} = 3 cm/s. Since the typical velocity of atoms in the beam is about 10^5 cm/s, then in order to stop the sodium atoms, it is necessary to produce an act of absorption-emission of a photon about $3 \cdot 10^4$ times. Since the atom can emit and absorb photons at a rate equal to half the rate of decay of the excited state (if we assume the atom to be a two-level quantum system), for Na atoms this means that photons can be emitted and absorbed approximately every 3.2 ns. In this connection, the atom in the beam can be stopped in approximately 0.1 ms.

When laser cooling of atoms in a certain volume, a problem arises with optical pumping. The problem is that the absorption of photons, in addition to slowing down the atoms, can also cause the acceleration of atoms if it occurs with atoms moving in a co-direction with the laser beam. Therefore, to slow down the atoms, it is necessary to provide a larger number of photon absorption in the case when the atom and photon fly towards each other. In practice, this is achieved by tuning the laser frequency (ω_L) just below the resonance absorption of stationary atoms (ω_A) (Fig. 2). [1] In the figure energy level $|g\rangle$ corresponds to the ground state of the atom, and the energy level $|g\rangle$ to its excited state, and Γ is the natural radiation width of the energy level, determined by the lifetime of the atom in the excited state.

According to [1], when $\omega_L < \omega_A$, for the atom moving toward the laser beam, the emission frequency of the laser photon is considered to be shifted upward to its resonance frequency by an amount of kV, and therefore such photons are more strongly absorbed than photons moving in the opposite direction (Fig. 3) [7].

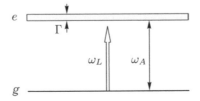

Fig. 2. Two-level model of energy transitions in an atom [1].

In the case where Na atoms are at room temperature, the frequency of the incoming photon, due to the Doppler effect, will be shifted by 0.97 GHz, and in order to get a collision of an atom and a photon in accordance with the resonance frequency, it will be necessary for the laser to be tuned below the resonance peak to this difference frequency.

As the atom, repeatedly absorbing photons, slows down, the Doppler frequency shift decreases, and the atom comes out of resonance with the light wave. The natural line width for the optical transition to Na is $\Gamma \sim 10$ MHz. The change in the velocity of an atom by a value of 6 m/s leads to practically the same frequency shift. Therefore, after the Na atom is absorbed by about 200 photons, the frequency of the laser radiation soon enough is far from resonance.

Nevertheless, this process of slowing down the atoms and leaving them out of resonance leads to their cooling and narrowing the peak of their velocity distribution. In an atomic beam, there is usually a fairly wide velocity distribution around their average thermal velocity. Therefore, atoms with the necessary speed for cooling rapidly absorb photons and also rapidly slow down. Too fast atoms first absorb photons more slowly, and then, when they get into resonance, they begin to absorb faster and, as they move further, they slow down more and more. Too slow atoms absorb photons from the very beginning slowly and slow down a little. Thus, atoms from a certain range around the resonant velocity are knocked into a narrower velocity range around a lower velocity (Fig. 4) [5]. As can be seen from Fig. 4, the atoms are redistributed into the region of lower velocities, that is, cooling of the gas.

Since as a result of the Doppler shift of the frequency, the atoms moving toward the laser beam strongly absorb photons and

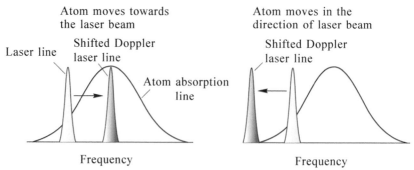

Fig. 3. Resonant scheme of interaction of a laser beam with colinearly moving atoms.

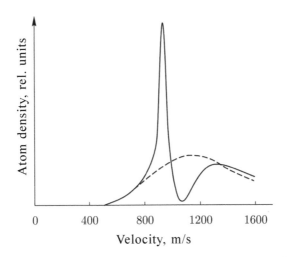

Fig. 4. The distribution of atoms in terms of velocities in an atomic beam cooled by laser with a fixed frequency (dashed curve – distribution up to cooling, solid curve – after cooling).

decelerate, this phenomenon can be interpreted as the result of the action of some viscous friction force. Since the intensity of the absorption of photons by an atom decreases with an increase in the deviation of the velocity of the atom from its resonant velocity, we can assume that the magnitude of this force will be proportional to the magnitude of the atom's velocity [5]: $F = -\eta V$, where η is the coefficient of viscous friction. Therefore, a medium consisting of actively retarding gas atoms of photons and gas atoms is called 'optical molasses' [1–5].

Thus, the viscous friction of gas atoms due to the Doppler effect leads to cooling of the gas. As noted in [1], the gas temperature achieved using the phenomenon of Doppler cooling will always be greater than a certain temperature T_D, called the Doppler temperature: $k_B T_D = \hbar\Gamma/2$, where k_B is Boltzmann's constant. For alkali atoms, $T_D \sim 100~\mu K$, and this temperature is the limiting temperature for Doppler cooling of atoms. In Chu's work, performed in 1985, sodium atoms were cooled from 500 K to 240 mK [5].

9.2. Zeeman cooling

According to the Doppler laser cooling method described above, the atomic beam interacts with an oppositely directed laser beam whose radiation frequency is tuned so as to be in accordance with the atomic transition. However, the gradual deceleration of atoms, due to the Doppler effect, removes the laser frequency from resonance, which limits the cooling process. As was shown in [5, 8], the Doppler frequency shift can be compensated for using the linear Zeeman effect. In this case, the distance between the energy levels of atoms changes with the help of a magnetic field, and thus they are held in resonance with a fixed laser frequency. Figure 5 illustrates this by showing the scheme of the energy levels of the Na atom in a magnetic field [5].

The scheme of cooling an atomic beam by laser radiation in a magnetic field, called the Zeeman slowing, is most often used to obtain slow atomic beams. The idea of the method is illustrated in Fig. 6.

As can be seen from Fig. 6 *a*, in the Zeeman slowing, the source directs a beam of atoms whose velocities vary over a wide range along the axis of a cone-shaped solenoid. The solenoid has a denser

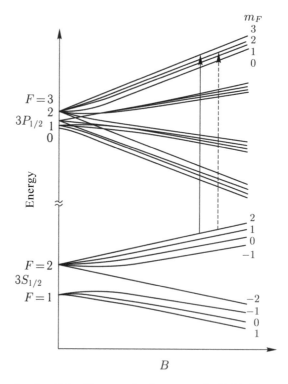

Fig. 5. Energy levels of the Na atom in the magnetic field. The energy transition used for laser cooling is indicated by the solid arrow, and one of the forbidden transitions which usually leads to undesirable optical pumping is indicated by the dashed line [5].

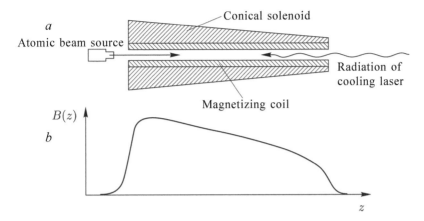

Fig. 6. Zeeman slowing of atomic beam (*a*) and the dependence of the distribution of the magnetic field along the solenoid axis (*b*) [6].

winding in the inlet (near the source of atoms), which gradually falls to its opposite end. As a result, a non-uniform distribution of the magnetic field induction $B(z)$ is created on the axis of the solenoid (Fig. 6 *b*). The cooling laser is tuned so that its frequency is equal to the frequency of the energy transition in the atom increased on the basis of the Doppler effect: $\omega_L = \omega_A + kV$. According to the quantum mechanical selection rules, only those transitions are allowed in an atom in a magnetic field for which the change in the magnetic quantum number is equal to $\Delta m = \pm 1$. The energy transitions to this selection rule correspond to emission or absorption of a circularly polarized electromagnetic wave. Therefore, the laser radiation must have a circular polarization σ^+ to ensure transitions between Zeeman sublevels differing in the magnetic quantum number by an amount of $\Delta m = +1$, and circular polarization σ^- to provide transitions between Zeeman sublevels differing in the magnetic quantum number by an amount of $\Delta m = -1$.

The Zeeman shift of the frequency of atomic transitions $\Delta\omega_Z$ is proportional to the magnetic field induction B: $\Delta\omega_Z = \alpha_Z B$, where α_Z is the constant of the linear Zeeman effect. Therefore, the condition for the resonance interaction of laser radiation with an atomic beam moving with velocity V in the magnetic field of the Zeeman slowing has the form

$$\Delta + kV - \alpha_Z B = 0, \qquad (9.1)$$

where $\Delta = \omega_L - \omega_A$, $k = 2\pi/\lambda$ is the wave number; λ is the wavelength of laser radiation.

Thus, if the magnitude of the induction of the magnetic field along the axis of the Zeeman slowing satisfies condition (9.1), the laser radiation will always be in resonance with the energy transitions in an atomic beam.

In other words, if the condition (9.1) is satisfied, the laser beam will exert pressure on the atomic beam, leading to deceleration of the atoms with acceleration α. As a result, the velocity of atoms along the axis of the solenoid will vary according to the law:

$$V(z) \approx \sqrt{V_0^2 - 2az}, \tag{9.2}$$

where V_0 is the initial velocity of the atom, and the acceleration of the atom can be calculated from the following equation

$$a = \frac{hk\gamma}{m} \frac{G}{1 + G + (\Delta + kV - \alpha_z B)^2 / \gamma^2}, \tag{9.3}$$

where $2\gamma = \Gamma$, m is the mass of the atom, $G = I/I_S$ is the saturation parameter of the laser beam, I is the intensity of the laser beam,

$$I_S = \frac{h\omega_A}{2\tau\sigma}, \tag{9.4}$$

where $\tau = 1/(2\gamma)$ is the spontaneous emission of the photon, σ is the absorption cross section of laser radiation.

Using (9.1), we can also obtain the optimum profile of the change in the induction of the magnetic field on the axis of the solenoid:

$$B(z) = B_0 \sqrt{1 - \frac{2az}{V_0^2}}. \tag{9.5}$$

Thus, if the laser is tuned so that when the atoms reach the point $z = 0$ (where the magnetic field has the maximum value B_0) changed as a result of the Zeeman shift, the transition frequency for their energy levels falls in resonance with the laser frequency, then such atoms begin to absorb light and because of the change in speed, the Doppler shift changes, but this in turn is compensated by the Zeeman shift, since the atoms move to the region of a weaker magnetic field. This process continues as the atomic beam moves along the solenoid, and initially the fast atoms, while decelerating, remain in resonance with the laser radiation, Initially, the slower atoms approaching this region also go into resonance and slow down. Therefore, in the final analysis, all the atoms with initial velocities below V_0 will have the same final velocity, which depends on the parameters of the magnetic field and the laser tuning. Figure 7 [5] schematically shows the distribution by the velocities of Na atoms, realized under Zeeman cooling. As can be seen, as a result of cooling, most of the atoms are grouped at low velocities.

The main advantage of Zeeman cooling of atoms is the ease with which problems of Doppler detuning of optical pumping can be avoided. The minimum temperature of the cooled atoms T_D, as in the case of Doppler cooling, is determined by the natural width of the energy level for their excited states Γ [3, 8]:

$$T_D = \frac{\hbar\Gamma}{k_B}. \tag{9.6}$$

For example, for ^{85}Rb atoms, the minimum achievable temperature is 141 μK, which corresponds to a velocity of 0.12 m/s. Nevertheless, a number of factors do not allow this temperature to be reached with Zeeman cooling. The main problem is the problem of collision of low-energy atoms in a cooled atomic beam. For this reason, it is not possible to reduce the velocity of the atoms below 50 m/s.

In practice, it is more difficult to cool molecules than individual atoms. This is due to the following factors: the molecules are heavier than individual atoms and therefore they are less sensitive

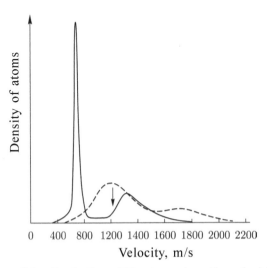

Fig. 7. Diagram of the distribution of Na atoms along the velocities up to (dashed line) and after (solid line) Zeeman cooling. The arrow indicates the maximum velocity of atoms in the beam, resonant with the frequency of the slowing laser [5].

to the action of laser radiation; unlike atoms, the energy received by molecules as a result of the action of the laser beam can go over into the vibrational and rotational energy of the molecule. Therefore, initially the ultracooled molecules were obtained by cooling individual atoms and then synthesized from them molecules. However, if we select molecules in which the vibrational degrees of freedom are unlikely, and the laser frequency does not cause rotational vibrations of its atoms, this makes it possible to simplify the cooling process. The cooling of SrF molecules to a temperature of 300 μK has already been reported.

9.3. Stopping and trapping atoms

As was experimentally established, with Zeeman cooling of atomic beams it is not possible to detect completely the resting atoms. In connection with this, several methods of trapping them were developed.

9.3.1. Doppler traps

The very first and simplest method was proposed and investigated in [1–3, 5, 9]. The method consists in using two laser waves directed towards each other (Fig. 8).

The cooling and stopping of the atom results from an imbalance induced by the Doppler effect between two opposing forces of radiation pressure. In this case, two laser waves propagating towards each other have the same (small) intensity and the same frequency (ω_L), slightly tuned to the red side of the resonant frequency of the atom (ω_A): $\omega_L < \omega_A$.

For a fixed atom, two radiation pressure forces exactly cancel each other, and the resultant force acting on the atom is zero. For a moving atom due to the Doppler effect, the effective frequencies of the two laser waves will be shifted relative to each other. The frequency of a wave propagating towards a moving atom becomes closer to its resonance frequency and, consequently, causes a greater radiation pressure force than a wave co-directed with an atom whose frequency is removed from resonance.

Fig. 8. Stopping and capturing an atom using two laser beams.

Thus, the resultant force of radiation pressure (friction) is directed towards the moving atom: $F = -\eta V$, and will slow it down to its full stop. If we use three pairs of lasers that radiate towards each other along three orthogonal axes, we can quickly quench the velocity of the atom, realizing the 'optical molasses' and localizing its position in space, thereby creating a trap for it.

Doppler radiation friction leading to cooling of atoms is inevitably accompanied by velocity fluctuations. These fluctuations are associated with fluorescence, during which excited atoms spontaneously emit photons in arbitrary directions and at random instants of time. These photons transmit to the atom a random recoil pulse $\hbar\mathbf{k}$, causing the diffusion of atoms in the space of pulses with a diffusion coefficient D [1, 5]. As in the ordinary Brownian motion, the competition between friction and diffusion leads to the establishment of a stationary state with a temperature proportional to ratio D/η [5]. This temperature always exceeds the Doppler limit of the cooling temperature T_{D}, which was determined earlier, depending on the natural width of the excited state of the atom.

9.3.2. Magneto-optical traps

As shown in [5], the disbalance between two oppositely directed forces of radiation pressure can be provided by the spatial one that in a magnetic field an atom with a nonzero magnetic moment has quantum states whose magnetic or Zeeman energy increases or decreases (depending on the orientation of the magnetic moments) with an increase in the induction of the magnetic field. Atoms with increasing energy can be captured by a magnetic field having a configuration with a local minimum of its induction. In this case, for a stable trapping of an atom, it is necessary that the orientation of its magnetic moment with respect to the field does not change.

The potential energy of a Zeeman energy sublevel with a magnetic quantum number m_F in the magnetic field of induction B is [1,2]

$$U_{m_F} = \mu_{\mathrm{B}} g_F m_F B, \tag{9.7}$$

where μ_{B} is the Bohr magneton, g_F is the Landé factor.

According to (9.7), the force acting on a neutral atom located in a magnetic field non-uniform in space will be

$$\mathbf{F}(r) = -\nabla U_{m_F}(r) = -\mu_{\mathrm{B}} g_F m_F \nabla B(r). \tag{9.8}$$

As seen from (9.8), an atom moving in an inhomogeneous magnetic field will be acted upon by a force directed in the direction opposite to the direction of increase in the induction of the magnetic field.

If we use a spherical magnetic quadrupole composed of two coils with the opposite flow of an electric current, which has a zero value of induction of the total magnetic field at the centre between these loops (Fig. 9) [5], then, since the field in it grows in any direction from it center, such a design will be a magnetic trap for cooled atoms.

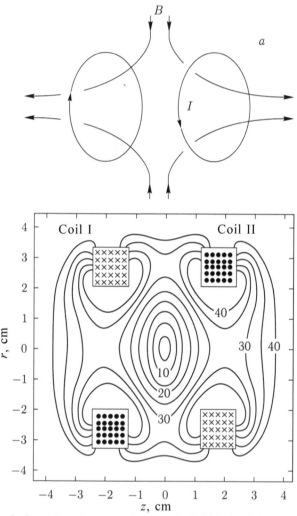

Fig. 9. Spherical quadrupole trap and magnetic field induction lines in it (*a*). Lines of equal intensity of magnetic field induction in the trap, demonstrating the presence of a region with zero induction value at its centre (*b*) [5].

As was shown above, the Doppler cooling of atoms is capable of ensuring the capture and cooling of atoms to a sufficiently low temperature, but it does not allow the trapping of atoms in the trap due to their diffusion into the surrounding space. Therefore, the combination of the principles of Doppler cooling of atoms with a magnetic trap makes it possible to create an effective magneto-optical trap for cooled atoms (Fig. 10) [7].

In this trap, the atomic beam is fed through the capillary inside the cooling region, six orthogonal laser beams create low-temperature optical molasses, and the two current coils form a spherical magnetic quadrupole that locks the cooled atoms, not allowing them to diffuse into the surrounding space. The efficiency of such a magneto-optical trap depends on the quality of evacuation of the volume, and also on the degree of isolation of the cooled atoms from the radiation heat fluxes.

Figure 11 show the experimental dependence of the change in the concentration of Na atoms in a magneto-optical trap with time [5]. The 'depth' of the trap is 17 mK, which corresponds to a velocity of Na atoms of 3.5 m/s. Since the vacuum in the chamber was $\sim 10^{-6}$ Pa, as a result of collision with the atoms of the residual gas the number of atoms in the trapped cooled optical molasses decreases exponentially. A separate, bright dot on the graph corresponds to the case when the vacuum in the chamber is several times worse. Improving the vacuum in the chamber makes it possible to ensure the decay times of cooled atoms in magnetic traps for about one minute or more.

9.4. Sisyphus cooling

The desire to achieve lower cooling temperatures of atoms led to the

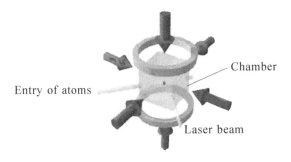

Fig. 10. Scheme of the magneto-optical trap [7].

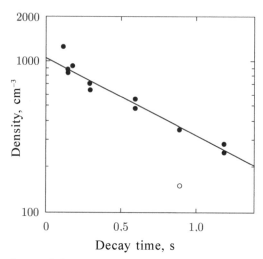

Fig. 11. Dependence of the number of trapped atoms in a magneto-optical trap on time [5].

search for new methods based on the modification of the Doppler cooling method. One of their varieties is the so-called 'Sisyphus cooling', in which the atom, like the mythical Greek hero, reaching the top of the periodic potential relief, periodically rolls down, losing energy and cooling.

If an atom simply moves in a periodic potential, its kinetic energy periodically changes into a potential and vice versa. In this case, the total energy of the atom does not change, which means that its average velocity remains unchanged and no cooling takes place. To cool the atom, it is necessary to take away its energy. For this, a combination of a number of physical processes is used: Doppler cooling in counter laser beams, Stark light shifts of the energy levels of atoms, and a gradient of light polarization.

It is known that an atom placed in an electric field of intensity **E** changes its energy. This phenomenon is called the 'variable Stark effect'. As a result, the energy levels of the atom are shifted by an amount [10]

$$\delta E = -c^2 \frac{\Omega^2}{2\Delta}, \tag{9.9}$$

where $\Delta = \omega_L - \omega_A$ is the detuning of the laser radiation frequency from atomic resonance, c is the speed of light in vacuum, and Ω is the Rabi frequency:

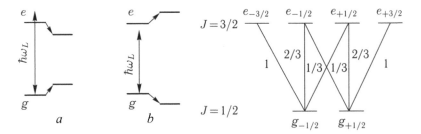

Fig. 12. Stark shift of the energy levels of an atom in a laser radiation field.

Fig. 13. Energy structure of levels in an atom and the probability of transitions between them [10].

$$\Omega = \frac{\mathbf{dE}}{\hbar}, \qquad (9.10)$$

d is the electric dipole moment of the transition in the atom.

As follows from expressions (9.9) and (9.10), the magnitude and sign of the energy shift depend on the tuning of the laser radiation (Fig. 12).

Therefore, the Stark splitting of the energy levels of the atom occurs in the laser radiation field. Each level with an orbital quantum number l splits into $l+1$ sublevels, in accordance with the number of possible values of the modulus of the magnetic quantum number. Figure 13 shows a model diagram of the energy levels of a two-level atom and denotes energy transitions between the ground state $|g\rangle$ (the total orbital number $J = 1/2$) and the excited $|e\rangle$ (the total orbital number $J = 3/2$) by energy levels under the action of laser radiation and the corresponding probabilities of transitions between levels.

According to Fig. 13, depending on the polarization of the light wave, the following transitions are possible between the energy levels [5]: – for linearly polarized radiation: transitions $(g_{-1/2} \rightarrow e_{-1/2})$ and $(g_{+1/2} \rightarrow e_{+1/2})$ with the probability 2/3;

– for circularly polarized radiation: transitions $(g_{-1/2} \rightarrow e_{-3/2})$ and $(g_{+1/2} \rightarrow e_{+3/2})$, with the probability 1 and also transitions $(g_{-1/2} \rightarrow e_{+1/2})$ and $(g_{+1/2} \rightarrow e_{-1/2})$ with the probability 1/3.

This feature of the interaction of atoms with light waves opens the possibility of manipulating them with the use of specially organized polarized laser radiation. To this end, we consider the interference of two oppositely directed linearly polarized plane waves of laser radiation with equal amplitudes E_0 propagating along the z axis (Fig. 14) [4]:

– the case of equally polarized waves

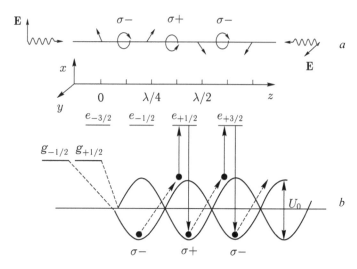

Fig. 14. Spatial distribution of the polarization state in the interference field of orthogonally polarized laser waves (*a*) and spatial modulation of the potential relief of the ground state of a two-level atom in an interference field with a polarization gradient (*b*).

$$\mathbf{E} = E_0\mathbf{e}\cos(\omega_L t - kz) + E_0\mathbf{e}\cos(\omega_L t - kz) = 2E_0\mathbf{e}\cos(kz)\cos(\omega_L t), \quad (9.11)$$

where **e** is the unit vector;
  – the case of orthogonally polarized waves

$$\mathbf{E} = E_0\mathbf{e_x}\cos(\omega_L t - kz) + E_0\mathbf{e_y}\cos(\omega_L t + kz) =$$
$$= E_0[(\mathbf{e_x} + \mathbf{e_y})\cos(kz)\cos(\omega_L t) + (\mathbf{e_x} - \mathbf{e_y})\sin(kz)\sin(\omega_L t)], \quad (9.12)$$

where $\mathbf{e_x}$ and $\mathbf{e_y}$ is the unit vectors in the direction of the appropriate axes.

As follows from (9.11) and (9.12), in the case of the addition of equally polarized laser waves, a standing wave is formed in space. In the case of the addition of orthogonally polarized waves at different points in space, different situations take place, for example,
  – at the starting point ($z = 0$, $kz = 0$):

$$\mathbf{E} = E_0\left(\mathbf{e_x} + \mathbf{e_y}\right)\cos(\omega_L t). \quad (9.13)$$

The wave will be linearly polarized and repeat its state at points that are multiples of $z = \lambda/4$, with each time the tension vector **E** undergoes a turn by 90°;

– at the point $z = \lambda/8$, when $kz = \pi/4$:

$$\mathbf{E} = E_0 \left[-\mathbf{e}_x \sin\left(\omega_L t + \frac{\pi}{4} \right) + \mathbf{e}_y \cos\left(\omega_L t + \frac{\pi}{4} \right) \right], \tag{9.14}$$

the wave has a circular (left-sided) polarization σ^-, which is periodically repeated through distances that are multiples of $\lambda/2$;
– at the point $z = 3\lambda/8$, when $kz = 3\pi/4$, the wave has an opposite circular (right-hand) polarization σ^+, the state of which also repeats through distances multiples of $\lambda/2$:

$$\mathbf{E} = E_0 \left[\mathbf{e}_x \cos\left(\omega_L t + \frac{\pi}{4} \right) - \mathbf{e}_y \sin\left(\omega_L t + \frac{\pi}{4} \right) \right]. \tag{9.15}$$

The result of interference of two orthogonally polarized plane waves is illustrated in Fig. 14 *a*.

Thus, as a result of interference of orthogonally polarized counterpropagating waves, a radiation polarization gradient is created in space. This gradient of polarization leads to the fact that the atom moving along the z axis will have different light shift of energy levels, which is illustrated in Fig. 14 *b*. Consider this phenomenon in more detail for the case when $\Delta < 0$ and the motion of the atom occurs from left to right. At the points $z = 0$, $\lambda/4$, $\lambda/2$, $3\lambda/4$,..., where the wave has linear polarization, the energy shift for the two ground-state energy levels $g_{-1/2}$ and $g_{+1/2}$ will be the same. At points at distances $z = \lambda/8 + n\lambda/2$, where $n = 0$, 1, 2..., the polarization wave σ^- results in the following transitions: $g_{-1/2} \rightarrow e_{-3/2}$ and $g_{+1/2} \rightarrow e_{-1/2}$. Since the first transition has a higher probability than the second (Fig. 13), its dipole moment is greater, and therefore the energy level $g_{-1/2}$ according to (9.9) and (9.10), undergoes a greater energy shift at these points (see Fig. 14 *b*). At the points $z = 3\lambda/8 + n\lambda/2$, the polarization wave σ^+ provides only the transitions of $g_{+1/2} \rightarrow e_{+3/2}$ and $g_{-1/2} \rightarrow e_{+1/2}$, of which the former has a greater probability and a greater dipole moment. Therefore, at these points already the energy level of the ground state $g_{+1/2}$ undergoes a greater energy shift (Fig. 14 *b*).

Thus, for an atom moving in an interference light field periodically changing the polarization state (a light field with a polarization gradient along the z axis), for the Zeeman energy sublevels ($g = \pm 1/2$) in the ground state (the total orbital momentum $J = 1/2$) the degeneracy is removed, and the energy of their splitting δE turns out to be modulated in space with period $\lambda/2$. Moving in such periodic energy potential, the atom located at the point $z = \lambda/8$ in the state $g_{-1/2}$ (Fig. 14 *b*), as it spreads along the *z* axis, loses its kinetic energy (slows), climbing upward to the 'potential hill' and thereby increases its potential energy. If we select the laser radiation frequency so that the atom absorbs the photon only at the maximum of the potential at the point $z = 3\lambda/8$, then, under the action of a circularly polarized light wave, it will go over to the state of $e_{+1/2}$ after the forced transition $g_{-1/2} \rightarrow e_{+1/2}$. Further, it spontaneously emits a photon and passes into the ground state energy, which corresponds to the energy of the level $g_{+1/2}$. In this case, the atom loses the energy equal to the difference between the energies of the absorbed and emitted photons, and is cooled. After that, the atom again begins to climb the next 'potential hill', located at the point $z = 5\lambda/8$ and the process will be repeated again. After each cycle, the motion of the atom slows down, i.e., it cools, and the total energy of the atom decreases by a magnitude of the order of the depth of the potential well U_0 (Fig. 14 *b*). This process can continue until the total energy of the atom is less than U_0 and the atom will be trapped in the potential well. Thus, Sisyphus cooling is capable of lowering the temperature of an atom to a T_S value, determined by the relation

$$k_B T_S \approx U_0. \tag{9.16}$$

With a low intensity of light radiation, the depth of the potential well will be close to the value of the stark energy splitting of the energy levels in the atom, which depends on the square of the Rabi frequency:

$$U_0 \propto -c^2 \frac{\Omega^2}{2\Delta}. \tag{9.17}$$

The value of U_0 will be less than $\hbar\Gamma$, which, according to (9.6), explains why the temperature of the atoms with Sisyphus cooling is lower than the temperature of the Doppler limit.

However, in the process of Sisyphus cooling, the temperature of the atoms can not be made arbitrarily low. The fundamental limitation for all mechanisms of laser cooling is the finite value of the recoil momentum of the photon re-emitted by the atom [4, 5]. It is known that in most laser cooling schemes the luminescence of atoms never stops. This means that an excited atom spontaneously emits a quantum of light in an arbitrary direction, which transfers the recoil momentum to the atom. In this case, it is impossible to reduce the spread of the atomic pulse by less than a certain value corresponding to the pulse of the emitted photon $\hbar k$, and consequently, to reduce the total energy of the atom below the value

$$E_R = \frac{\hbar^2 k^2}{m}.$$ (9.18)

So, when U_0 becomes equal to or less than E_R, Sisyphus cooling will become less effective than heating due to recoil photons. This means that the extremely low temperature of the atoms, which can be achieved by means of Sisyphus cooling, amounts to several values of E_R/K_B. Using the Sisyphus cooling effect, very low temperatures have been experimentally attained: up to 35 mK for sodium and 3 mK for cesium.

9.5. Laser cooling below the recoil level

As already noted above, since it is impossible to control random recoil pulses transmitted to an atom by spontaneously emitted photons, it is impossible to reduce the scatter of the atomic momentum below the value corresponding to the pule of the emitted photon, $\delta p = \hbar k$. This condition determines the limit of the 'recoil temperature' for the atom:

$$T_R = \frac{\delta p^2}{k_B m}.$$ (9.19)

9.5.1. Cooling of atoms based on the selective coherent trapping of their populations based on their velocity

The values for T_R, depending on the type of atoms, lie in the range of several nano- to several micro-Kelvins. It is possible to reach the temperature of the atoms below this limit if we go from the resonant interaction of light waves with two-level atoms to the use of their resonant interaction with multilevel atoms [5, 12–14]. In this case, it is possible to create a situation where the rate of absorption of photons or the rate of atomic jumps in the event of a random walk in the velocity space turns out to depend on the velocity of the atom and tends to zero for $V = p/m \rightarrow 0$ (Fig. 15 *a*). For such an atom, there will be no spontaneous reradiation of light, and therefore no recoil momentum will be observed. Thus, a cooled atom, for which $V \approx 0$, will be protected from harmful, warming-up effects of light.

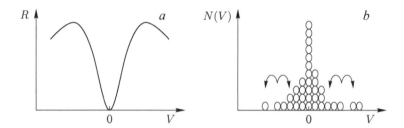

Fig. 15. Illustration of the process of cooling of atoms below the recoil level: *a* – dependence of the frequency of atomic jumps (*R*) for their random wandering in the velocity space (*V*); *b* – process of the inhomogeneous random wandering of atoms, leading them to a neighborhood where $V \approx 0$, in which the atoms cease to absorb light and accumulate [5].

Since other atoms for which $V \approx 0$ will continue to absorb and emit light waves, a random change in their velocities (wandering in the velocity space) can randomly transfer them to a state with $V \approx 0$, where they are trapped and will accumulate (Fig. 15 *b*).

At present, various cooling schemes for atoms below the recoil level have been proposed and implemented, realizing the process of Fig. 15 [5]. One of them is called the *rate-selective coherent population trapping* (SPT). In this scheme, the decrease to zero of the rate of jumps at $V \approx 0$ is achieved due to the destructive quantum interference of oppositely directed light waves in the flow of three-level atoms, which have the so-called Λ-scheme of energy levels (Fig. 16) [5].

Consider an atom with a three-level energy system consisting of two sublevels g_1 and g_2 of the ground energy state and a singlet excited energy level e_0. Suppose that this quantum system is excited by two laser beams with frequencies ω_{L1} and ω_{L2}, which cause the corresponding transitions ($g_1 \leftrightarrow e_0$) and ($g_2 \leftrightarrow e_0$). Let $\Delta = \omega_{L1} - \omega_{L2}$ be the detuning from resonance for the stimulated Raman process, consisting in the absorption of one photon with a frequency of ω_{L1} and the subsequent stimulated emission of one photon with a frequency of ω_{L2}. In the case of when $\Delta = 0$, the rate of the recoil process in the atom turns out to be zero and $R = 0$. In this case, the dependence $R(V)$ will be similar to the one shown in Fig. 15 *a*. Interpretation of this effect consists in the fact that in the process of two-wave optical pumping the atoms are transferred to a state that is a linear combination of states g_1 and g_2 from which transitions to the state e_0 do not occur due to the destructive interference of two excitation channels ($g_1 \rightarrow e_0$) and ($g_2 \rightarrow e_0$).

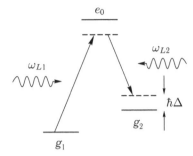

Fig. 16. Diagram illustrating the coherent population trapping in a three-level energy system (Fig. 16) [5].

The basic idea of the process of cooling atoms based on the selective velocity of coherent trapping of populations in atoms is to make the frequency of the stimulated Raman process in the three-level energy system (Fig. 16), which is proportional to the velocity of the atom, be tuned off using the Doppler effect. This can be realized by directing circularly polarized in opposite directions laser beams with frequencies ω_{L1} and ω_{L2} along the z-axis towards each other. As a result, if the frequencies of the laser beams are chosen so that for a stationary atom ($V = 0$) $\Delta = 0$, then for an atom moving along the z axis with a velocity V, the opposite Doppler frequency shifts for the colliding beams will give a detuning of the frequency of the Raman cooling process:

$$\Delta = (k_{L1} - k_{L2})V, \qquad (9.20)$$

where k_{L1} and k_{L2} are the magnitudes of the wave vectors of the laser beams. But this means that at the point of the velocity space of atoms, when $V = 0$, the atoms cease to absorb radiation and for them the recoil process ceases to work, leading to their heating. That is, for such a cooling scheme, the process illustrated in Fig. 15 will operate. As shown by the results of the studies [5], this scheme of cooling atoms allows reaching record low temperatures $\sim T_R/800$, i.e., ~ 5 nK.

We note that the effect of coherent population trapping selective in terms of the rates is very common for multi-level interaction schemes and can occur in all cases when conditions for the destructive interference of excitation channels are fulfilled.

9.5.2. Evaporative cooling of atoms

Evaporative cooling is used to obtain lower cooling temperatures of atoms [16]. Evaporation is a well-studied physical phenomenon that

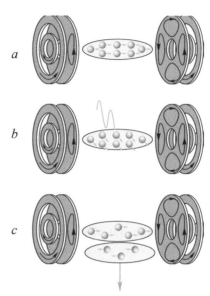

Fig. 17. An illustration of the process of evaporative cooling of atoms in a magneto-optical trap [17]: *a* – in all the retained atoms the spins are parallel to the stationary magnetic field; *b* – short pulse of a high-frequency magnetic field tilts the spins of the atoms; *c* – by selecting the frequency of the high-frequency electromagnetic field, the spins of high-energy atoms unfold against the direction of the stationary magnetic field and are pushed out of the trap.

describes the transformation of a substance from a liquid state to a gaseous one. In the process of evaporation, particles with energy above the binding energy leave the system, and this depletion by high-energy particles leads to cooling of the system. Evaporative cooling is a universal process both in the microworld (evaporation of a neutron from the nucleus) and in the macroworld (evaporation of stars from clusters).

To realize evaporative cooling, a gas preliminarily cooled by a laser method and locked in a magneto-optical trap is irradiated with a high-frequency magnetic field. If an atom with a nonzero magnetic moment oriented against a stationary magnetic field (Fig. 17 *a*) acts as a pulse of a high-frequency magnetic field with a magnetic induction vector rotating with a certain frequency in the (x, y) plane (Fig. 17 *b*), then the magnetic moment of the atom can be rotated so that it no longer holds in the trap (Fig. 17 *c*). By selecting the frequency of the high-frequency magnetic field, it is possible to achieve that the magnetic moment is reversed only at warmer atoms. Then warm atoms will begin to leave the trap, and colder ones will accumulate in it. By analogy with the process, when the evaporation of a liquid from its volume leaves more energy molecules, this process of cooling atoms in a trap is also called evaporative cooling. Evaporative cooling of neutral atoms, previously localized in magneto-optical traps, makes it possible to obtain a temperature in the laboratory conditions of up to 100 nK [16]. The main conditions for evaporative cooling of atoms are the sufficient decay times of the atoms in the trap and the sufficient density of atoms to begin the process of effective evaporation.

Since the decay time of the atoms in the trap is limited by inelastic collisions of atoms that determine the thermalization of the atomic ensemble, in order to realize the evaporative cooling of the atoms, it is necessary that the thermalization time be shorter than the decay time of the atoms in the trap. We also note that in the process of evaporative cooling, a significant part of the atoms of the original ensemble are lost, up to 0.1%.

9.6. The physics of cold atoms and its applications

Ultra-low cooling of atoms with the help of methods of the rate-selective coherent population trapping, evaporation of atoms, etc., make it possible to reach a temperature below the recoil limit T_R.

Figure 18 [18] shows the temperature scale of atoms experimentally achieved using different cooling methods and the de

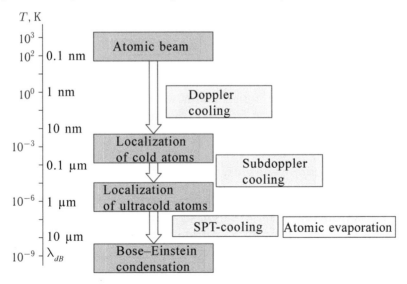

Fig. 18. The scale of temperatures attained during cooling of atoms, and the corresponding de Broglie wavelength [18].

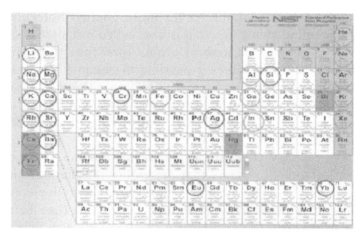

Fig. 19. Elements of the periodic table, for which laser cooling of atomic beams is realized (marked by circles) [18].

Broglie wavelengths of atoms corresponding to these temperatures are given.

As can be seen, with ultralow cooling, the width of the atomic wave packet becomes comparable with the wavelengths of the light acting on atoms: 1–5 μm. Such deep cooling of atoms makes it possible to advance significantly in such areas as atomic interferometry, the creation of frequency standards, the spectroscopy of cold atoms in traps, the study of quantum-statistical effects under Bose condensation of atoms, and so on.

Significant progress has been achieved by cooling atoms of the most diverse chemical elements (Fig. 19) [18], which opens new horizons for the development of a fundamentally new direction of research, called atomic optics, and new technological methods in the field of constructing new structures by nanolithography methods. In this section, we briefly review some of them.

9.6.1. One-component plasma

The trapping of ionized atoms of matter in a magneto-optical trap can be considered as the creation of a 'one-component' plasma [2].

Cooling this plasma to an ultralow temperature can lead to the appearance of liquid and crystalline plasma states of matter (Fig. 20).

The crystals formed in the trap can be considered as the classical limit of Wigner crystallization, when the wave functions of the ions do not overlap and quantum statistics do not play a decisive role.

Fig. 20. Photograph of an ionic crystal [2].

On the other hand, the crystallization process can be regarded as establishing a balance between the ion trapping force in the trap and their strong long-range Coulomb repulsion. In the above photograph, the crystal is formed by approximately 2000 laser-cooled beryllium ions. The distance between the ions in the crystal is about 15 μm.

9.6.2. Bose–Einstein condensation of atoms

Consider a gas of identical atoms interacting with themselves only through elastic collisions. If the number of atoms in thermodynamic equilibrium N is sufficiently large, then their behaviour is described by Maxwell's statistics for an ideal gas [2, 16]. However, according to the quantum theory, every freely moving atom with momentum **p** has wave properties characterized by the de Broglie wavelength inversely proportional to its momentum, $\lambda_{dB} = h/p$. The classical treatment will be valid as long as the quantum mechanical restrictions associated with the Heisenberg uncertainty principle on the localization of an atom in space are unimportant. That is, until the de Broglie wavelength for an atom becomes comparable with the interatomic distance. If the gas atoms occupy a cubic volume $v = Na^3$, where a is the side of the cubic cell per atom, the condition for the applicability of the classical treatment will be

$$\lambda_{dB} \ll a = \left(\frac{v}{N}\right)^{1/3}.$$

Taking into account that the mean square velocity of atoms at the temperature T is equal to $\bar{V} = (3k_B T/m)^{1/2}$, the condition for applicability of the classical representation will be written in the following form:

$$\frac{\lambda_{dB}}{a} = \frac{h}{m\bar{V}a} = \frac{h}{(3mk_BT)^{1/2}}\left(\frac{N}{v}\right)^{1/3} \ll 1. \tag{9.21}$$

As can be seen, the present inequality can be violated for the cases of very light atoms, their high concentration and very low temperatures.

From the consideration of (9.21), we can introduce the notion of a certain temperature, called the degeneracy temperature T_0:

$$T_0 = \frac{h^2}{3mk_B}\left(\frac{N}{v}\right)^{2/3}, \tag{9.22}$$

which allows us to represent condition (9.21) in the form

$$T \gg T_0. \tag{9.23}$$

Thus, for $T \gg T_0$, the gas is well described by the laws of classical physics, and for $T < T_0$ the Maxwell statistics ceases to work. If the particles (atoms) have an integer spin value, then they begin to obey the Bose–Einstein statistics (the particles in this case are called bosons), and the Fermi–Dirac statistics work for atoms with half-integral spins (the particles are called fermions).

The phenomenon of Bose–Einstein condensation (BEC) consists in the fact that at $T < T_0$ for atoms with an integer total spin in a state with a zero momentum value a very large number of atoms N_0 (such a state is called degenerate) accumulates, and the fraction of these atoms in the total number of atoms is [17]

$$\frac{N_0}{N} = 1 - \left(\frac{T}{T_0}\right)^{3/2}. \tag{9.24}$$

This phenomenon is called the phenomenon of Bose–Einstein condensation in analogy with the condensation of vapour into a liquid, although the distribution particles in space remains the same, and condensation arises in the space of velocities or momenta.

For most gases, the degeneracy temperature T_0 turns out to be so small that the substance passes into a solid state much earlier than the process of its Bose–Einstein condensation may occur. Therefore, until recently BEC substances were associated only with the superfluidity of helium and the appearance of exciton droplets in semiconductors. Nevertheless, the usual condensation of gas into a liquid can be 'circumvented' if one works at very low gas densities – of the order of one hundred-thousandth of the density of atmospheric air.

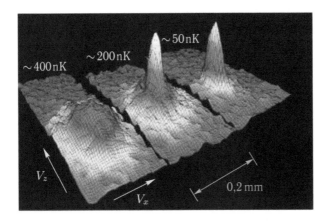

Fig. 21. Dynamics of the Bose–Einstein condensation of Na atoms [2].

Under such conditions, the time of formation of molecules or clusters as a result of three-particle inelastic collisions, the probability of which is proportional to the square of the gas density, is prolonged to several seconds and even minutes. The velocity of the two-particle elastic collisions decreases in proportion to the first degree of gas density, so such collisions will occur much more often. As a result, the thermal equilibrium for the translational motion of atoms is established faster than chemical, and quantum degeneracy (BEC) can occur in the gas phase, which in this case is metastable. However, at such extremely low gas densities, the temperature of the BEC decreases to nano- and micro-Kelvins. Therefore, only the process of laser cooling of atoms, which makes it possible to provide extremely low atomic temperatures at low densities, made it possible to realize in practice the production of a Bose condensate of atomic gas long before the atoms are in the solid phase. In this connection, although the phenomenon of Bose–Einstein condensation of gases was predicted in 1924, the first atomic condensates were obtained only in 1995, first for vapours of very rarefied alkali metals (rubidium, sodium and lithium), and two years later for atomic hydrogen [2, 16].

Figure 21 shows the experimental results illustrating the dynamics of the BEC process of Na vapour cooled to different temperatures: 400, 200 and 50 nK, respectively, using a combination of laser and evaporative cooling processes. As can be seen, a decrease in temperature leads to an increasing concentration of the velocities of the atoms near its zero value. At a temperature of 50 nC, almost 90% of the Na atoms are collected in the condensate.

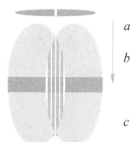

Fig. 22. Scheme of observation of the interference of atomic condensates: *a* – atomic condensate in a magneto-optical trap, cut into two halves by a laser beam; *b* – after the magnetic field is turned off, the atomic condensates fall under the action of gravitational forces, expand and begin to overlap; *c* – a high-contrast interference pattern consisting of atoms is formed in the region of overlapping clouds of condensate.

9.6.3. Atomic laser

Since the de Broglie wave is associated with each atom, we can draw an analogy between atoms and electromagnetic waves. This is often used in atomic optics, when they consider the processes of reflection, diffraction, and interference of atomic beams. The presence of magneto-optical traps for cooled atoms and potential magnetic mirrors strengthened this analogy, which served as the basis for investigating the coherence of atomic beams. The coherence of the atomic beam was experimentally demonstrated in [16]. The scheme of the experiment is illustrated in Fig. 22.

The idea of checking coherence was to create two identical sources of atomic beams that could interact with each other. For this purpose, in a magneto-optical trap, a cloud of trapped atoms was cut by a laser beam into two halves separated by a distance *d*. After that, the trap was turned off, and the atomic condensates began to move in the field of gravity of the Earth.

As a result of ballistic expansion, the condensates increased their volumes and began to overlap. If the atomic beams are coherent, then after the time *t* their interference occurs and in the overlapping region interference fringes are formed from the atoms, which should be spaced apart by a de Broglie wavelength: $\lambda = h/mv$. Figure 23 shows an experimentally observed interference pattern in the region of overlapping of condensates from Na atoms. The observed interference pattern corresponds to an atom temperature of 0.5 nK. The presence of an interference pattern was a clear indication of the coherence of atomic beams emanating from a Bose condensate.

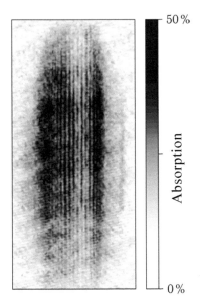

Fig. 23. The interference pattern of two expanding atomic condensates, recorded after 40 ms of atomic flight. The period of the interference atomic bands is 15 μm. The interference bands are visualized by the absorption of the laser radiation radiating through the region [16].

Fig. 24. Illustration of the work of a laser on Na atoms. Coherent clusters of atoms leave the trap every 5 ms [16].

Further [16], to create a coherent atomic beam, it was proposed to extract atoms from a condensate using the spins of atoms induced by a radio-frequency magnetic field in a manner analogous to that for evaporative cooling of atoms. As a result, the atoms go into a trap-free state and are extracted from it by gravity. The result of the operation of such a laser operating at a frequency of 200 Hz is illustrated in Fig. 24. The pulses of coherent Na atoms are derived from a trap with a condensate. The radio frequency pulse transfers a part of the trapped atoms every 5 ms to an unsealed state. These atoms are accelerated by the gravitational field, and their cloud expands due to repulsion. Each pulse contains 10^5 to 10^6 atoms. The only drawback of such a laser is the exhaustibility of the number of atoms that can be extracted from the trap. In addition, in order to realize the true laser effect, it is necessary to create conditions for increasing the flux of atoms. Amplification of the atomic beam is more complicated than amplification of electromagnetic waves, because external fields can only change the state of atoms, but one can not generate their number. Schematically this difference

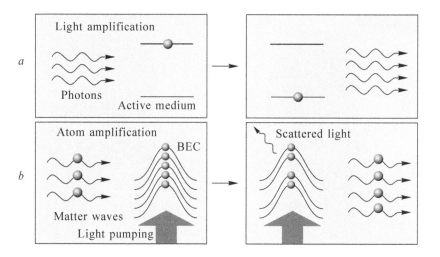

Fig. 25. Schemes of light (*a*) and atom (*b*) amplifiers. In the optical laser light amplifies when an electromagnetic wave passes through an excited inverse-populated environment. In an atomic amplifier, the incident beam of atoms passes through a Bose–Einstein condensate irradiated with laser light [16].

is illustrated in Fig. 25 [16]. Therefore, the atomic amplifier must convert the atoms of the active medium (Bose–Einstein condensate) into a stream of atoms that are in exactly the same state as the atoms of the atomic beam incident on the condensate.

For an atomic amplifier, it is required to create a 'reservoir' of ultracold atoms with a very narrow velocity distribution, which can be converted into an atomic beam.

Such a 'reservoir' can be represented by a magneto-optical trap with the Bose–Einstein condensate of atoms contained in it. The atomic amplifier should also provide some mechanism capable of transferring atoms from the 'reservoir' to atoms that have the state of the atoms of the incident beam, with observance of the laws of conservation of energy and momentum.

Figure 26 shows photographs describing the result of the enhancement of the Na atomic beam. The output atomic beam is not simply amplified – the condensate atoms pass exactly into the same quantum-mechanical state in which the atoms in the incident beam are. This was verified by interference of the amplified output atomic beam and observation of phase coherence [16].

The creation of an atomic amplifier has added a new element to the optics of atoms in addition to the passive elements used. Coherent atomic amplifiers can contribute to the improvement of

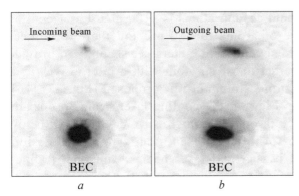

Fig. 26. The result of experimental amplification of an atomic beam. *a* – atomic the beam passes through the condensate without amplification; *b* – amplification of an atomic beam after activation of a condensate by laser radiation. The typical beam gain is from 10 to 100.

atomic interferometers and can also be used as sensitive sensors for acceleration, gravity and rotation.

9.6.4. Atomic fountain and atomic clock

Beginning in 1967, the international system of SI units defines the time interval of one second as 9192631770 periods of electromagnetic radiation that occurs during the transition between two hyperfine levels of the ground state of the isotope of the cesium-133 atom.

The fundamental limitation of the accuracy of measuring frequency and time in an atomic clock is the time of measurement of the frequency of the atomic working transition. In the frequency standards for cesium atoms, the accuracy of measuring the transition frequency is determined by the time of flight of the cesium atom through the probe microwave field.

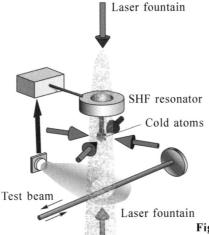

Fig. 27. Diagram of an atomic clock based on an atomic fountain.

The velocity of thermal cesium atoms is ~100 m/s. Laser cooling lowers the temperature of atoms to ~10 nK and thereby reduces their thermal velocity to several centimeters per second. To provide a long time for the interaction of an atom with a microwave field, the construction of an atomic fountain was proposed (Fig. 27) [16].

As can be seen from Fig. 27, the cooling laser beams propagate vertically and pass twice through the microwave cavity: from below upwards and from top to bottom. In an atomic fountain, cesium atoms are first cooled by vertical laser beams, and then trapped by a magneto-optical trap in which they are further cooled. The resulting cloud of ultracold atoms serves as a source of slow atoms in the fountain. A cooled cloud of cesium atoms is directed from below by a laser pulse, vertically upwards, as a result of which the atoms begin to move along ballistic trajectories in the gravitational field of the Earth. Moving upwards, the atoms cross the probing field of the RF resonator and rise to a height of about one meter. Due to the gravity field, the velocity of atoms gradually decreases to zero, and they begin to fall down, passing through the resonator again. Thus, a cloud of cesium atoms twice cuts off the resonator, where the microwave frequency is used to read the frequency of oscillations of their atomic transition. At the same time, there are always atoms in the resonator moving only in one direction, which makes the transition line more specific. As a result, the duration of the frequency measurement in the atomic fountain increases by two orders of magnitude in comparison with the frequency standard on the thermal atomic beam and is ~1 s. The microwave transition is registered by recording the fluorescence of cesium atoms initiated by a probe laser beam. The large interaction time makes it possible to realize the relative accuracy of frequency measurement in atomic fountains ~$6 \cdot 10^{-16}$. If a cesium atom fountain is placed in a cryogenic chamber, it is possible reduce the effect of thermal radiation from the walls of the chamber on the heating of atoms. This made it possible to increase the accuracy of the frequency measurement by an order of magnitude. Improved frequency control leads to the fact that the atomic clock based on the atomic fountain becomes one of the most accurate hours of the world.

9.6.5. Atomic optics

Atomic optics is the optics of material particles and deals with the problems of formation of ensembles and beams of neutral atoms, their control, and also questions of their application. Atomic optics was formed into an independent discipline in the mid-1980s as a result of studies of the effect of laser radiation pressure on the translational motion of atoms.

The invention of the laser gave the researchers a fundamentally new light source with high spectral brightness, monochromaticity and high directivity of radiation. With the first use of laser radiation, light pressure from a barely observed phenomenon has become an effective means of influencing the motion of atoms. The development of atomic optics is closely connected with the development of methods for laser cooling and the localization of neutral atoms [18, 19]. Laser cooling of atoms and their spatial localization make it possible to form atomic ensembles and beams with given parameters. The localization of neutral atoms opens the possibility of working with both single atoms localized with nanometer accuracy, and with macroscopic ensembles of cold atoms having a high phase density.

Laser methods for cooling neutral atoms, which significantly increased the de Broglie atomic wavelength, led to the emergence of atomic wave optics. On the basis of various configurations of laser light fields and micro- and nanostructures (zone plates, multi-gap diaphragms, etc.), coherent divisors of atomic beams, atomic interferometers and waveguides are created.

The possibilities of atomic optics are much wider than the capabilities of optics of other types of material particles (electrons and neutrons) due to the presence of an internal structure in the atom. At a temperature close to absolute zero, when the de Broglie wavelength becomes comparable with the distance between atoms, the behaviour of the atomic ensemble begins to depend appreciably on the internal quantum characteristic of the spin-atom.

9.6.5.1. Methods of constructing elements of atomic optics

Although atomic and light optics have similar mathematical justifications, their 'technical means' are different. The basis for creating the means of optical optics is the technique of grinding and polishing surfaces of the required shape from various reflective and transparent materials. In atomic optics, the main 'technical means' are electromagnetic fields. The use of various configurations of laser light fields, electric and magnetic fields, as well as micro- and nanostructures, made it possible to construct the basic elements of atomic optics, similar to those of ordinary optics – atomic lenses, mirrors, deflectors and modulators of atomic beams, coherent divisors of atomic beams, atomic interferometers, waveguides.

Control of atomic beams using material structures. One of the first methods of controlling atomic beams was the mirror reflection and diffraction of an atomic beam from the surface of a solid [18, 19]. For mirroring, the following two conditions must be met.

1. The projection of the average height of the surface irregularities on the direction of the atomic beam should be less than the de Broglie wavelength of the atom or molecule. If δ is the average height of the irregularities of the surface, and φ is the angle of the sliding incidence of the beam, then these requirements can be expressed as

$$\delta \cdot \sin \varphi < \lambda_{dB}. \tag{9.25}$$

2. To ensure that the state of the reflected atom is the same as that of the incident atom on the surface, the mean residence time of the atom on the surface should be small.

The irregularities of well mechanically polished surfaces usually have a level of the order of 10^{-5} cm. Therefore, for example, for molecular hydrogen, for which at a temperature of 300 K $\lambda_{dB} \sim 10^{-8}$ cm, in accordance with (9.25), for mirror reflection of molecules from the surface, the angle $\varphi < 10^{-3}$ rad.

The cleavage surfaces of the crystals are much smoother. The thermal vibrations of the crystal lattice limit the smoothness of the surface to a level of 10^{-8} cm. In this case, the atomic beam is reflected specularly at the angles of incidence $\varphi \sim 20°-30°$, which significantly exceeds the value of the angle when reflected from the polished surface. This was confirmed in experiments with He and LiF-crystals [18]. The temperature dependence of the angle of specular reflection has a pronounced singularity, indicating a transition from specular reflection of atoms to diffuse reflection, which indicates the influence of thermal vibrations on the smoothness of the surface of the crystal.

The first experiment to observe the diffraction of atoms on the cleavage surface of a crystal acting as a two-dimensional plane lattice was performed by Stern. The results of a detailed study of this phenomenon are presented in [19], in which the diffraction of atoms on an artificial periodic structure (slits in the membrane) with a much longer lattice period was observed.

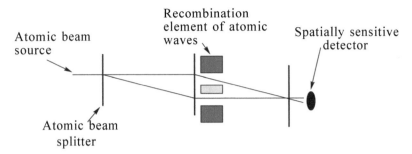

Fig. 28. Generalized scheme of an atomic interferometer.

The effect of quantum reflection of ^{4}He and ^{3}He atomic beams from the surface of liquid helium in vacuum has been successfully used to focus atoms on a concave surface and to focus He atoms on a zone plate [18].

Atomic interferometry, based on microstructures, was implemented in two elegant devices: a Young atomic interferometer with two slits and an atomic Michelson interferometer [18].

To create an atomic interferometer (Fig. 28), several elements are needed: the source of the atomic beam, the coherent divisor of the de Broglie atomic wave, the mirror for recombination (addition) of waves, and the interfering wave detector [2]. The simplest source of diatomic beams is an atomic beam formed by two collimating slits. The disadvantage of this source is its small phase density. Therefore, a source prepared by laser cooling and subsequent trapping of atoms in electromagnetic traps is preferred. The next step is coherent splitting of the atomic wave. One such method is based on the phenomenon of diffraction of atoms. For diffraction, both material nanogratings and light gratings are used. The drawback of the first is that a significant part of the atomic beam incident on them is blocked. Light gratings are a standing light wave, that is, they are phase gratings. They pass all the atoms of the beam and are therefore more effective for atomic interferometers.

To observe the interference pattern, it is necessary to recombine two atomic waves. In an atomic interferometer, as in the optical interferometer, recombination is achieved through the use of mirrors. In this case atomic mirrors are used. The first demonstration of the atomic mirror was the reflection of atoms from the surface light wave. With spatial modulation of such a wave, it can serve as a coherent splitter of an atomic beam. A laser beam can also act as an atomic mirror. In this case, after reflection, the atom completely passes into an excited state and its deviation by an angle determined by the momentum of the absorbed photon occurs.

The scheme of an atomic interferometer based on the use of laser pulses is today the most common. It should also be noted that such a simple control of atomic beams became possible only thanks to the development of methods of laser cooling and atomic trapping, which made it possible to form, first, slow atomic beams and, secondly, to create beams with a narrow velocity distribution. The first circumstance makes it possible to deflect the atomic beam noticeably even when a single photon pulse is transmitted to it, and the second is to efficiently separate atomic beams in space.

By analogy with optics, interference fringes in an atomic interferometer can be observed as oscillations of the intensity of an atomic beam directly in space using a spatially sensitive detector.

The great importance of atomic interferometry in both fundamental studies and in numerous applications is that it is possible to measure the phase shift caused by extremely small potentials of physical fields. For example, a beam of sodium atoms acquires a phase shift of one radian at the potential $6 \cdot 10^{-12}$ eV with an interaction length of 10 cm. The measurement of the phase shift by 10^{-3} radians corresponds to the measurement of the potential with a relative accuracy of 10^{-14}. If cold atoms are used, the accuracy of the measurements is increased by a factor of 10^3. With the help of atomic interferometers, ultra-precise measurements of rotation, gravity of the earth's field, atomic polarizability, etc., are carried out, and the accuracy of the measurements made exceeds the accuracy of all the preceding methods.

Control of atomic beams using static electric and magnetic fields. Some elements of the optics of atoms and molecules based on the interaction of spatially inhomogeneous magnetic or electric fields with magnetic or electric dipole moments of particles have been known for a long time and are successfully used in experimental physics and have been considered in part by us above.

In the presence of a magnetic or electric field, the spatial position of the atom or molecule varies, and the magnitude of the displacement depends on the initial quantum state of the particle and the magnitude of the field (Zeeman and Stark effects). In the adiabatic approximation (the fields vary little in time and space, and the particles move rather slowly), the internal state of the particles follows the changes in the field strength or, in other words, the particles remain on the same quantum sublevel whose energy depends on the intensity of the corresponding field.

Magnetic interaction

For an atom or a molecule with a constant magnetic moment μ the effective potential energy U in an external magnetic field with induction **B** is expressed as $U = -\mu B$. As a result, the force acting on an atom or molecule that has a potential energy U is $F = -\nabla U = \nabla(\mu B)$ (9.8). That is, a particle in a non-uniform magnetic field is acted upon by a force directed along the gradient of its induction. On the basis of this, an inhomogeneous magnetic field was proposed to be used for focusing atomic and molecular beams.

Electrical interaction

Since the energy of an atom or a molecule possessing a dipole moment **d** in an electric field depends on the field strength $E: U = -dE$, then $F = -\nabla U = \nabla(dE)$.

Thus, the force acting on an atom (molecule) with a nonzero dipole moment in an inhomogeneous electric field is also directed along the gradient of the electric field strength, which can also be used to control atomic or molecular beams.

Control of atomic beams by means of laser radiation. Atoms and molecules that do not have a static magnetic or electric dipole moment can not change their trajectory in a constant magnetic or electric field. However, an atom in a quasiresonant laser field has a high-frequency polarizability and, if the intensity of the laser field is spatially inhomogeneous, a gradient (dipole) force acts on the atom [18, 20].

The control of atoms by means of light fields is based on the recoil effect. In the optical region of the spectrum for low-energy light beams, the recoil effect that occurs in an atom upon emission or absorption of light is very small. Intensive laser radiation tuned to resonance with any allowed dipole transition of an atom can cause in the atom reradiation of a million photons per second and, therefore, significantly affect the velocity and trajectory of the atom's motion.

So a new direction in the development of atomic physics arose, based on the effects of resonance interaction of laser light with atoms using the entire arsenal of well-known effects of atomic physics: the recoil effect, the Doppler, Stark, Zeeman effects and the Raman effect.

9.6.5.2. Atomic optical nanolithography

Of particular interest are the methods of atomic optics for

nanolithography. Nanolithography is often associated with the achievement of a high density of elements on a chip, described by Moore's law. Today, methods of optical lithography with the use of far vacuum ultraviolet radiation, lithography methods through electron, ion and X-ray beams are successfully developing. These methods allow one to create nanostructures with a resolution of several nanometers. Other methods are also known that make it possible to form nanostructures, these are methods: 1) scanning nanoprobe, 2) imprinting, and 3) 'self-assembly' of nanostructures.

At the same time, an extensive search for alternative lithographic methods is under way. It is from the point of view of searching for nanolithography methods of the future that atomic-optical nanolithography should be considered, which is represented by two main methods: *direct deposition of atoms on the surface and lithography with the help of excited (metastable) atoms and chemically active atoms (alkali metals)*. The main achievements in this field are briefly discussed below [18, 19].

Direct deposition nanolithography
A typical scheme of direct deposition nanolithography suitable for the use of many types of atoms is shown in Fig. 29 [18].

An atomic beam from a thermal source, passed through material collimating apertures, undergoes transverse laser cooling, which is necessary to reduce its angular divergence. Thus, in atomic optics, the barrier inherent in light optics due to the diffraction divergence of beams is overcome. Further, a highly collimated atomic beam passes through an intense standing light wave whose frequency is shifted to the blue spectral region by several hundred MHz with respect to

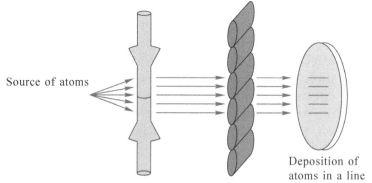

Source of atoms

Deposition of atoms in a line

Transverse laser cooling Standing wave

Fig. 29. Scheme of direct deposition nanolithography [18].

Fig. 30. An image of nanostructures obtained by depositing Cr atoms using one-dimensional (*a*) and two-dimensional (*b*) photon lenses [18]. Period of lines and points on the surface is 213 nm.

the frequency of the atomic resonance, which allows the atoms to be drawn by the gradient force into the nodes of the standing wave, ie, to the region of the minimum potential energy of the atom in the light field . This structure serves as a one-dimensional photon lens and allows focusing atoms in the line of nanometer width. In the case of atoms with a magnetic structure ($J \neq 0$), it is necessary to control the population of the magnetic sublevels, ensuring the population of the state $|m| = J$ by optical pumping by circularly polarized light.

The experiment with the deposition of Na atoms on a silicon substrate, performed according to this scheme, showed that the atomic beam forms a 249.5-nm line system on the surface, which corresponds to half the wavelength of the radiation producing the interference field [43].

A similar experiment was carried out for the deposition of Cr atoms in the form of one-dimensional and two-dimensional structures (Fig. 30). Transverse laser cooling of the atomic beam made it possible to make its angular divergence 0.1 mrad. The thickness of the deposited layers was ~22 nm, and the width of the nanolines at half-height was 50 nm.

As the results of the conducted studies show, a standing light wave is ideally suited for obtaining 1D (lines) and 2D (points) nanostructures on the substrate surface. By varying the wavelength, you can control the period of the lattice being created. The complexity of the structures created on the surface depends on the configuration of the light field produced by the superposition of many laser beams. Complex configurations of the created structures can also be obtained with the help of holographic reconstruction of the light field.

Nanolithography on resist. In traditional methods of lithography, a resist (a thin film on a substrate) is used, in particular a photoresist sensitive to UV or VUV radiation. The photoresist applied to the substrate is exposed through a light mask to create the desired topological pattern on the surface. Then, the processing (manifestation) of the photoresist and subsequent etching or application of a new layer of material takes place. The same method can be used for nanolithography using a collimated atomic beam passed through a light mask. The light mask creates a spatially inhomogeneous distribution of excited (metastable) or chemically active atoms, which modify the resist applied to the substrate. Further etching of the exposed resist is performed by standard lithographic methods. For this approach, substrates of any materials that can be etched, including such important magnetic materials as Ni and Fe, are suitable in nanofabrication. Methods of nanofabrication using excited (metastable) atoms of noble gases (He*, Ne*, Ar*) and chemically active alkali metal atoms (Na, Cs) are demonstrated.

Metastable atoms of noble gases. The first resist used in nanolithography is a self-assembled monolayer (1.5 nm thick) of dodecanethiol on a gold-coated substrate. The molecules of a highly ordered monolayer form a hydrophobic surface, which protects the substrate from chemical etching in aqueous solution. Metastable atoms with a large internal energy (up to 20 eV for He*) or chemically active atoms destroy the local ordering of organic molecules, allowing subsequent local etching. This technique was demonstrated for the case of a standing light wave. In the experiments nanostructures with a size of 65 nm, determined by the wave nature of the atoms, were obtained [19].

Instead of local destruction of the resist film from self-assembling molecules, destruction in the film of the background oil molecules that are deposited on the resist surface during the experiment due to oil pumping can be used. The created spatial structure of local destruction of this background film can be transferred to a substrate for subsequent etching by an ion beam [18].

Chemically active atoms of alkaline elements. They can be focused by means of the gradient force of the light field by radiation from both continuous and pulsed lasers and, using the high chemical activity of these atoms, to modify the surfaces on a nanometer scale.

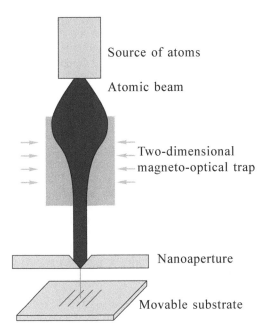

Source of atoms

Atomic beam

Two-dimensional
magneto-optical trap

Nanoaperture

Movable substrate

Fig. 31. Scheme of the atomic-optical pen for 'nanodrawing' using an atomic beam.

Atomic nanopen lithography. The transportation of atoms in a hollow fiber waveguide and their focusing in the near field of a nanopulse formed the basis for the idea of an atomic optical 'pen', the scheme of which is shown in Fig. 31 [19].

A hollow fiber or a screen with a nanoaperture can be moved in the transverse direction with the help of a cantilever. Laser cooling, collimation and the concentration of an atomic beam can be obtained in a tapering (like a horn) hollow optical waveguide [20]. Using an atomic pen can be a universal way of drawing, although there are obvious limitations to the performance of such a process due to the slow scanning process. In principle, 'painting' can also be done with the help of spatial scanning of the de Broglie atomic wave by laser light. Such an atomic scanner has already been demonstrated experimentally.

References

1. Dalibard J., Cohen-Tannoudji C., Laser cooling and trapping of neuratral atoms.//in book Atomic and Molecular Beams, ed. R. Campargac. Springer Verlag Heidelberg. 2001. P. 1–40.
2. Wieman C.E., et al., Reviews of Modern Physics. 1999. V. 71, No. 2. P. 253–262.
3. Adams, C.S., Riis E., Prog. Quant. Electr. 1997. V. 21, No. 1. P. 1–79.

4.   Cohen-Tannoudji C., Usp. Fiz. Nauk. 1999. Vol. 169, No. 3. p. 292–304.
5.   Phillips W.D., *ibid*, 1999. Vol. 169, No. 3. P. 306–322.
6.   Balykhin V.I., et al., *ibid,* 1985. P. 147, No. 1. P. 117–156.
7.   Vieira D.J., Zhao X., Experiments on Cold Trapped Atoms, Los Alamos Science. 2002. No. 27. P. 168–177.
8.   Melentiev P.N., et al., J. Experimental and Theoretical Physics. 2004. V. 98, No. 4. P. 767–677.
9.   Schmidt P.O., et al., J. Opt. Soc. Am. 2003. V. 20, No. 5. P. 960–967.
10.   Li Yo., et al., Laser Physics. 1994. V. 4, No. 5. P. 829–834.
11.   Castin Y., et al., Phys. Rev. 1994. V. A50, No. 6. P. 5092–5115.
12.   Mazets I.E., Matisov B.G., Pis'ma Zh. Eksper. Teor. Fiz. 1994. V. 60, P. 686–690.
13.   Minogin V.G., Ninhaus G., Zh. Eksper. Teor. Fiz. 1998. V. 114, No. 2. P. 511–525.
14.   Cognet L., Clairon A., Phys. Rev. A. 1996. V. 53, No. 6. P. R3734–R3737.
15.   Kazantsev L.I., Usp. Fiz. Nauk. 1978. V. 124, No. 1. P. 113–145.
16.   Ketterle B., *ibid,* 2003. V. 173, No. 12. P. 1339–1358.
17.   Gorokhov A.V., Soros. Obrazov. Zh. 2001. V. 7, No. 1. P. 71–76.
18.   Balykin V.I., Usp. Fiz. Nauk. 2009. V. 179, No. 3. P. 297–305.
19.   Balykin V.I., Kvant. elektronika. 2011. V. 81, No. 4. P. 291–315.
20.   Meschede D., Metcalf H., J. Phys. 2003. V. D36. P. R17–23.

10

Photonics of Nanostructures

Introduction

The current stage of the development of photonics is characterized by the fact that the main structural objects of its study are not structures from massive crystalline materials, but thin films, multilayer thin-film structures, threads, nanoparticles and nanodots. The small size (a) of these structures in any of the geometric directions, comparable in magnitude to the de Broglie wavelength $a \sim \lambda_{dB}$, according to the laws of quantum mechanics, leads to a change in them in the energy spectrum of charge carriers. The energy spectrum of the charge carriers becomes discrete as they move along the direction in which there is a restriction on the size of the nanostructure. The presence of this 'size' quantization under certain conditions can significantly affect the physical properties of the quantum-dimensional structures under consideration and lead to the emergence of a set of unique optical and electrical properties that are different from those observed in bulk single-crystal materials [1].

10.1. Energy spectrum of nanoscale structures

In solids, the quantum constraint can be realized in three spatial directions. The number of directions in the solid structure in which the effect of quantum limitation is observed is used as a criterion for the classification of elementary nanostructures in three groups. Thus, among the low-dimensional nanostructures, three basic elementary structures can be distinguished. These are quantum wells, quantum threads, and quantum dots (Fig. 1). These elementary structures are a crystalline material, spatially limited in one, two and three dimensions, respectively.

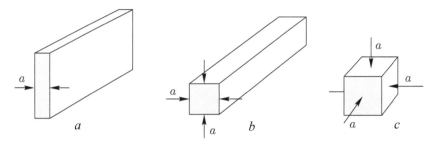

Fig. 1. Schematic representation of quantum wells (*a*), quantum threads (*b*) and quantum dots (*c*).

The unique physical properties of substances in the nanocrystalline state are due to the wave nature of particles, for example, electrons whose behavior obeys the laws of quantum mechanics.

To describe the energy spectrum and wave functions of low-dimensional systems, one or another form of the effective mass method (**kP**-perturbation theory), which has been developed in detail for bulk solid bodies, is most often used. The attractiveness of this approach is based on the fact that in a number of cases it allows one to obtain analytical results that explicitly take into account the boundary conditions and the shape of nanoscale structural elements. In addition, it is relatively easy to take into account the interactions of the electronic subsystem of low-dimensional systems with lattice vibrations, static deformations, and external fields within the framework of the **kP**-perturbation theory. This approach allows explaining many qualitative regularities inherent in low-dimensional systems, even on the basis of the simplest two-band semiconductor model, which explicitly allows for only one conduction band (E_c) and one valence band (E_v).

The basic idea of the **kP**-perturbation theory is that the wave function of an electron (hole) is a linear combination of products rapidly oscillating in the region of the unit cell of a crystal of Bloch amplitudes $u(r)$ and slowly varying in the scale of the unit cell of the envelopes of the wave functions $\psi(r)$.

According to the band theory of a solid body, the energy spectrum and envelopes of the wave functions of charge carriers in an external field are a solution of the stationary Schrödinger equation in the effective mass approximation [2, 3]:

$$\left[-\frac{h}{2m^*}\nabla^2 + V(r))\right]\psi(r) = E\psi(r), \tag{10.1}$$

where $V(r)$ is the potential energy of the charge carrier in an external field (a potential well bounding the motion of a charged particle); E is the total energy of the charge, measured from the edge of the charge carrier zone in the absence of an external field; $m*$ is the effective mass of the charge carrier in the vicinity of the edge of the corresponding energy zone; $\psi(r)$ is the wave function (envelope).

10.1.1. Bulk crystal structure

10.1.1.1. Energy spectrum of charge carriers in the bulk crystal structure

A free electron moving in a three-dimensional system has a potential energy $V(r) = 0$ and its corresponding wave function is described by the expression

$$\psi(r) = \exp(ikr) = \exp(ipr / \hbar), \qquad (10.2)$$

where $\mathbf{k}$ is the quasiwave vector.

Then the total energy of an electron is equal to its kinetic energy, whose value in accordance with the spatial components of its quasimomentum $\mathbf{p} = (p_x, p_y, p_z)$ is

$$E = \frac{1}{2m^*}\left(p_x^2 + p_y^2 + p_z^2\right) = \frac{\hbar^2}{2m^*}(k_x^2 + k_y^2 + k_z^2). \qquad (10.3)$$

As a result, for a direct-band crystal structure, the dependence of the electron energy near the edges of the conduction band and the valence band will be described by the expressions:

$$E_c(p) = E_g + \frac{p^2}{2m_e^*} = E_g + \frac{\hbar^2 k^2}{2m_e^*}, \qquad (10.4)$$

$$E_v(p) = -\frac{p^2}{2m_h^*} = -\frac{\hbar^2 k^2}{2m_h^*}, \qquad (10.5)$$

where E_g is the energy of the bandgap; m_e^* and m_h^* are the effective masses of the electron and hole, respectively.

Figure 2 shows the dispersion dependence for a direct-gap semiconductor.

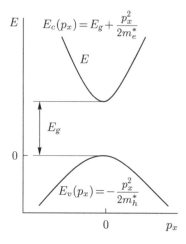

$$E_c(p_x) = E_g + \frac{p_x^2}{2m_e^*}$$

$$E_v(p_x) = -\frac{p_x^2}{2m_h^*}$$

Fig. 2. Schematic diagram of the band diagram for a straight-gap semiconductor.

10.1.1.2. Density of electron states in the energy band

When studying the optical spectral characteristics of crystalline substances, the important value is the density of states of carriers in one or another energy zone. Suppose that the free electrons occupy a conditional cubic volume (box) with the side of a face of length L. Assuming the volume of the crystal to be infinitely large, we introduce the boundary conditions for the wave function (10.2):

$$\psi(x+L) = \psi(x), \tag{10.6}$$

which are similarly satisfied for other directions y and z.

Taking into account (10.2) and (10.4), we have: $\psi(x + L) = e^{ik_x x} e^{ik_x L} = e^{ik_x x}$, from which we get:

$$k_x = \pm n_x \frac{2\pi}{L}, \tag{10.7}$$

where n_x is an integer.

According to (10.3), the electron energy depends quadratically on the magnitude of its quasi-wave vector **k** and does not depend on the direction of its motion, and to each of its states there corresponds a point in the quasimomentum space (Fig. 3) [4]. Moreover, each point corresponds to the presence of two electrons having oppositely directed pulses.

Each state of an electron in the quasimomentum space is separated from each other by a distance of $2\pi/L^3$. Therefore, we can assume that each electron occupies in this space a cell with an elementary

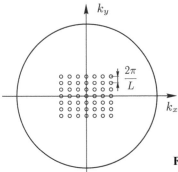

Fig. 3. Schematic representation of the state of an electron in the space of quasimomenta.

volume $(2\pi/L^3)$. Then the number of electrons in the volume of a quasisphere bounded by a sphere of radius k is equal to

$$N = 2\frac{4\pi k^3/3}{(2\pi/L)^3} = \frac{V}{3\pi^3}k^3, \qquad (10.8)$$

the factor 2 takes into account the fact that in an elementary volume there can be two electrons with opposite spin directions.

From (10.3) we find the value for $k = (2m^*E/h^2)^{1/3}$. Substituting this expression in (10.8), we get the dependence for the electron density function:

$$D_{3D}(E) = \frac{dN}{dE} = \frac{d}{dE}\left[\frac{V}{3\pi^2}\left(\frac{2m^*E}{\hbar^2}\right)^{3/2}\right] = \frac{V}{2\pi^2}\left(\frac{2m^*}{\hbar}\right)^{3/2}E^{1/2}. \qquad (10.9)$$

Figure 4 shows the dependence of the electron density in a direct-gap semiconductor as a function of its energy normalized to the value of the Fermi energy (E_F) [4].

10.1.2. One-dimensional isolated quantum well and quantum thread

A quantum limitation occurs when the free motion of electrons, at least in one of the directions, is limited by the potential barriers that form in the nanostructure in which these electrons are located. This restriction introduces new regularities in the spectrum of allowed energy states and in the process of transfer of charge carriers through the nanostructure.

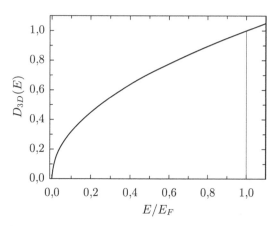

Fig. 4. Dependence of the density function of states on the electron energy for a direct-gap semiconductor [4].

10.1.2.1. One-dimensional isolated quantum well

For a one-dimensional isolated quantum well (Fig. 1 *a*) with the *z* axis perpendicular to the layer, equation (10.1) takes the form

$$\left[-\frac{h}{2m^*}\nabla^2 + V(z)\right]\psi(r) = E\psi(r). \qquad (10.10)$$

The solution of this equation is sought by the method of separation of variables in the form

$$\psi_{k_{x,y}n} = \frac{1}{\sqrt{S}}\exp\left[i\left(k_x x + k_y y\right)\right]\varphi_n(z), \qquad (10.11)$$

where *S* is the area of the cross section of the quantum well along the normal to the *z* axis; k_z and k_y are the projections of the wave vector onto the corresponding axes *x* and *y* for a charged particle moving freely in the (*x*, *y*) plane; *n* is an integer.

For the case of a one-dimensional rectangular infinitely deep quantum well, the particle potential has the form

$$V(z) = \begin{cases} 0, -\dfrac{a}{2} < z < \dfrac{a}{2}, \\[2mm] \infty, |z| \ge \dfrac{a}{2}. \end{cases} \qquad (10.12)$$

The boundary conditions for the given problem have the form

$$\varphi_n\left(-\frac{a}{2}\right) = \varphi_n\left(-\frac{a}{2}\right) = 0. \tag{10.13}$$

Taking into account (10.12), (10.13) we obtain a solution of equation (10.2) for $\varphi_n(z)$, which has the form of a standing wave:

$$\varphi_n(z) = \begin{cases} \sqrt{\dfrac{2}{a}}\cos\left(n\dfrac{\pi}{a}z\right) \text{ for odd values of n,} \\ \sqrt{\dfrac{2}{a}}\sin\left(n\dfrac{\pi}{a}z\right) \text{ for even values of } n. \end{cases} \tag{10.14}$$

This standing wave at the boundaries of the potential well has nodes, which requires that the condition that an integer number of half-waves be laid on the length of the potential well:

$$a = n\frac{\lambda_n}{2}. \tag{10.15}$$

From this condition it follows that the projection of the wave vector of a particle on the direction of the z axis can take only discrete values:

$$k_z = \frac{\sqrt{2m^* E_m}}{\hbar} = \frac{\pi}{a}n. \tag{10.16}$$

As a result, the total energy of a particle in a one-dimensional quantum well with infinitely high walls will be:

$$E = \frac{\hbar^2}{2m^*}\left(k_x^2 + k_y^2 + k_z^2\right) = \frac{\hbar^2}{2m^*}\left(k_x^2 + k_y^2\right) + E_n = $$
$$= \frac{\hbar^2}{2m^*}\left[\left(k_x^2 + k_y^2\right) + \frac{\pi^2}{a^2}n^2\right], \tag{10.17}$$

where the allowed values of the energy of a particle moving across a quantum well assume discrete values:

$$E_n = \frac{\pi^2 \hbar^2}{2\,m^* a^2}n^2, \tag{10.18}$$

where $n = 1, 2, 3, \ldots, \infty$.

402 Modern Optics and Photonics of Nano- and Microsystems

Thus, expression (10.17) describes a discrete spectrum of the size quantization of the energy of a charged particle moving in a rectangular infinitely deep quantum well. The allowed energy values for a particle are a set of discrete energy levels, into which the charge carrier band splits as a result of the size quantization. The quantum number n is the number of the corresponding energy level.

The total energy of an electron in the conduction band

$$E_e(\mathbf{p}) = \frac{p_x^2 + p_y^2}{2\,m_e^*} + \frac{\pi^2\hbar^2}{2m_e^*a^2}n^2 = E_c(p_x, p_y) + E_{en}, \qquad (10.19)$$

and the total hole energy in the valence band for a direct-band semiconductor

$$E_h(\mathbf{p}) = -\left[\frac{p_x^2 + p_y^2}{2\,m_h^*} + \frac{\partial^2\hbar^2}{2m_h^*a^2}n^2\right] = -E_v(p_x, p_y) + E_{vn}. \qquad (10.20)$$

Figure 5 shows the energy spectrum of electrons and holes in a one-dimensional infinitely deep quantum well. As can be seen from the figure, the energy levels for the bottom of the conduction band and the top of the valence band are shifted to the depth of the corresponding bands. As a result, the width of the forbidden band due to the process of size quantization increases and is equal to

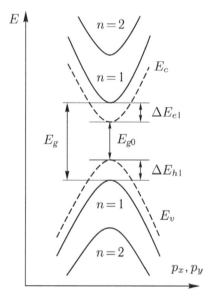

Fig. 5. Energy spectrum of the electrons and holes in the one-dimensional quantum well [4].

$$E_g = E_{g0} + (\Delta E_{e1} + \Delta E_{h1}) = E_{g0} + \frac{\pi^2 \hbar^2}{2\mu^* a^2}, \tag{10.21}$$

where $\Delta E_{e1} = \frac{\pi^2 \hbar^2}{2m_e^* a^2}$, $\Delta E_{h1} = \frac{\pi^2 \hbar^2}{2m_h^* a^2}$ and $\mu^* = \left[\frac{m_e^* + m_h^*}{m_e^* m_h^*} \right]^{-1}$ is the reduced effective mass.

As follows from (10.21), the quantum-size additive to the width of the forbidden band varies inversely with the square of the width of the quantum well.

10.1.2.2. Quantum thread

For quantum threads (Fig. 1 *b*), we assume that the direction of motion of the particle in the direction of the *z* axis is free, and the potential energy of the particle in (10.1) is a function of the coordinates (x, y): $V(x, y)$. The solution of equation (10.1) in this case is the envelope function:

$$\psi_{k_z n} = \frac{1}{\sqrt{L}} \exp(ik_z z) \varphi_n(x, y), \tag{10.22}$$

where k_z is the projection of the quasi-wave vector onto the direction of free motion of the particle z; L is the length of the quantum thread; $\varphi_n(x, y)$ is the envelope function that describes the motion of a particle in the plane of the cross section of a quantum thread, which is the solution of the equation:

$$\left[-\frac{\hbar^2}{2m^*} \left(\frac{d^2}{dx^2} + \frac{d^2}{dy^2} \right) + V(x, y) \right] \varphi_n(x, y) = E_n \varphi_n(x, y), \tag{10.23}$$

n is the quantum number.

The boundary conditions for this problem

$$\varphi_n(x, y) = 0 \text{ at } x, y \to \infty. \tag{10.24}$$

As in the previous case for a one-dimensional potential well, the energy spectrum of charge carriers in a quantum thread consists of two parts: the energy of free motion in the *z* direction and the discrete levels of size quantization E_n:

$$E = \frac{\hbar^2}{2m^*} k_z^2 + E_n. \tag{10.25}$$

In the simplest case, when the potential energy of a particle in a quantum thread can be represented as two rectangular infinitely deep potential quantum wells of width a along the x axis and the width b along the y axis:

$$V(x,y) = V_a(x) + V_b(y), \tag{10.26}$$

the solution of equation (10.23) will be the envelope function

$$\varphi_n(x,y) = \varphi_{n_x}(x) \varphi_{n_y}(y). \tag{10.27}$$

The energy of the quantization levels corresponding to these functions is

$$E_n = E_{n_x n_y} = \frac{\pi^2 \hbar^2}{2m^*} \left[\left(\frac{n_x}{a} \right)^2 + \left(\frac{n_y}{b} \right)^2 \right], \tag{10.28}$$

where n_z, n_y = 1, 2, 3,...n,

Thus, according to (10.24) and (10.28), the allowed energy values for a particle in a quantum thread, like in a quantum well, split into a sequence of discrete energy levels E_n, where n is a combination of the integers n_x and n_y.

As a result, the total energy of the electron and hole in the corresponding energy bands, respectively, will be described by the expressions:

$$E_e(p) = \frac{p_z^2}{2m_e^*} + \frac{\pi^2 \hbar^2}{2m_e^*} \left[\left(\frac{n_x}{a} \right)^2 + \left(\frac{n_y}{b} \right)^2 \right] = E_c(p_z) + E_{e n_x n_y}. \tag{10.29}$$

$$E_h(p) = -\left\{ \frac{p_z^2}{2m_h^*} + \frac{\pi^2 \hbar^2}{2m_h^*} \left[\left(\frac{n_x}{a} \right)^2 + \left(\frac{n_y}{b} \right)^2 \right] \right\} = E_v(p_z) + E_{h n_x n_y}. \tag{10.30}$$

The edges of the energy bands will be shifted by an amount

$$\Delta E_{e11} = \frac{\pi^2 \hbar^2}{2m_e^*} \left[\left(\frac{n_x}{a} \right)^2 + \left(\frac{n_y}{b} \right)^2 \right]$$

into the conduction band and by an amount

$$\Delta E_{e11} = \frac{\pi^2 \hbar^2}{2 m_e^*} \left[\left(\frac{n_x}{a} \right)^2 + \left(\frac{n_y}{b} \right)^2 \right]$$

into the valence band. This will lead to an increase in the width of the forbidden band for a quantum thread

$$E_g = E_{g0} + \Delta E_{e11} + \Delta E_{h11}. \qquad (10.31)$$

in the same way as shown in Fig. 5.

10.1.2.3. The density of electron states for an isolated one-dimensional quantum well

Suppose that an electron moves freely in a one-dimensional quantum well (Fig. 1 *a*) in the directions of the axes *x* and *y*. In the direction of the *z* axis its motion is bounded and the wave function satisfies condition (10.13). Since the motion of the electron in the directions of *x* and *y* occurs in the periodic potential field of a crystal with the characteristic dimensions L_x and L_y, then the projections of the quasi-wave electron vector on these axes must take on discrete values:

$$k_x = \frac{2\pi}{L_x} n_x, \; k_y = \frac{2\pi}{L_y} n_y, \qquad (10.32)$$

where n_x, n_y = 1.2.3, ...
   Thus, each electron in the *k*-space occupies a site of size

$$S_k = \frac{2\pi}{L_x} \frac{2v}{L_y} = \frac{4\pi^2}{L_x L_y} = \frac{4\pi^2}{S}, \qquad (10.33)$$

where $S = (L_x L_y)$ is the area of the transverse surface of the quantum well (plate).
   Then, taking into account the electron spin, the number of quantum states for an electron in a circle of radius *k* is defined as:

$$N(k) = 2 \frac{\pi k^2}{(4\pi^2 / S)} = \frac{k^2 S}{2\pi}. \qquad (10.34)$$

The number of states in the interval from *k* to *k* + *dk* is defined as

$$dN(k) = \frac{dN}{dk} dk = \frac{k}{\pi} S \, dk. \qquad (10.35)$$

Similarly, the number of states in the energy gap from E to $E + dE$ can be calculated as

$$\frac{dN}{dE}dE = \frac{dN}{dk}\frac{dk}{dE}dE. \tag{10.36}$$

The number of states per unit area in the unit energy interval, with allowance for (10.35), is defined as

$$\frac{1}{S}\frac{dN}{dE} = \frac{1}{S}\frac{dN}{dk}\frac{dk}{dE} = \frac{k}{\pi}\frac{dk}{dE}. \tag{10.37}$$

Because for a free electron $E = \hbar^2 k^2/2m$, then

$$k = \frac{\sqrt{2mE}}{\hbar} \tag{10.38}$$

and, consequently,

$$\frac{dk}{dE} = \frac{1}{2\hbar}\sqrt{\frac{2m}{E}}. \tag{10.39}$$

Using (10.36) and (10.37) for the density of surface states in a quantum well (plate), we obtain expression

$$\rho_{2D}(E) = \frac{1}{S}\frac{dN}{dE} = \frac{k}{\pi}\frac{dk}{dE} = \frac{1}{\pi}\frac{\sqrt{2mE}}{\hbar}\frac{1}{2\hbar}\sqrt{\frac{2m}{E}} = \frac{m}{\pi\hbar^2}. \tag{10.40}$$

As follows from (10.40), the density function of the surface states for a one-dimensional quantum well does not depend on the energy E. Nevertheless, ρ_{2D} depends on the number of levels of spatial quantization in the zone and, thus, is the sum of the contributions of each of the discrete energy levels in the conduction band:

$$\rho_{2D}(E) = \sum_{n=1}^{\infty}\frac{m}{\pi\hbar^2}H(E - E_{cn}), \tag{10.41}$$

where E_{cn} is the minimum value of the electron energy at the corresponding level in the conduction band (10.28), and $H(E-E_{cn})$ is the Heaviside's step function:

$$\begin{cases} H(x) = 0, \text{ at } x < 0, \\ H(x) = 1, \text{ at } x > 0. \end{cases} \tag{10.42}$$

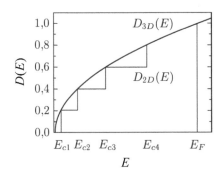

Fig. 6. Density of states for electrons in a three-dimensional medium $(D_{3D}(E))$ and in a medium with a one-dimensional quantum well $(D_{2D}(E))$ [4].

Thus, the density of electron states in a one-dimensional quantum well is independent of the energy and the number of the energy level. Figure 6 shows the stepwise dependence of the density of states of an electron on energy, determined by the formula $D_{2D}(E) = S \cdot \rho_{2D}(E)$ [4]. A characteristic feature of this dependence is the same step height which takes place in the approximation of the independence of the effective mass of the electron from the energy level number.

10.1.2.4. The density of states for a quantum thread

The quantum thread is an extended object, the motion of electrons in which is bounded in two directions by a two-dimensional potential well (Fig. 1 *b*). If the length of the quantum thread in the direction of free motion of the electron z is L_z, then the projection of its quasi-wave vector onto this axis must take on discrete values:

$$k_z = \frac{2\pi}{L_z} n_z, \qquad (10.43)$$

where $n_z = 1, 2, 3,....$

With the one-dimensional motion of an electron taken into account, the size of the minimum linear region occupied by an electron in the k-space is $2\pi/L_z$. This means that the possible number of electron states is

$$N(k) = 2\frac{k}{2\pi / L_z} = \frac{kL_z}{\pi}. \qquad (10.44)$$

The density of the electron states in a quantum thread is calculated as

$$D_{1D}(E) = \frac{dN}{dE} = \frac{dN(k)}{dk}\frac{dk}{dE}. \qquad (10.45)$$

Using (10.28) and (10.29) for the one-dimensional motion of the

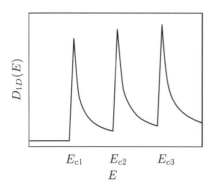

$D_{1D}(E)$

E_{c1} E_{c2} E_{c3}

E

Fig. 7. Dependence of the state density function for electrons in an isolated quantum thread [4].

electron in a quantum thread, we obtain

$$k = \frac{1}{h}\sqrt{2m_e^*(E - E_{cn})},$$ (10.46)

where E_{cn} is the minimum energy of an electron at a quantum-well level with a quantum number n.

As a result, using (10.44)–(10.46), for the density of states of electrons in a quantum thread, we obtain

$$D_{1D}(E) = \frac{L_z}{\pi\hbar}\sqrt{\frac{m_e^*}{2(E - E_{cn})}}.$$ (10.47)

Figure 7 shows the qualitative dependence of the density function of states for electrons for an isolated quantum thread is presented [4].

10.1.3. Quantum dots and the density of states of electrons in them

In quantum dots (Fig. 1 c), the motion of charge carriers is limited in all three coordinates, and therefore quantum dots are called zero-dimensional objects. The envelopes of the wave functions of charge carriers are solutions of equation (10.1) which describe the localized motion and depend on three quantum numbers. At its core, the quantum dot is a three-dimensional quantum potential hole with dimensions along the axes L_x, L_y, L_z and infinitely high potential walls. If the potential energy of charge carriers in a quantum dot is represented as the sum of the potential energies of three rectangular infinitely deep potential wells:

$$V(x,y,z) = V_{L_x}(x) + V_{L_y}(y) + V_z(z), \tag{10.48}$$

then the envelope function will have the form of a product of three functions, which depend on three quantum numbers (n_x, n_y, n_z)

$$\psi_n(x,e,z) = \varphi_{n_x n_y n_z} = \varphi_{n_x}(x)\varphi_{n_y}(y)\varphi_{n_z}(z), \tag{10.49}$$

each of which is a standing wave of the type

$$
\begin{cases}
\varphi_{n_x}(x) = \sqrt{\dfrac{2}{L_x}}\sin(k_x x) = \sqrt{\dfrac{2}{L_x}}\sin\left(\dfrac{2\pi}{L_x}n_x x\right), \\[2mm]
\varphi_{n_y}(y) = \sqrt{\dfrac{2}{L_y}}\sin(k_y y) = \sqrt{\dfrac{2}{L_y}}\sin\left(\dfrac{2\pi}{L_y}n_y y\right), \\[2mm]
\varphi_{n_z}(z) = \sqrt{\dfrac{2}{L_z}}\sin(k_z z) = \sqrt{\dfrac{2}{L_z}}\sin\left(\dfrac{2\pi}{L_z}n_z z\right).
\end{cases} \tag{10.50}
$$

This function will correspond to energy levels of size quantization, equal to the sum of the energy of motion along each of the axes

$$E_n = E_{n_x n_y n_z} = \frac{\pi^2\hbar^2}{2m^*}\left[\left(\frac{n_x}{L_x}\right)^2 + \left(\frac{n_y}{L_y}\right)^2 + \left(\frac{n_z}{L_z}\right)^2\right], \tag{10.51}$$

where m^* is the effective mass of the charge carrier (electron or hole); n_x, n_y, $n_z = 1, 2, 3,...,\infty$.

Thus, the total energy of the electron and the hole in the corresponding energy bands will be described by the expressions:

$$E_e(\mathbf{p}) = E_c + \frac{\pi^2\hbar^2}{2m_e^*}\left[\left(\frac{n_x}{L_x}\right)^2 + \left(\frac{n_y}{L_y}\right)^2 + \left(\frac{n_z}{L_z}\right)^2\right], \tag{10.52}$$

$$E_h(\mathbf{p}) = E_v - \frac{\pi^2\hbar^2}{2m_h^*}\left[\left(\frac{n_x}{L_x}\right)^2 + \left(\frac{n_y}{L_y}\right)^2 + \left(\frac{n_z}{L_z}\right)^2\right]. \tag{10.53}$$

As can be seen, the energy spectrum of the motion of electrons and holes is discrete and has a form analogous to the spectrum of a solitary atom. As a result, the quantum dot can be considered as

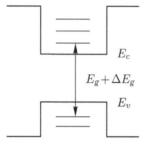

Fig. 8. Energy spectrum of electrons and holes in a quantum dot.

an artificial point atom. The peculiarity of the energy spectrum of charges in the quantum dot lies in the fact that the edges of the forbidden band are displaced into the depths of the conduction and valence band (Fig. 8) by an amount

$$\Delta E_c = \frac{\pi^2 \hbar^2}{2m_e^*}\left[\left(\frac{1}{L_x}\right)^2 + \left(\frac{1}{L_y}\right)^2 + \left(\frac{1}{L_z}\right)^2\right], \qquad (10.54)$$

$$\Delta E_v = \frac{\pi^2 \hbar^2}{2m_h^*}\left[\left(\frac{1}{L_x}\right)^2 + \left(\frac{1}{L_y}\right)^2 + \left(\frac{1}{L_z}\right)^2\right], \qquad (10.55)$$

which, accordingly, leads to an increase in the width of the forbidden zone by

$$\Delta E_g = \frac{\pi^2 \hbar^2}{2\mu}\left[\left(\frac{1}{L_x}\right)^2 + \left(\frac{1}{L_y}\right)^2 + \left(\frac{1}{L_z}\right)^2\right]. \qquad (10.56)$$

Since at each quantum level there can be two electrons, the function of their density of states in the quantum dot has the form

$$D_{0D}(E) = 2\sum_{n_x, n_y, n_z=1}^{\infty} \delta(E - E_{n_x, n_y, n_z}) = 2\sum_{n=1}^{\infty}\delta(E - E_n). \qquad (10.57)$$

Figure 9 shows the form of the function of the density of states of electrons in a quantum dot, constructed in accordance with the expression (10.57) [4].

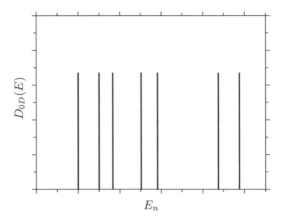

Fig. 9. The function of the density of states of electrons in a quantum dot.

10.2. Exciton states in semiconductor and dielectric materials

10.2.1. Free excitons or Wannier–Mott excitons

The simplest excitation of the electronic system of a semiconductor or dielectric consists in the transition of an electron from the valence band to the conduction band. In this case, a positively charged hole is formed in the valence band. Electrons and holes formed during the generation process undergo a Coulomb interaction. This interaction leads to the fact that nonequilibrium electrons and holes should be considered in the coordinate space as a bound electron–hole pair – exciton. An exciton is a quasiparticle that arises during excitations in semiconductors and dielectrics. Depending on the nature of the bond, there are two types of excitons. The first type is free large-radius excitons (Wannier–Mott excitons), whose characteristic dimensions reach dozens of interatomic distances (Fig. 10).

The second type is bound small-radius excitons (Frenkel excitons) (Fig. 10), whose dimensions do not exceed one interatomic distance.

The exciton, transferring energy, does not transfer the charge and is similar in its structure to a neutral hydrogen atom, in which the hole plays the role of the atomic nucleus, i.e., the place in which the electron was before excitation.

The corresponding Schrödinger equation for the exciton motion in the field of the Coulomb potential will have the form

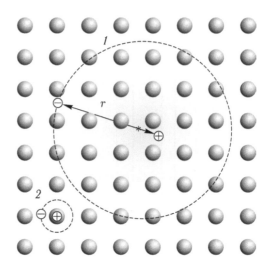

Fig. 10. Wannier–Mott (1) and Frenkel (2) excitons formed in the crystal lattice of a semiconductor or dielectric [5].

$$\hat{H}\,\Psi\left(r_e, r_h\right) = (E - E_g)\Psi\left(r_e, r_h\right), \tag{10.58}$$

where is the Hamilton operator [1]:

$$\hat{H} = -\sum_{i=e,h} \frac{\hbar^2 \nabla_{r_i}^2}{2m_i} - \frac{e^2}{\varepsilon \left|r_e - r_h\right|}, \tag{10.59}$$

$\psi(\mathbf{r}_e, \mathbf{r}_h)$ is the wave function of the exciton; $\mathbf{r}_e$, $\mathbf{r}_h$ and m_e, m_h are the radius vectors and effective masses corresponding to an electron and a hole; e is the charge of an electron; ε is the electric permittivity of the medium; E is the total energy of the exciton; E_g is the maximum energy of the forbidden band. The eigenvalues of the exciton energy E are reckoned from the maximum energy of the forbidden band of the crystal E_g.

The exciton motion can be represented as two components: the slow translational motion of an exciton as a free particle with mass M and relatively fast rotation of an electron with reduced mass μ around a fixed hole [6].

Given the independent and different nature of these two components of the exciton motion in a crystal, it is natural to introduce new coordinates separating the slow translational motion of the centre of mass of the exciton (with the coordinates $\mathbf{R} = (m_e \mathbf{r}_e +$

$m_h \mathbf{r}_h) / (m_e + m_h))$ from the rapid motion of the electron with respect to the hole (with the radius vector $\mathbf{r} = \mathbf{r}_e - \mathbf{r}_h$), representing the wave function of the exciton in the form

$$\Psi(\mathbf{r}_e, \mathbf{r}_h) = \Phi(\mathbf{R})\psi(\mathbf{r}). \tag{10.60}$$

In this case the Hamiltonian operator (10.59) can be regarded as consisting of two terms

$$\hat{H} = \hat{H}_R + \hat{H}_r. \tag{10.61}$$

where

$$\hat{H}_R = -\frac{h^2 \nabla_{\mathbf{R}}^2}{2M}, \ \hat{H}_r = -\frac{h^2 \nabla_r^2}{2i} - \frac{e^2}{\varepsilon |\mathbf{r}|}, \tag{10.62}$$

and $M = m_e + m_h$; $\mu = m_e + m_h$; $(m_e + m_h)$.

Substituting (10.60)–(10.62) into (10.58), we obtain two independent expressions:

$$-\frac{\hbar^2}{2M} \nabla_{\mathbf{R}}^2 \Phi(\mathbf{R}) = E_R \Phi(\mathbf{R}), \tag{10.63}$$

$$-\frac{\hbar^2}{2\mu} \nabla_r^2 \psi(\mathbf{r}) - \frac{e^2}{\varepsilon |\mathbf{r}|} \psi(\mathbf{r}) = E_r \psi(\mathbf{r}), \tag{10.64}$$

where the sum of the eigenenergies in equations (10.63) and (10.64) are equal to energy in (10.58):

$$E_R + E_r = E - E_g. \tag{10.65}$$

Equation (10.63) describes the motion of a free particle (an exciton) with a mass M whose kinetic energy is

$$E_R = \frac{\hbar^2 k^2}{2M}, \tag{10.66}$$

where k is the wave vector of the exciton as a whole particle.

Equation (10.63) is equivalent to the Schrödinger equation for the hydrogen atom with the only difference being that in the corresponding atom around the nucleus with charge $+e$ a quasiparticle with charge $-e$ and the mass μ rotated and the field of Coulomb attraction between which is weakened ε times (Fig. 10). The potential energy of an electron in such an 'atom' $U(r) = -e^2/|\mathbf{r}|$ depends only on the distance r electron from the hole and, as a result, has a spherical symmetry. Possible values of the electron energy follow from the exact solution of equation (10.64). Therefore, the binding energy of an exciton is expressed in this model in the same way as for a hydrogen atom, and has the form

$$E_r(n) = -\frac{e^4\mu}{2\varepsilon^2\hbar^2}\frac{1}{n^2} = -\frac{e^2}{2\varepsilon a_{ex}}\frac{1}{n^2}, \tag{10.67}$$

where n is the principal quantum number of the exciton and a_{ex} is the efficient exciton radius

$$a_{ex} = \frac{\varepsilon\hbar^2}{e^2\mu}. \tag{10.68}$$

The total energy of an exciton consists of the sum of its kinetic energy $E_R = \hbar^2 k^2/2M$ and the potential energy $E_g + E_r(n)$:

$$E_{ex}^n(k) = E_g + E_r(n) + E_R = E_g - \frac{e^2}{2\varepsilon a_{ex}}\frac{1}{n^2} + \frac{\hbar^2 k^2}{2M}. \tag{10.69}$$

According to (10.67) for $n = 1$ we obtain the binding energy for the main states of an exciton

$$E_{ex} = |E_r(n)| = \frac{e^4\mu}{2\varepsilon^2\hbar^2} = \frac{e^2}{2\varepsilon a_{ex}}. \tag{10.70}$$

The minimum energy $E_{gx} = E_g - E_{ex}$, which is necessary for creating an exciton, is called the exciton band gap width.

Multiplying and dividing the right-hand side of expression (10.68) by the free electron mass in vacuum, we transform the expression for the radius of the exciton (10.68) to the form

$$a_{ex} = \frac{\hbar^2}{me^2}\frac{m}{\mu}\varepsilon = 0.53\frac{m}{\mu}\varepsilon \text{ Å}, \tag{10.71}$$

where $\hbar^2/me^2 = 0.53$ Å is the radius of the first Bohr orbit in the hydrogen atom.

The energy of the ground state of the exciton (10.70) is also called the dissociation energy of the exciton, or, by analogy with Rydberg for the hydrogen atom ($R_y = 13.6$ eV), the Rydberg exciton. The value of the dissociation energy of a free exciton, calculated according to the relation (10.70) at an effective mass of $m_r = 0.0667\, m_0$, the dielectric constant ε equal to 12, is 5.6 meV, and the value of the exciton radius is $a_{ex} = 10.6$ nm. As follows from the above estimates, the binding energy of an electron in Wannier–Mott excitons is thousands of times smaller than in a hydrogen atom. And the effective radius of free excitons is hundreds of times greater than the Bohr radius of the hydrogen atom.

A direct optical transition (Fig. 11) from the exciton state with emission of a photon $\hbar\omega$ requires the fulfillment of the laws of conservation of energy and quasimomentum:

$$\hbar\omega = \frac{\hbar^2 k^2}{2M} + E_g + E_r(n), \tag{10.72}$$

$$\hbar k_{phot} = \hbar k. \tag{10.73}$$

Since the quasimomentum of a photon is much smaller than the quasimomentum of an electron and a hole, optical transitions are

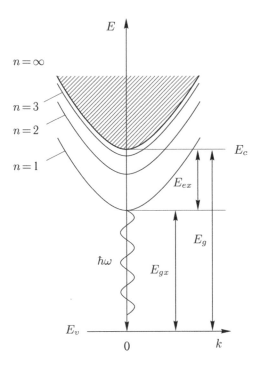

Fig. 11. Structure of exciton energy levels: E_c – the energy of the bottom of the conduction band, E_v – the energy of the top of the valence band, E_{gx} – exciton width of the forbidden band [6].

possible only in the state $k = 0$. Then the third term on the right-hand side can be neglected in (10.69). As a result, the optical spectrum of an exciton can be expressed as

$$E_{ex}^n(k) = E_g - \frac{e^2}{2\varepsilon a_{ex}} \frac{1}{n^2},$$ (10.74)

Then for the values of the quantum number $n = 1, 2, 3,...$ the exciton spectrum will be analogous to the Lyman series in the hydrogen atom. It is in this sense that the spectrum of the exciton is 'hydrogen-like', since it represents a series of spectral lines that 'thicken' as $1/n^2$ as it approaches the edge of dissociation, beyond which a continuous spectrum (continuum) begins. Thus, the presence of exciton levels leads, in addition to the smooth spectrum caused by transitions between the continua of states in the valence band and the conduction band, to the appearance of a line spectrum localized near the bottom of the conduction band in the forbidden band region. Figure 12 shows the pattern of formation of the exciton spectrum.

Under fundamental absorption, indirect exciton transitions are also possible. The conservation laws in this case will have the form

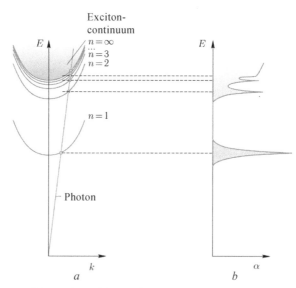

Fig. 12. Scheme of formation of the exciton spectrum at the edge of the fundamental absorption of the material: a – energy scheme demonstrating the conservation of energy and momentum; b – spectral dependence of the radiation absorption coefficient of the material (α) [5].

$$\hbar\omega = \frac{\hbar^2 k^2}{2M} + E_g + E_r(n) - \hbar\Omega, \qquad (10.75)$$

$$\hbar\mathbf{k}_{phot} = \hbar\mathbf{k} - \hbar\mathbf{k}_{phon}, \qquad (10.76)$$

where Ω, $\mathbf{k}_{phon}$ is the eigenfrequency and quasimomentum of the phonon of the crystal lattice.

Taking into account the real band structure of semiconductors essentially modifies the energy spectrum of excitons. Table 1 shows the values of the effective masses of electrons and holes, the dielectric constant, the ionization energy of the exciton and the exciton radius for various semiconductor materials [6]. Bold letter in Table 1 denote those values of the ionization energy of the exciton that exceed the value of thermal energy at room temperature (T_{room}): $k_B T_{room} = 26$ meV, k_B is the Boltzmann constant. It follows from the Table 1 that, for the main spectrum of semiconductor materials, the dissociation energy of free excitons is much less for most materials than thermal energy.

Table 1. Parameters of excitons in various semiconductor materials

Material	m_n in units of m_0	m_p in units of m_0	ε	E_{ex}, meV	a_{ex}, nm
GaN	0.20	0.80	9.3	25.2	3.1
InN	0.12	0.50	9.3	15.2	5.1
GaAs	0.063	0.50	13.2	4.4	12.5
InP	0.079	0.60	12.6	6.0	9.5
GaSb	0.041	0.28	15.7	2.0	23.2
InAs	0.024	0.41	15.2	1.3	35.5
InSb	0.014	0.42	17.3	0.6	67.5
ZnS	0.34	1.76	8.9	**49.0**	1.7
ZnO	0.28	0.59	7.8	**60.0**	2.2
ZnSe	0.16	0.78	7.1	**35.9**	2.8
CdS	0.21	0.68	9.4	24.7	3.1
ZnTe	0.12	0.6	8.7	18.0	4.6
CdSe	0.11	0.45	10.2	11.6	6.1
CdTe	0.096	0.63	10.2	10.9	6.5
HgTe	0.031	0.32	21.0	0.87	39.3

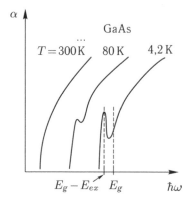

Fig. 13. An example of exciton absorption spectra in a GaAs bulk semiconductor obtained at different temperatures [6].

Consequently, exciton recombination for Wannier–Mott excitons at room temperature does not make significant contributions to luminescence because of the high probability of thermal dissociation of the exciton. It is precisely such small values of the binding energy of excitons that prevent the observation of exciton transitions at room temperature of the samples.

In addition, in the case of indirect-gap semiconductors, the need for interacting with a phonon in radiative recombination significantly reduces the intensity of exciton recombination.

Figure 13 shows an example of absorption spectra for samples from the GaAs semiconductor compound obtained at different temperatures. As can be seen, the observation of exciton absorption peaks in the volume of the semiconductor is possible only at low temperatures, when the binding energy of the exciton exceeds the value of the energy of the thermal vibrations.

Thus, the low binding energy of excitons and the decay of excitons at room temperature for most types of semiconductor and dielectric materials hamper the development of semiconductor devices based on them.

10.2.2. Bound excitons (Frenkel excitons)

Under certain conditions, nonequilibrium electrons and holes in crystals can form not only an exciton freely moving along the crystal and called free (the Wannier–Mott exciton), but also a small-radius exciton localized on the defect, which is bound (Frenkel exciton) [5].

In most cases, bound excitons are formed at neutral centres, although under certain conditions their formation on charged defects is possible. Bound excitons are formed with high probability on *isoelectronic* traps. An isoelectronic impurity is an atom of such an element that is in one group of a periodic table with a substitutable atom. However, not every isoelectronic impurity can bind an exciton. Isoelectronic substitutional impurities form bound excitons when impurity and substitutable atoms differ substantially in both electronegativity and covalent radii. In this case, the impurity atom distorts the potential energy of the electron in the lattice, thus forming a deep potential well that does not exceed the interatomic distance in size. In this potential well, an electron or hole is trapped near the isoelectronic impurity. After the carrier of one sign (for example, an electron) is captured-localized, the isoelectronic center acquires a charge and then quite easily captures the carrier of the opposite sign (in our case, a hole). Thus, a bound electron-hole pair is formed in the form of an exciton strongly localized in space. The Frenkel exciton has two significant differences in comparison with the Wannier–Mott exciton:

1. The energy level of the ground state of the Frenkel exciton is below the analogous level of the Wannier–Mott exciton, i.e., the dissociation energy E_{ex} for a bound exciton is greater than for a Wannier–Mott exciton. For some materials, the dissociation energy can be equal to the fraction of the electron volt. Therefore, the Frenkel exciton is more stable at room temperature in comparison with a large-radius exciton.

2. Since the Frenkel exciton is localized in the coordinate space, according to the Heisenberg uncertainty relation, it is completely delocalized in k-space. Using the uncertainty relation: $\Delta p \cdot \Delta r \geq \hbar$, we estimate the quasi-pulse Δp.

If the region of electron localization in the coordinate space $\Delta r \sim a$, where a is the dimension of the interatomic distance in the lattice, then the change in the quasimomentum of the pulse:

$$\Delta p = \hbar \Delta k \geq \frac{\hbar}{\Delta r} \sim \frac{\hbar}{a}. \qquad (10.77)$$

It follows from (10.77) that the uncertainty of the magnitude of the wave vector of the bound exciton is $\Delta k \geq 1/a$. This means that the wave vector of the Frenkel exciton can take on any values within the Brillouin zone of the crystal. This fact is quite important for indirect-gap semiconductors, since it does not require the participation of phonons in the bound excitons in radiative recombination.

The mechanism of radiative recombination of bound excitons is very effective in semiconductors with an indirect structure of energy bands (silicon, germanium, and gallium phosphite), since the probability of radiative recombination through such a centre is much greater than the probability of indirect interband transitions. A typical example of an isoelectronic trap can be a nitrogen atom N in a gallium phosphide (GaP) – semiconductor with an indirect structure of energy bands. The nitrogen atom N replaces the phosphorus atom P at the lattice sites. Nitrogen N and phosphorus P have the same external electronic external configuration ($1s^2 2s^2 2p^3$ and $1s^2 2s^2 2p^6 3s^2 3p^3$), since both belong to the group V of the elements of the periodic system, and the structures of their inner shells are very different.

The spectrum of radiative recombination of bound excitons is narrower than the spectrum of free excitons, since the bound exciton is localized in the coordinate space and its kinetic energy is small compared to that of a free exciton of large radius.

10.3. Influence of the form of nanoparticles on the energy subsystem of charge carriers

Nanoparticles are usually called objects consisting of atoms or molecules having a size of less than 100 nm. Due to the small size, these particles differ in their properties from both atoms and bulk material. Therefore, nanoobjects are characterized not only by small dimensions, but also by special physical and chemical properties.

In connection with the fact that nanoparticles contain a limited number of atoms, their properties differ from the properties of crystals, which contain a huge number of atoms. First of all, this is due to the special role of the surface. The atoms lying on the surface have more energy, since the number of neighbours is less than for the atoms lying in the depth of the crystal lattice. The gradual decrease in particle size leads to an increase in the total surface area, and thus to an increase in the fraction of atoms on the surface (Fig. 14) and an increase in its role in determining their physical and chemical properties [8].

The dependence of the physical and chemical properties of nanoparticles on their geometric dimensions is called dimensional effects. For example, the size effect is reflected in the difference between the optical properties of nanoparticles and the properties of bulk samples. First of all, this is due to the difference in the energy characteristics of the charge carrier for nanoparticles and bulk materials. Indeed, as was shown in Sec. 10.1 and 10.2.

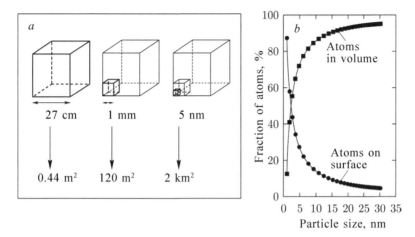

Fig. 14. Schematic representation of the growth of the total surface area with a decrease in the particle size constituting a quartz cube weighing 50 kg (*a*) and the dependence of the fraction of atoms in the volume and on the surface of the nanoparticle on its size (*b*).

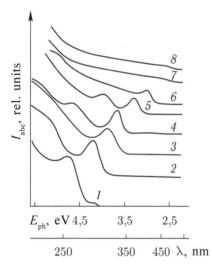

Fig. 15. Dependence of light absorption by colloid solutions of CdS nanoparticles. The size of nanoparticles: 1-6,4; 2-7.2; 3-8.0; 4-9.3; 5-11.6; 6-19.4; 7 28.0 and 8 48.0 nm. E_{ph} is the energy of photons [9].

The appearance of nanoparticle sizes leads to the appearance of levels of quantization in the valence and conduction bands, an increase in the band gap, and the formation of stable exciton states whose energy spectrum depends on the size of the nanoparticles, as shown in Fig. 15.

Another reason is the existence of defects and impurity centers on the surface and in the bulk of nanoparticles, which also affect the energy spectrum as well as the dynamics and mechanisms of relaxation processes of charge carrier states and other elementary excitations [8].

Another reason that causes the appearance of peculiarities of optical properties of nanoparticles is in the form of nanoparticles, which directly affects the energy spectrum of charge carriers.

The effect of size quantization leads to a radical transformation of the quasicontinuous spectrum and wave functions of the charge carriers. Therefore, when describing the states of charge carriers, one should keep in mind not only the size, but also the shape and size of the surface of the nanoparticles, which must determine the boundary conditions for the corresponding wave functions.

10.3.1. Single-particle states in the nanoparticles of the complicated shape

Modeling the processes of the effect of the nanoparticle shape on the energy spectrum of charge carriers and other elementary excitations is extremely difficult. Nevertheless, for some particular cases it is possible to find a solution to this problem. In particular, a variational method was developed in [10], which makes it possible to estimate the effect of the axially symmetric deformations of the shape of the surface of spherical nanoparticles on the energy spectrum and the density of states of charge carriers.

We represent the energy of charge carriers as a functional $E[s(\mathbf{r})]$, defined on the space of functions describing the shape of the surface of the nanoparticle $s(\mathbf{r})$. Because of the continuity of the wave function describing the state of charge carriers, this functional is continuous and can be decomposed into a variational series around a certain initial volume of the nanoparticle V_0 bounded by the surface $s_0(\mathbf{r})$:

$$E[s(\mathbf{r})] = E_0[s_0(\mathbf{r})] + \int_{V(\xi)} \frac{\delta E[s(\mathbf{r})]}{\delta s(\xi)}\bigg|_{s=s_0} \delta s(\xi)\, d\xi +$$

$$\frac{1}{2!}\iint_{V(\xi)} \frac{\delta^2 E[s(\mathbf{r})]}{\delta(s(\xi_1))\delta(s(\xi_2))}\bigg|_{s=s_0} \delta(s(\xi_1))\delta(s(\xi_2)) d\xi_1\, d\xi_2 + ..., \tag{10.78}$$

where V is the nanoparticle volume bounded by the surface $s(\mathbf{r})$; $\mathbf{r}$ is the radius vector; $\dfrac{\delta}{\delta s(\xi)}$ is the variation derivative; $\delta s(\xi) = s(\xi) - s_0(\xi)$ is the variation of the nanoparticle surface.

Thus, if we know the value $E_0[s_0(\mathbf{r})]$, using the variational series (10.78) and the variational corrections to the value of $E_0[s_0(\mathbf{r})]$ resulting from the deviation of the shape of the nanoparticle from the original, we can find the value of $E_0[s(\mathbf{r})]$.

Assuming that the deviations from the original form are insignificant, we take into account only the first two terms of the series (10.78). Then, using the transformation

$$\frac{\delta E[s(\mathbf{r})]}{\delta s(\xi)} = \frac{dE[s(\mathbf{r})]}{ds(\xi)} \delta(\mathbf{r}-\xi)$$

we obtain for the charge energy functional in the nanoparticle E

$$[s(\mathbf{r})] \approx E_0[s_0(\mathbf{r})] + \frac{dE[s(\mathbf{r})]}{ds(\mathbf{r})}\bigg|_{s=s_0} (\mathbf{r}). \qquad (10.79)$$

Suppose that the initial nanoparticle has a spherical surface s_0 of radius R_0, which makes the problem symmetric and allows us to use simple expressions for the energy spectrum of a charge moving in a spherically symmetric potential well with infinitely high walls. This also makes it possible to use the spherical coordinate system $(\mathbf{r},\Theta,\phi)$. Let us assume that the unperturbed surface of a sphere in a spherical coordinate system is described by the expression: $s_0(\mathbf{r}) = R_0^2$. In this case, the distance between the centre of the sphere and the point on its surface $R(\Theta,\phi)$ will satisfy the condition: $R_0^2(\Theta,\phi)$ R_0^2. In this case, the variation in the shape of the surface, describing the deviation of the shape of the nanoparticle from the sphere, will be written as

$$\delta s(R,\Theta,\phi) = s(R,\Theta,\phi) - s_0(R,\Theta,\phi) = \delta R^2(\Theta,\phi) =$$
$$= R^2(\Theta,\phi) - R_0^2, \qquad (10.80)$$

where $R^2(\Theta,\phi)$ is the square of the distance between the centre of the sphere and the point on the surface of a nanoparticle with a perturbed form $s(R,\Theta,\phi)$.

As is known [11], in the effective mass approximation, the energy of a charged particle in a spherical potential well with infinitely high walls is described by the expression

$$E_{n,l}^0(R_0) = \frac{\alpha_{n,l}^2 \hbar^2}{2R_0^2 m^*}, \tag{10.81}$$

where m^* is the effective mass of a charged particle; $\alpha_{n,l}^2$ are the roots of the spherical function of the number (n, l)

We introduce the notation $E_{n,l}(R_0) = E_{n,l}^0$, then using equation (10.79) one obtains

$$E_{n,l}(\Theta, \varphi) = E_{n,l}^0 \left(1 - \frac{\delta R^2(\Theta, \phi)}{R_0^2} \right). \tag{10.82}$$

The angle Θ in equation (10.82) can be considered as the angle between the angular momentum vector of the charged particle and the selected direction of the Z-axis of the nanoparticle. In this case, the value of the angle is quantized and satisfies the relation

$$\cos^2(\Theta) = \frac{m^2}{l(l+1)}, \tag{10.83}$$

where l and m are the orbital and magnetic quantum numbers of a charged particle, respectively.

Thus, unlike the unperturbed state, the energy value for a charged particle in a nanoparticle with a perturbed surface turns out to depend on three quantum numbers: n, l and m. This means that in a nanoparticle with a perturbed surface shape, the degeneracy is removed from the magnetic quantum number, which should lead to the splitting of the energy levels.

In the general case, the perturbed axially symmetric surface can be described analytically by the expression [12]

$$R^2(\Theta) = R_0^2[1 + p^2 \cos^\gamma(k\Theta)], \tag{10.84}$$

where $p^2 = \| \delta R^2(\mathbf{r})/R^2 \|$.

Let us consider several particular cases of deformation of the surface of a spherical nanoparticle.

1. Let: $\gamma = 2$, $k = 3$, $p^2 = 0.2$, then the expression (10.84) takes the form

$$R^2(\Theta) = R_0^2[1 + 0.2\cos^2(3\Theta)]. \tag{10.85}$$

The shape of the nanoparticle whose surface is described by Eq. (10.85) is shown in Fig. 16 *a*.

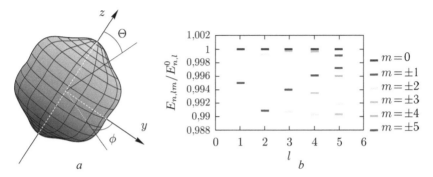

Fig. 16. *a* – form of the deformed surface of a nanoparticle; *b* – diagram of the energy spectrum of electrons [10, 12].

Using (10.82), (10.83) and (10.85), we can obtain an expression for describing the energy of charged particles in a deformed nanoparticle:

$$E_{n,l,m} \approx E_{n,l}^0 \left\{ 1 - 0.2 \left[4 \left(\frac{m^2}{l(l+1)} \right)^{3/2} - 3 \left(\frac{m^2}{l(l+1)} \right)^{1/2} \right]^2 \right\}. \qquad (10.86)$$

Figure 16 *b* shows the distribution of energy levels in a deformed nanocrystalline particle calculated in accordance with (10.86).

2. Suppose that: $\gamma = 3$, $k = 2$, $p^2 = 0.2$, then the expression (10.84) takes the form

$$R^2(\Theta) = R_0^2 [1 + 0.2 \cos^3(2\Theta)]. \qquad (10.87)$$

The shape of the surface of such a nanoparticle is shown in Fig. 17 *a*.

Using the equations (10.82), (10.83) and (10.87), we obtain an expression for calculating the energy of charged particles in a deformed nanoparticle:

$$E_{n,l,m} \approx E_{n,l}^0 \left\{ 1 - 0.2 \left[\frac{2m^2}{l(l+1)} - 1 \right]^3 \right\}. \qquad (10.88)$$

Figure 17 *b* shows the energy distribution diagram for the deformed nanoparticle calculated in accordance with (10.86).

The obtained results show that the deformation of the shape of the nanoparticle leads to the removal of the degeneracy by the magnetic quantum number, which leads to a change in the energy spectrum of the electrons. In turn, these changes lead to the formation of

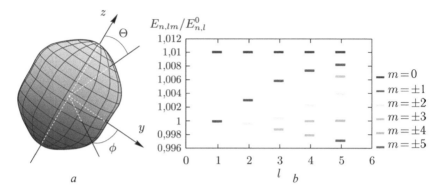

Fig. 17. *a* – shape of the deformed surface of a nanoparticle; *b* – the diagram of the energy spectrum of the electrons [10, 12].

additional allowed mini-energy zones, which may lie below (Fig. 16 *b*) or higher (Fig. 17 *b*) than the corresponding unperturbed energy levels in nanoparticles of spherical shape. This can also lead to a corresponding change in the density of allowed states, which may be greater or less than the baseline state density in the unperturbed nanoparticle, which may affect the optical properties of the nanoparticles.

10.3.2. Two-particle (exciton) states in nanoparticles with irregular shape geometry

Above we considered single-particle states of charge carriers, however, the formation of the optical properties of dielectric and semiconductor nanoparticles depends to a large extent on the states of excitons in nanoparticles. The contribution of exciton states to the optical properties of nanoparticles is most significant when their Bohr radius turns out to be comparable with the size of nanoparticles. Unlike single-particle states, an accurate description of exciton states in nanoparticles is impossible even for nanoparticles with a regular spherical shape. Therefore, approximate approaches are used, which are based on an approximate model of nanoparticles in the form of a deep spherical potential well.

As a rule, the most common approximate geometric shapes for nanoparticles can be considered a sphere, an ellipsoid of rotation and a pyramid.

Since the ellipsoid of rotation is an intermediate form between the two previous ones, an approximate model of a nanoparticle in the form of a deep potential well was used to estimate the effect of

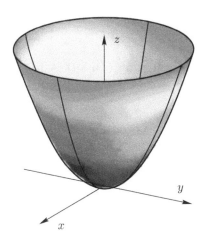

Fig. 18. Graphical representation of a nanoparticle in the form of a paraboloid of revolution.

the shape of nanoparticles on the energy spectrum of excitons, the surface of which is a paraboloid of revolution oriented along the z axis (Fig. 18). The relationship of the parabolic coordinates (ξ,η,ϕ) to the Cartesian coordinates (x, y, z) is given by the transformations:

$$x = \sqrt{\xi\eta}\cos\phi, \; y = \sqrt{\xi\eta}\sin\phi, \; z = \frac{(\xi - \eta)}{2}.$$

Then the equations (10.82)–(10.85) for the wave function of the exciton in a parabolic coordinate system tied to the center of mass of the exciton, $\Psi(\xi,\eta,\phi)$, according to [13], transform to the form

$$\frac{4}{\xi+\eta}\left[\frac{\partial}{\partial\xi}\left(\xi\frac{\partial\Psi}{\partial\xi}\right)+\frac{\partial}{\partial\eta}\left(\eta\frac{\partial\Psi}{\partial\eta}\right)\right]+\frac{1}{\xi\eta}\frac{\partial^2\Psi}{\partial\phi^2}+2\left(E+\frac{2}{\xi+\eta}\right)=0, \qquad (10.89)$$

in the given equation, the values of Planck's constant ($\hbar$), the effective mass of the exciton (μ) and the charge of the electron (e) are taken equal to unity, and the charge interaction is assumed to be only of the Coulomb type, and all coordinates are measured in units of the Bohr radius of the exciton.

We assume that the maximum size of the paraboloid (nanoparticle) in the direction of the z axis is z_0. The maximum section of the paraboloid (nanoparticles at the base) in the (x, y) plane is given by the parameter $\xi = 2z_0 + \eta$. Since the changes are in an infinitely deep potential well, the boundary conditions for the wave function must satisfy the condition:

$$\Psi\left(\xi(z,\eta_0),\eta_0,\phi\right) = 0. \tag{10.90}$$

Taking this into account, the solution of equation (10.89) is found in the form of a set of degenerated hypergeometrical functions $F(\alpha, |m| + 1, \rho)$ [13] and has the form

$$\Psi(\xi,\eta,\phi) = C \frac{1}{n^2 |m|!} \sqrt{\frac{(\alpha+|m|)!(\beta+|m|)!}{\alpha!\beta!}} \left(\frac{\xi\eta}{n^2}\right)^{|m|} \times$$
$$F\left(-\alpha, |m|+1, \frac{\eta}{n}\right) \exp\left[-\left(\frac{\xi}{2n} + \frac{\eta}{2n}\right) + im\phi\right], \tag{10.91}$$

where C is a constant.

In equation (10.91) n is the quantum number which determines the energy level of the exciton

$$E_n = -\frac{\mu e^4}{2\hbar^2 \varepsilon^2 n^2} = -\frac{\mu}{m_e \varepsilon^2 n^2} \; 13.6 \text{ eV}. \tag{10.92}$$

In expression (10.91), there are two sets of numbers α and β for which the wave function satisfies the boundary conditions (10.90) on the lateral surface of the paraboloid of revolution [12].

1. For a state where $\alpha = 0$ and $\beta = 1$, the boundary condition is satisfied if

$$n = n_1 = \frac{|m|+1}{2} \pm \sqrt{\frac{(|m|+1)^2}{4} + \frac{\eta_0}{|m|+1}}. \tag{10.93}$$

2. For a state where $\alpha = 1$ and $\beta = 0$, the boundary condition will be satisfied when

$$n = n_2 = \frac{|m|+1}{2} \pm \sqrt{\frac{(|m|+1)^2}{4} + \frac{2z_0+\eta^2}{|m|+1}}, \tag{10.94}$$

where the values η vary from 0 to η_0.

The value $m = 0$ in (10.93) corresponds to the motion of the exciton in a plane parallel to the $0z$ axis.

Dependence (10.94) reflects the influence of the shape of the surface of the potential well on the energy spectrum of excitons.

As can be seen, the inhomogeneity of the surface of the nanoparticle leads to an increase in the interaction of electrons and

Table 2. Calculated values of the energy of exciton states

m		0	1	2	3	4
E, eV	I	−0.04	−0.05	−0.03	−0.04	−0.02
	II	−0.07	−0.21	−0.58	−1.36	−4.06

holes, which leads to the removal of degeneracy by the magnetic quantum number and the splitting of the energy levels.

Table 2 shows the calculated values of the exciton energy spectrum obtained using expressions (10.92) and (10.93), assuming that the kinetic energy of the excitons is small, for model nanoparticles from Al_2O_3 having $z_0 = 40$ nm, $\eta_0 = 13$, $m_e^* = 0.04\, m_e$ and m. The use of (10.92) and (10.94) gives practically analogous dependences.

The values shown in the table show that the energy spectrum of the excitons in a nanoparticle with a variable shape contains two regions: one – quasi-continuous (row I), lying near the bottom of the conduction band, and the other consists of discrete energy levels (row II) lying in the depth of the forbidden band.

The energy states of excitons with quantum numbers satisfying the expression (10.94) introduce an additional splitting of the energy levels, which also leads to the smearing of the energy levels into quasi-continuous energy bands lying in the forbidden band.

The solution of equation (10.89) for particles moving freely in a parabolic potential well has the form [11]

$$\Psi(\xi,\eta,\phi) = CJ_{m/2}\left(\frac{k}{2}\xi\right)J_{m/2}\left(\frac{k}{2}\eta\right)\exp(m\phi), \qquad (10.95)$$

where $J_{m/2}()$ is the Bessel function, C is a constant, k is the wave number.

As a result, the energy spectrum of the size-quantization levels for charged particles is determined as

$$E_{n,m}^{e,h} = \frac{2\hbar^2}{m_{e,h}}\alpha_{n,m}^2\left[\frac{1}{\eta_0^2}+\frac{1}{(2z_0+\eta)^2}\right], \qquad (10.96)$$

where $\alpha_{n,m}$ is the n-th root of the Bessel function of order $m/2$; $m_{e,h}$ are the masses of an electron or hole; the coordinates ξ and η are measured in nanometers.

As seen from (10.96), the spectrum of the quantum-size levels of the charge energy also turns out to depend on the shape of the nanoparticles.

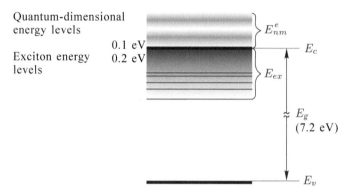

Fig. 19. Diagram of the energy levels of excitons in nanoparticles with a nonuniform geometric shape.

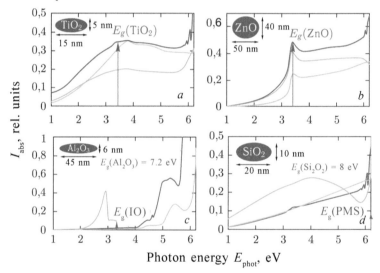

Fig. 20. Absorption spectra of solutions of nanoparticles of wide-gap semiconductors and dielectrics: green curves – nanoparticles in polymethylsiloxane PMS or immersion oil (IO), red – in isopropanol, blue – in water [14–16].

Thus, taking into account the foregoing, one can say that the variation of the shape of nanoparticles from Al_2O_3 leads to the fact that, taking into account the temperature broadening, the energy spectrum of the excitons formed in them consists of a quasi-continuous energy band adjacent to the conduction band of width about 0.1 eV and lying within the forbidden zone at a depth of more than 0.2 eV conductivity (from the bottom of the conduction band) of discrete energy levels (Fig. 19). A similar energy spectrum of excitons will be observed for nanoparticles from other materials. The presence of such a specific energy state for excitons and free

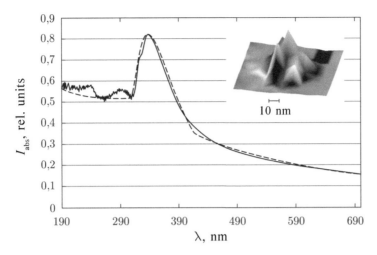

Fig. 21. Absorption spectrum of a solution of Al_2O_3 nanoparticles with an average size of 40 nm (dashed line – calculated values) [17].

electrons should lead to broadening of the luminescence spectrum and erosion of the edge of the absorption band for nanoparticles, which must have a dependence on the shape of the nanoparticles.

Figure 20 shows the experimental spectral dependences of the radiation absorption efficiency obtained for nanoparticles of wide-band semiconductors and dielectrics of various shapes dissolved in various liquids (hereinafter $I_{abs} = \log (I_0 / I)$, where I_0 is the intensity of the radiation incident on the sample, I is the intensity of the transmitted radiation).

The dependences shown by red arrows in Fig. 20 show the values of the energy of the bandgap for bulk materials. The dependences in Fig. 20 *a* and *b* show the asymmetry of the absorption bands of visible and UV radiation for dielectric nanoparticles from Al_2O_3 and ZnO caused by exciton effects. As can be seen, in all cases, the edge of the absorption band is blurred, which depends both on the size and shape of the nanoparticles.

Figure 21 shows the absorption spectrum of a solution of Al_2O_3 nanoparticles with an average size of 40 nm in the immersion oil. The inset shows the photograph of the nanoparticle obtained using an atomic force microscope. Nanoparticles have a complex defect form due to their production in a detonation manner.

The complex form of nanoparticles made it possible to more clearly distinguish the effect of the surface on the energy spectrum of charge carriers in nanoparticles. Two shallow absorption bands appeared in the spectrum of the solution of the nanoparticle array

– in the spectral ranges 220–225 nm and 265–307 nm, as well as a relatively deep absorption band in the range 308–400 nm.

The value of the photon energy corresponding to the maximum absorption wavelength is $\lambda = 337$ nm. The energy of photons corresponding to this wavelength is 3.7 eV, which is much less than the width of the forbidden band for a bulk crystal $E_g = 7$ eV, which is due to the exciton absorption of radiation. The photon energies corresponding to the wavelengths of the other two absorption maxima are 4.1 eV and 5.2 eV. Such features are explained by the significant change in the electronic structure of Al_2O_3 crystals caused by the size effect, which leads to the formation of additional allowed energy bands within the forbidden band. The appearance of these zones can be due to the presence of a significant number of defects near the surface of the crystal structure of the nanoparticle, which is facilitated by their complex shape.

10.4. Influence of the environment on the energy spectrum of excitons in nanoparticles

As noted above nanoparticles (or quantum dots, see section 1.3) are characterized by maximally developed interphase surfaces and possess an excess of energy, compared to bulk materials, as a result of which they are often called artificial atoms (or superatoms). Such terminology can be qualified if one considers the similarity of the discrete spectra of the electronic states of quantum dots and atoms, as well as their chemical activity. When nanoparticles are placed in media with a permittivity ε_1 different from the dielectric constant of ε_2 nanoparticles, effects associated with changes in the energy spectrum are observed [18].

Consider, according to [19], a simple model of an artificial atom (Fig. 22): a neutral spherical nanoparticle of radius a, which contains in its volume of a semiconductor (or dielectric) material with a permittivity ε_2 surrounded by a dielectric matrix with a permittivity ε_1. We assume that the relative permittivity $\varepsilon_2/\varepsilon_1 \gg 1$. In the volume

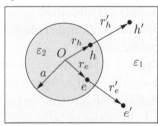

Fig. 22. The image of an exciton in a spherical nanoparticle placed in a medium with an excellent dielectric permittivity [19].

of a nanoparticle a hole h moves with an effective mass of M_h and an electron e with an effective mass of m_e, which are at distances r_h and r_e from the centre of the nanoparticle. Suppose that the valence band in the nanoparticle has a parabolic shape, and an infinitely high potential barrier exists on the spherical interface (nanoparticle-dielectric matrix). Typical sizes in this task are: $a_e = \varepsilon_2 \hbar^2 / m_e e^2$; $a_p = \varepsilon_2 \hbar^2 / m_p e^2$ and: $a_{ex} = \varepsilon_2 \hbar^2 / \mu e^2$ are the Bohr radii of the electron, hole, and exciton, respectively.

It is assumed that the characteristic dimensions are: a, a_e, a_h, $a_{ex} \gg a_0$, where a_0 is the interatomic distance, which allows us to consider the motion of an electron and a hole in a nanoparticle in the approximation of the effective mass.

As can be seen from Fig. 22, the energy of the polarization interaction $U(r_e, r_h, a)$ of the electron and hole with a spherical interface (nanoparticle–dielectric matrix) with relative permittivity $\varepsilon \gg 1$ can be represented as an algebraic sum of the energies of the interaction of a hole and an electron with its own (self-action energy) $V_{hh'}(\mathbf{r}_h, a)$, $V_{ee'}(\mathbf{r}_e, a)$, and 'alien' $V_{he'}(\mathbf{r}_e, \mathbf{r}_h, a)$, $V_{eh'}(\mathbf{r}_e, \mathbf{r}_h, a)$ images respectively:

$$U(\mathbf{r}_e, \mathbf{r}_h, a) = V_{hh'}(\mathbf{r}_h, a) + V_{ee'}(\mathbf{r}_e, a) + V_{he'}(\mathbf{r}_e, \mathbf{r}_h, a) + V_{eh'}(\mathbf{r}_e, \mathbf{r}_h, a), \quad (10.97)$$

where

$$V_{hh'}(\mathbf{r}_h, a) = \frac{e^2}{2\varepsilon_2}\left(\frac{a^2}{a^2 - r_h^2} + \varepsilon\right), \quad (10.98)$$

$$V_{ee'}(\mathbf{r}_e, a) = \frac{e^2}{2\varepsilon_2}\left(\frac{a^2}{a^2 - r_e^2} + \varepsilon\right), \quad (10.99)$$

$$V_{he'}(\mathbf{r}_e, \mathbf{r}_h, a) = V_{eh'}(\mathbf{r}_e, \mathbf{r}_h, a) =$$

$$= -\frac{e^2 \beta}{2\varepsilon_2} \frac{a}{\left[(r_e r_h / a)^2 - 2 r_e r_h \cos\Theta + a^2\right]^{1/2}}, \quad (10.100)$$

where $\mathbf{r}_e$, $\mathbf{r}_h$ are the radius-vectors that determine the position of the electron e and the hole h relative to the center of a nanoparticle with a radius; a; $e' = (a/r_e)e$ and $h' = (a/r_h)h$ are the charges of images located at distances $r_e' = (a_2/r_e)$ and $r_h' = (a_2/r_h)$ from the centre of the nanoparticle, which represent point charges of the image of an electron and a hole in the surrounding medium, respectively.

In expression (10.100), parameter β is defined as

$$\beta = \frac{\varepsilon - 1}{\varepsilon + 1}, \tag{10.101}$$

and Θ is the angle between the vectors $\mathbf{r}_e$ and $\mathbf{r}_h$.

The Hamiltonian of the exciton moving in the nanoparticle has the form [6,8,9]:

$$\hat{H}(\mathbf{r}_e, \mathbf{r}_h, a) = -\frac{h^2}{2m_e}\Delta_e - \frac{h^2}{2m_h}\Delta_h + E_g + U(\mathbf{r}_e, \mathbf{r}_h, a) + \\ + V_{eh}(\mathbf{r}_e, \mathbf{r}_h) + V_e(\mathbf{r}_e, a) + V_h(\mathbf{r}_h, a), \tag{10.102}$$

where the first two terms are operators of the kinetic energy of an electron and a hole; E_g is the width of the forbidden band in a bulk material with a permittivity ε_2, the energy of the Coulomb interaction between an electron and a hole is described by expression

$$V_{eh}(\mathbf{r}_e, \mathbf{r}_h) = -\frac{e^2}{\varepsilon_2 |\mathbf{r}_e - \mathbf{r}_h|}. \tag{10.103}$$

In equation (10.102) the potentials

$$V_e(\mathbf{r}_e, a) = V_h(\mathbf{r}_h, a) = \begin{cases} 0, & r_e, r_h \le a, \\ \infty, & r_e, r_h > a \end{cases} \tag{10.104}$$

describe the motion of an electron and a hole in the volume of a nanoparticle using the model of an infinitely deep potential well.

When the condition

$$a_h \le a \le a_e \approx a_{ex} \tag{10.105}$$

is fulfilled, it is possible to use the adiabatic approximation (in which the effective mass of the hole considerably exceeds the effective electron mass $m_e/m_h \ll 1$). In view of the weak mobility of the hole, the electron kinetic energy in the nanoparticle will be the largest value in (10.102):

$$T_{n_e, l_e=0}^e(a) = \frac{\pi^2 n_e^2}{(a/a_{ex})^2}. \tag{10.106}$$

Using the first order of perturbation theory, it is possible to obtain the binding energy of an exciton of radius a in the ground state

($n_e = 1$, $l_e = 0$; $n_h = 0$, $l_h = 0$), where n_e, l_e, n_h and l_h are the principal and orbital quantum numbers of the electron and holes:

$$E_{ex}^{1,0;0,0}(a,\varepsilon) = \bar{V}_{eh}^{1,0;0,0}(a) + \left[\bar{V}_{eh'}^{1,0;0,0}(a,\varepsilon) + \bar{V}_{he'}^{1,0;0,0}(a,\varepsilon) \right]. \qquad (10.107)$$

The average values of the energy of the Coulomb interaction of an electron with a hole $\bar{V}_{eh}^{1,0;0,0}(a)$, and also the electron and hole interaction energies with 'foreign' images $\bar{V}_{eh'}^{1,0;0,0}(a,\varepsilon)$ and $\bar{V}_{he'}^{1,0;0,0}(a,\varepsilon)$, obtained by averaging the energies from (10.100) and (10.104) for a potential well of infinite depth take the form

$$\bar{V}_{eh}^{1,0;0,0}(a) = -\left[\frac{2}{S}\left(\ln(2\pi) + j - \text{Ci}(2\pi) - \frac{3}{2}\omega(S, n_e = 1) \right) \right] R_{y_{ex}}, \qquad (10.108)$$

$$\omega(S, n_e = 1) = \frac{2(1+(2/3)\pi^2}{S^{3/2}}\left(\frac{\mu}{m_h} \right)^{1/2}, \qquad (10.109)$$

$$\bar{V}_{eh'}^{1,0;0,0}(a,\varepsilon) + \bar{V}_{he'}^{1,0;0,0}(a,\varepsilon) = -\frac{2\beta}{S}R_{y_{ex}}, \qquad (10.110)$$

where the binding energy of the electron is

$$E_{ex}^{0} = R_{y_{ex}} = \frac{(\mu/m_0)}{\varepsilon_2^2}R_{y_0}. \qquad (10.111)$$

In an unbounded bulk material with a permittivity ε_2: $R_{y_0} = 13.61$ eV – Rydberg; $S = a/a_{ex}$; Ci() – the integral cosine; $j = 0.577$ is the Euler constant.

According to (10.107)–(10.111), the effect of increasing the binding energy of the exciton $E_{ex}^{1,0;0,0}(a, \varepsilon)$ in a nanoparticle placed in a dielectric medium is determined by two factors:

1. With the renormalization of the energy of the Coulomb interaction of the electron with the hole $\bar{V}_{eh}^{1,0;0,0}(a)$, associated with a purely spatial limitation of the quantization region in the nanoparticle, since starting from the radius $a \sim 30$ nm, the binding energy of the exciton ground state is almost two orders of magnitude higher than the binding energy of the exciton in bulk crystalline media;

2. With a significant increase in the energy of the Coulomb interaction of an electron and a hole, due to the inclusion of the mechanism of interaction of an electron and a hole with 'foreign'

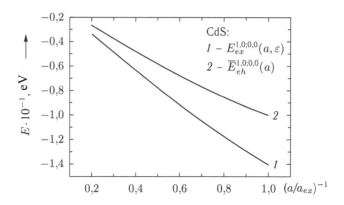

Fig. 23. Dependence of the binding energy of the exciton (1) and the energy of the Coulomb interaction (2) in a CdS nanoparticle on the reduced radius $(a/a_{ex})^{-1}$ of a nanoparticle placed in a borosilicate glass matrix $(a_{ex} = 2.5$ nm) [19].

images of $\overline{V}_{eh'}^{1,0;0,0}(a,\varepsilon) + \overline{V}_{he'}^{1,0;0,0}(a,\varepsilon)$, a nanoparticle–dielectric matrix arising on a spherical interface.

In the interaction between an electron and a hole, which move in the bulk of a nanoparticle with a radius a satisfying the relation (10.105), comparable to the Bohr radius of these quasiparticles, a significant contribution is made by the polarization interaction of these particles, which is described by the interaction of an electron and a hole with 'foreign' images. In this case, the polarization interaction (10.100) depends not only on the permittivity of the material of the nanoparticle ε_2, but also on the permittivity of the matrix ε_1 into which this nanoparticle is immersed. This dependence is related to the penetration of the electric field created by the particles beyond the nanoparticle surface into the matrix region. If the dielectric permittivity of the matrix is much less than the dielectric constant of the nanoparticle, the interaction between the electron and the hole is much larger than in an unbounded medium with a dielectric permittivity ε_2. This effect was called 'dielectric amplification' [10]. The last circumstance is due to the fact that in this case an appreciable role is played by the electric field created by quasiparticles in the matrix.

Figure 23 shows the calculated dependences of the binding energy of the exciton and the energy of the Coulomb interaction of an electron and a hole in a CdS nanoparticle grown in a matrix of borosilicate glass. As can be seen, the highest binding energy of the exciton $E_{ex}^{1,0;0,0}(a,\varepsilon) \approx 1.4 \cdot 10^{-1}$ is observed in a nanoparticle with radius $a = a_{ex} = 2.5$ nm. The binding energy of an exciton in a bulk semiconductor of CdS is $E_{ex}^0 = 31 \cdot 10^{-3}$ eV. Thus, the binding energy of an exciton in a nanoparticle placed in a dielectric matrix is much larger than the energy of excitons in a massive single crystal.

The main contribution to the binding energy of the exciton is the average value of the energy of the Coulomb interaction of the $\overline{V}_{eh}^{1,0;0,0}(a)$ between the electron and the hole. The average value of the interaction energy of an electron and a hole with 'foreign' images $\overline{V}_{eh'}^{1,0;0,0}(a,\varepsilon) + \overline{V}_{he'}^{1,0;0,0}(a,\varepsilon)$ gives a smaller but significant contribution to the binding energy of the exciton (see Fig. 23).

In [20, 21] a model of a quasi-atomic (QA) nanosized heterostructure consisting of a spherical core of radius a is described with dielectric permittivity ε_2, in the volume of which there is a semiconductor or dielectric material, selectively doped with donors in an unrestricted semiconductor (or dielectric) matrix with permittivity ε_1 (with a band gap less than the band gap of the nanoparticle material). Electrons of donors flow into the matrix, while in the nucleus there is a positive charge, determined by the number of donors (in this case, heavy holes, whose effective mass is much larger than the effective mass of electrons, remain in the bulk of the nanoparticle). In such a nanostructure, the lowest electronic level is located in the matrix, and the lowest hole level is in the volume of the nanoparticle. A large shift of the valence band (of the order of 700 meV) causes the localization of holes in the volume of the nanoparticle. A large shift in the conduction band (about 400 meV) is a potential barrier for electrons (electrons move in the matrix and do not penetrate the nanoparticle volume) [22].

Since the dielectric constant ε_2 of the nanoparticle material that contains a semiconductor (or a dielectric, in particular aluminum oxide) in its volume greatly exceeds the permittivity of the ambient matrix, the energy of the polarization interaction of the electron with the interface (nanoparticle–matrix) causes the localization of the electron in a polarization well near the outer surface of a nanoparticle [17, 23].

This causes the probability of the electron flowing from the volume of the nanoparticle into the matrix, and the localization of the electron in the polarization well near the outer surface of the nanoparticle, with the hole continuing to move in the volume of the nanoparticle [24].

Since there is an infinitely high potential barrier on the spherical interface (nanoparticle matrix), the hole h can not get out of volume, but the electron e can not penetrate into the nanoparticle [20, 21]. This can lead to a significant increase in the lifetime of the exciton states.

Of special interest is the localization of charge carriers in nanostructures on macroinhomogeneities, which in a number of cases can be considered as dielectric nanoparticles immersed in a medium with a different dielectric constant. The formation of such macroinhomogeneities, due in particular to doping, composition fluctuations, growing conditions in multicomponent crystals, plays a special role in nanophysics and nanochemistry, where the formation of artificial atoms (as well as macrocluster atoms) is accompanied by the formation of acceptor levels, the nature of which is in most cases unknown, and a sharp drop in the mobility of charge carriers. The localization of charge carriers and large-radius excitons in isoelectronic solid solutions A_2B_6 has been experimentally studied in [25]. In particular, it was established in [25] that in solid solutions $ZnSe_{1-x} Se_x$, $CdS_{1-x} Se_x$ on macroscopic clusters of atoms (which can be regarded as artificial atoms) forming a narrow-band component of the solution, bound states of the hole were formed.

From the models of artificial atoms described above, it follows that they have the ability to attach to their electronic orbitals N electrons (where N can vary from one to several tens), means that such 'superatoms' will be N-valent. This new effect causes high physical and chemical activity and opens up new possibilities for artificial atoms associated with the appearance of new optical properties, the appearance of strong oxidizing abilities, the possibility of a significant increase in the intensity of photochemical reactions during catalysis and absorption, and with their ability to form set of new chemical compounds with unique properties (in particular quasimolecules and quasicrystals (quasi-one-dimensional and quasi-two-dimensional)).

A detailed study of the influence of the matrix on the energy spectrum of excitons in nanoparticles was carried out in [26–30]. To this end, nanoparticles were used from wide-band semiconductors and dielectrics: Al_2O_3, TiO_2, ZnO, SiO_2, having different shapes and sizes. As the dielectric matrix into which the nanoparticles were placed, polar: water, isopropanol, and nonpolar: polymethylsiloxane and immersion oil, liquids were used. The concentration of nanoparticles in liquid matrices was chosen not more than 0.1% of the mass of the solution, which made it possible to exclude the process of collective

Table 3. Parameters of liquids and bulk materials of oxides

Material	Static dielectric permittivity, ε_0 (SDP)	Bandgap width E_{gap} (eV)	Viscosity, (cSt)	Temperature coefficient of refractive index, dn/dT, (K^{-1})
Matrix				
Water* (H_2O)	81	6.5	101	$<10^{-6}$
Isopropanol* (C_3H_8O)	24	5.9	~10	~10^{-4}
Polymethylsiloxane (PMS)	2.5	6.2	350	$<10^{-6}$
Immersion oil (IO)	2.5	3.3	300	$<4 \cdot 10^{-4}$
Polar fluids – *				
Nanoparticles				
Al_2O_3	10	7.2–10		
TiO_2	86	3.3		
ZnO	8.8	3.4		–
SiO_2	4.5	8–11		

interaction of nanoparticles. The characteristics of the materials used are shown in Table 3.

Figure 24 shows the experimentally measured absorption spectra of visible radiation by Al_2O_3 nanoparticles. The measured absorption spectra of an array of non-interacting Al_2O_3 nanoparticles with a ruler of medium sizes from 10 to 50 nm and shapes from spheroid to scaly form the following:

- the process of absorption of radiation by an array of spheroidal nanoparticles 50 nm in diameter (Fig. 24 *a*) is characterized only by a monotonic increase in absorption with a decrease in the wavelength of optical radiation from 700 to 200 nm, regardless of the type of matrix;
- The absorption of radiation by an array of ellipsoidal nanoparticles with an average main diameter of 13 nm (Fig. 24 *b*) is also characterized by a monotonic increase in absorption with decreasing radiation wavelength. In this case, an absorption peak in the UV region of 205–210 nm ($E_{phot} \sim 6$ eV) is observed for nanoparticles surrounded by strongly polar molecules (water, isopropanol);
- The spectrum of radiation absorption by an array of flaky

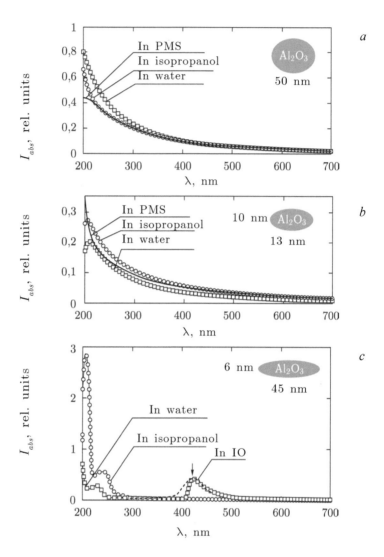

Fig. 24. Absorption spectra of visible radiation by an array of non-interacting Al_2O_3 nanoparticles of spherical (*a*), ellipsoidal (*b*) and scaly (*c*) shapes surrounded by dielectric matrices (water, isopropanol, polymethylsiloxane (PMS) and immersion oil (IO)).

nanoparticles (Fig. 24 *c*) surrounded by polar media (water, isopropanol) in the UV region contains several absorption bands in the vicinity of 210 and 250 nm, while for an array of these nanoparticles surrounded by a nonpolar medium (IO), an absorption band is observed in the visible region of the spectrum from 410 to 520 nm.

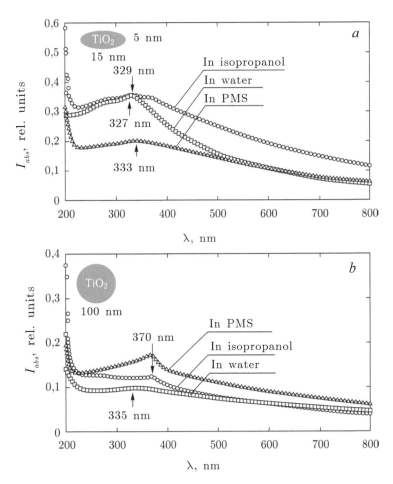

Fig. 25. Absorption spectra of visible radiation by an array of noninteracting TiO_2 nanoparticles of ellipsoidal (*a*) and spherical (*b*) forms surrounded by dielectric matrices (water, isopropanol and PMS (polymethylsiloxane)).

Figure 25 shows the absorption spectra of visible radiation by an array of non-interacting nanoparticles made of TiO_2 of ellipsoidal and spherical forms surrounded by dielectric matrices (water, isopropanol, and PMS). As can be seen from the presented dependences, in the spectra for arrays of wide-band semiconductor TiO_2 nanoparticles, both ellipsoidal and spherical in the 320–370 nm region, the fundamental absorption edge is observed, blurred by exciton states. For nanoparticles of smaller size (Fig. 25 *a*), this edge lies in the region 3.7–7 eV, the position of which varies within 0.08 eV, depending on the type of matrix. For larger TiO_2 particles (Fig. 25 *b*), the fundamental absorption edge is more pronounced, and its

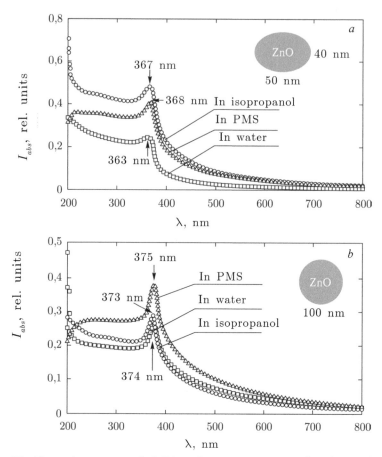

Fig. 26. Absorption spectra of visible radiation by an array of noninteracting ZnO nanoobjects of ellipsoidal (*a*) and spherical (*b*) forms surrounded by dielectric matrices (water, isopropanol and PMS).

position practically coincides with the width of the forbidden band of the bulk oxide and is ~3.35 eV, with a weaker dependence of the shift of the fundamental absorption edge upon a change in the type of the matrix.

Figure 26 shows the absorption spectra of visible radiation by an array of non-interacting ZnO nanoparticles of ellipsoidal and spherical forms surrounded by dielectric matrices (water, isopropanol, and PMS).

As can be seen from the presented dependences, the behaviour of the curves is similar to the character of the dependences in Fig. 25. Nevertheless, ZnO nanoparticles show a more pronounced edge

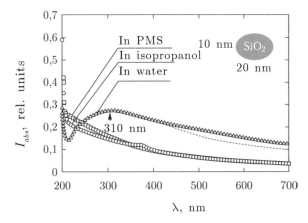

Fig. 27. Absorption spectra of visible radiation by an array of noninteracting nanoparticles of SiO_2 of scaly form surrounded by dielectric matrices (water, isopropanol and PMS).

of fundamental absorption is observed, which is also diffused by exciton transitions.

In this case, the position of the fundamental absorption edge for nanoparticles practically coincides with the position of the fundamental absorption edge for bulk ZnO crystals. There is also a dependence of the position of the fundamental absorption edge on the size of the ZnO nanoparticles: for smaller particles (Fig. 26 *a*), it is located in the vicinity of 3.4 eV, and for larger ones (Fig. 26 *b*) in the region of 3.31 eV. The effect of the same type of matrix on the nature of absorption is practically not detected.

Figure 27 shows the spectral dependences of the absorption of visible radiation by an array of noninteracting nanoparticles of flake-shaped SiO_2 surrounded by dielectric matrices: water, isopropanol, and PMS. Apparently, for nanoparticles from a material with a large band gap ($E_{gap} > 8$ eV) in a polar liquid medium (water, isopropanol), a monotonic dependence of absorption with a decrease in the wavelength of the radiation is observed. Replacing the matrix with a nonpolar one (PMS) leads to the appearance of a band of nonlinear absorption of radiation in the region from 230 to 500 nm.

Thus, the presented experimental data demonstrate the influence of the environment on the energy spectrum of charge carriers in nanoparticles. For nanoparticles suspended in nonpolar matrices, when the static permittivity of the matrix is less than that of the nanoparticle material, the jump in the permittivity at the nanoparticle-matrix interface leads to the concentration of the positively charged component of the polarized matrix molecules near the outer surface of the nanoparticle, and to an increase in the concentration electrons nanoparticles near its surface. In this case, because of the polarization interaction of the electron and hole with the interface, a high density of stable exciton states is formed in the defect surface layer with high binding energy of excitons, which leads to the formation of a wide band of their energies inside the forbidden band.

For nanoparticles suspended in polar matrices, the negative charges of the matrix molecules are attracted to the outer surface of the nanoparticles. This leads to an increase in the potential barrier for near-surface electrons excited to exciton states, which prevents the formation of exciton bands.

In the case when the shape of the nanoparticles becomes different from the spherical shape, the complex energy structure of the exciton energy levels arising in them leads to a noticeable erosion of the edge of the fundamental absorption band, even if these nanoparticles are enclosed in nonpolar matrices.

10.5. Low-energy optical nonlinearity of liquid nanocomposite media based on nanoparticles

The process of interaction of radiation with matter is determined by the behaviour of the charge carriers in it and, therefore, directly depends on their energy spectrum. The study of the passage of low-intensity laser radiation through weakly concentrated solutions of

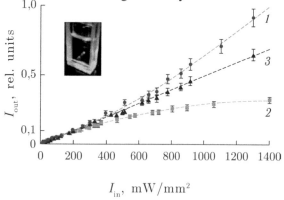

Fig. 28. Experimental dependences of the intensity of the radiation (I_{out}) transmitted through a cuvette with a solution of Al_2O_3 nanoparticles in vacuum oil on the radiation at the input (I_{in}): $1 - \lambda = 633$ nm; $2 - \lambda = 532$ nm; $3 -$ passage of radiation through pure vacuum oil at $\lambda = 633$ nm. The sidebar shows a photo of a cell with a solution [17].

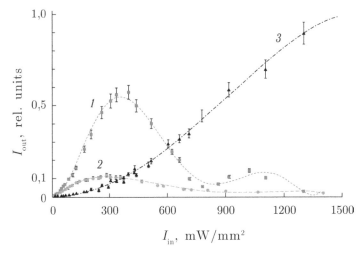

Fig. 29. The experimental dependences of the radiation intensity transmitted through a cuvette with a solution of Al_2O_3 nanoparticles in immersion oil (I_{out}) on the radiation at the input (I_{in}). 1 – λ = 633 nm; 2 – λ = 532 nm; 3 – passage of radiation through pure immersion oil with λ = 633 nm [17].

nanoparticles of wide-band semiconductors and dielectrics made it possible to detect low-threshold optical nonlinearity in these media.

Figure 28 shows the experimental dependences of radiation passing through a liquid nanocomposite, which is a solution of Al_2O_3 nanoparticles in vacuum oil, on the intensity of the incident laser radiation. The AFM photograph of nanoparticles is shown in Fig. 21. The concentration of nanoparticles was 0.3% of the total mass of the nanocomposite. When measuring the intensity of the transmitted radiation from the light beam, only its central part was cut out at the exit from the cuvette by a point diaphragm.

Figure 29 shows similar dependences obtained for a liquid nanocomposite prepared on the basis of Al_2O_3 nanoparticles and immersion oil.

As can be seen (Fig. 28), the dependences *1* and *2* demonstrate the phenomenon of self-action of low-energy laser beams, manifested in their self-focusing or self-defocusing, the magnitudes of which depend on the radiation intensity at the entrance to the cell. Since the pure matrix material does not exhibit this property (Fig. 28, curve *3*), this indicates that the nonlinear optical properties of the nanocomposite are determined by the characteristics of the nanoparticles and their interaction with the matrix.

The dependence of the nonlinear properties of the nanocomposite on the matrix material can be easily observed by comparing the

Table 4. Results of investigation of low-energy nonlinear response refractive index and absorption coefficient of liquid nanocomposites in the visible range of the radiation spectrum

Matrix		Water*	Isopropanol*	PMS	IO
TiO$_2$ — 5 nm, 15 nm	$E_g = 3.8$ eV	–	–	–	Not studied
TiO$_2$ — 100 nm	– " – " – "	–	–	–	
ZnO — 50 nm, 15 nm	$E_g = 3.4$ eV	–	–	–	
ZnO — 100 nm	– " – " – "	–	–	–	
SiO$_2$ — 5 nm, 15 nm	$E_g = 8$ eV	–	–	+	
Al$_2$O$_3$ — 10 nm, 13 nm	$E_g = 7.2$ eV	–	–	–	
Al$_2$O$_3$ — 50 nm	– " – " – "	–	–	–	
Al$_2$O$_3$ — 6 nm, 45 nm		–	–	Not studied	+
Al$_2$O$_3$ — 45 nm, 45 nm					+

+ – indicates the manifestation of the nonlinear optical properties of nanocomposites; * – notation of polar liquid matrices

curves in Figs. 28 and 29. As can be seen, the replacement of the matrix material not only leads to a change in the behaviour of the curves, but also changes the threshold for the appearance of nonlinear optical properties. For the dependences in Fig. 29 the nonlinearity was essentially low-threshold and was observed at radiation intensities ~ 40 mW/mm². Moreover, the matrix material also possessed nonlinear optical properties, which are explained by the temperature dependence of the refractive index of immersion oil [17].

Further studies of liquid nanocomposite materials based on 0.1% solutions in polar and nonpolar liquids of nanoparticles from wide-band semiconductors and dielectrics: Al_2O_3, TiO_2, ZnO, SiO_2, having different shapes and sizes, additionally showed that the nonlinearity of liquid nanocomposite materials depends on complex of parameters, which include material, geometric dimensions and shape of nanoparticles, as well as matrix material. The results of experimental studies are summarized in Table 4. Investigations of nonlinear changes in the refractive index and the absorption coefficient of nanocomposite media were carried out using the Z-scan method [31] using low-intensity sources ($I < 0.5$ kW/cm²) and highly stable continuous laser radiation with wavelengths of 442 nm (E_{ph} = 2.81 eV) and 532 nm (E_{ph} = 2.33 eV).

As can be seen from Table 4, in the visible range of the emission spectrum, nonlinear optical properties are not observed in nanocomposites based on polar liquid-phase matrices. In nanocomposites based on nonpolar liquids, the nonlinearity of optical properties is observed in those cases when the shape of nanoparticles differs significantly from spherical.

Figure 30 shows the dependence of the nonlinear changes in the absorption coefficient and the refractive index of nanocomposite media, based on the SiO_2 and Al_2O_3 nanoparticles, measured in the Z-scan, dissolved in nonpolar matrices: immersion oil and PMS, respectively.

As follows from the presented dependences, nonlinear processes in the above-mentioned nanocomposite media are observed in cases when the photon energy is much less than the forbidden band widths of the nanoparticle and matrix matrices, and the radiation power is so low that the flow of multiphoton absorption processes is impossible. In this case, the amplitude of changes in the absorption coefficient and the refractive index rapidly increases with increasing intensity and then decreases to zero. This indicates the presence of a developed

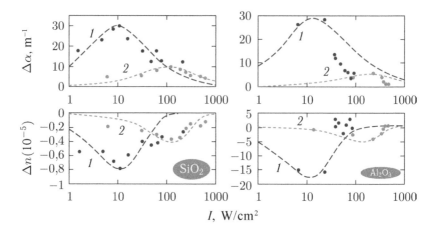

Fig. 30. Non-linear dependences of changes in absorption coefficients $\Delta\alpha$ and refractive indices Δn of liquid nanocomposite media based on SiO_2 nanoparticles in PMC and flaky Al_2O_3 nanoparticles in IO on the intensity of visible laser radiation (442 nm – *1* and 532 nm – *2*). Points – experimental data; dashed lines – calculated dependences.

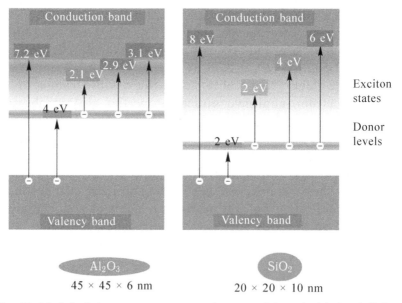

Fig. 31. Model of the energy spectrum of nanoparticles of wide-band dielectrics (Al_2O_3 and SiO_2) in nonpolar matrices,

system of energy levels and energy subbands within the forbidden band of nanoparticle materials in dielectric media.

Based on the results of an experimental study of spectral absorption by liquid nanocomposite media (Figs. 24 and 27) and using the results of the sections 10.3.1–10.3.3, a model of the energy spectrum for dielectric nanoparticles (Al_2O_3 and SiO_2) in nonpolar matrices, shown in Fig. 31.

The model was based on the following facts that were found the reflection in Fig. 31:

- The band gap of nanoparticles contains bands of allowed energy states for electrons, including: defective energy levels and exciton energy levels;
- The edge of the fundamental absorption band of arrays of dielectric nanoparticles is blurred by size-quantization levels and exciton states with different binding energies;
- The width of the exciton states depends on the size and shape of the nanoparticle, which limit the spatial motion of the exciton and increase its binding energy;
- The matrix surrounding the nanoparticles affects their energy spectrum: if the dielectric constant of the nanoparticle material is larger than that of the matrix material, the dielectric enhancement of the interaction of electrons with the surface occurs, which increases in the case of a nonpolar medium and decreases for the polar one and thereby increases/decreases the binding energy of excitons and the density of their states in the surface layer and,

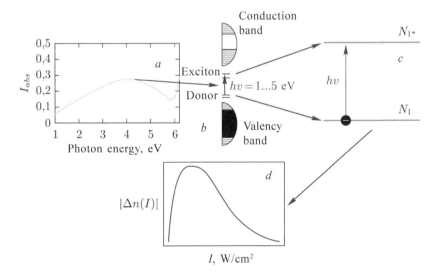

Fig. 32. Scheme of the appearance of the phenomenon of low-threshold nonlinearity in nanocomposites based on wide-band dielectric nanoparticles in nonpolar matrices.

as a consequence, leads to broadening of the exciton band inside the band gap;
- Defective states of surface atoms of dielectric nanoparticles are donor due to the strong binding energy of outer shell electrons with the nucleus [32–35].

Using the remarks made, the scheme of the energy band of dielectric nanoparticles in nonpolar matrices, the appearance of a low-threshold and low-energy nonlinearity can be explained using the diagram shown in Fig. 32. In this case, the nonlinearity is due to electron transitions from the donor energy levels to the quasicontinuous system of exciton levels (Fig. 32 *b*) due to the absorption of low-energy quanta of visible radiation (Fig. 32 *c*). Initially, the transfer of electrons to bound exciton states reduces the number of freely moving electrons, and thus increases the refractive index of the medium with increasing radiation intensity (Fig. 32 *d*). Due to the fact that the density of electronic states at the donor levels is limited, when a certain level of radiation intensity is reached, saturation of the transitions occurs and a further increase in intensity leads to the breakdown of the bound exciton states, which leads to a decrease in the refractive index of the nanoparticles.

The components of the polarization vector of a nanoparticle, located in the field of an electromagnetic light wave with electric field strength **E** can in a general case be represented in the form [35]

$$\mathbf{P} = \sum_{i=1}^{3} \alpha_i E_i = \chi^{(1)} \mathbf{E}, \qquad (10.112)$$

where $\alpha = \{\alpha_i\}$ is the polarizability of the nanoparticle.

If we use the general model of a nanoparticle in a nonpolar dielectric medium (Fig. 31), then low-intensity optical radiation in such an energy system should provide three main energy transitions for electrons from the initial state $|n\rangle$ to the state $|g\rangle$: on quasicontinuous states in the conduction band, on discrete quantum-size states in the conduction band, and on exciton states localized in the forbidden band.

As is known, the i-th component of the polarizability of a nanoparticle when it interacts with electromagnetic radiation with a frequency ω is defined as

$$\alpha_i = \frac{1}{V} \sum_{n<g} \sum_g \frac{(D_{ng}^i)^2}{\hbar(\omega - \omega_{ng} + i\Gamma_{ng})} \rho_{ng}(\hbar\omega), \qquad (10.113)$$

where D_{ng}^i is the i-th component of the matrix element of the dipole moment for transition between the states $|n\rangle$ and $|g\rangle$, described by the expression

$$D_{ng}^i = \int_g \psi(er_i)\psi_n d\mathbf{r}, \qquad (10.114)$$

$\rho_{ng}(\hbar\omega)$ is the joint density of states, depending on the difference in the density of states in the initial n and the final g energy states:

$$\rho_{ng} = \rho_{nn} - \rho_{gg} - \rho_{ng}^0;$$

ω_{ng} is the frequency of the transition between the energy levels of $|n\rangle$ and $|g\rangle$; Γ_{ng} is the half-width of the transition line; ρ_{nn} and ρ_{gg} are the densities of states of charge carriers at the corresponding energy levels; ρ_{ng}^0 is the equilibrium (thermal) joint density of states, and

$$|\mathbf{D}_{ng}|^2 = \left(\sum_{i=1}^{3}(D_{ng}^i)\right)^2. \qquad (10.115)$$

The joint density of states of charge carriers arising during the interaction of electromagnetic radiation with a nanoparticle is a function of the intensity of the irradiating radiation (I) [36]:

$$\rho_{ng}(\omega, I) = \left(1 - \sum_n \sum_g \frac{I/I_s}{(\omega - \omega_{ng})^2 + \Gamma_{ng}^2(1 + I/I_s)}\Gamma_{ng}^2\right)\rho_{ng}^0, \qquad (10.116)$$

where I_s is the intensity of saturation at which half of the charge carriers pass from the state $|n\rangle$ to the state $|g\rangle$.

Then the refractive index of the nanoparticle can be defined as

$$n(\omega, I) \approx n_0 + \frac{2\pi}{n_0}\text{Re}(\chi^{(1)}) =$$

$$= n_0 + \sum_n \sum_g \rho_{ng}(\omega, I)\left[\frac{\omega - \omega_{ng}}{(\omega - \omega_{ng})^2 + \Gamma_{ng}^2}\right]A_{ng}, \qquad (10.117)$$

where n_0 is the refractive index of a nanoparticle in the absence of radiation,

$$A_{ng}(Q_1, Q_2) = \frac{2\pi N}{\hbar n_0} \left[\frac{1}{3} |D_{ng}|^2 + Q_1 \left(|D_{ng}^1|^2 - |D_{ng}^3|^2 \right) + \right.$$
$$\left. Q_2 \left(|D_{ng}^2|^2 - |D_{ng}^3|^2 \right) \right],$$

(10.118)

where N is the number of nanoparticles, $Q_1 = (\cos^2 \theta_1 - 1/3)$; $Q_2 = (\cos^2 \theta_2 - 1/3)$ are the orientation parameters, and θ_1 and θ_2 are the angles that determine the orientation of the nanoparticle in the field of the light wave.

Taking into account (10.116), the expression for the refractive index (10.117) will have the form [10]:

$$n(\omega, I) = n_0 + \sum_n \sum_g \left[1 - \frac{I / I_s}{(\omega - \omega_{ng})^2 + \Gamma_{ng}^2 (1 + I / I_s)} \Gamma_{ng}^2 \right] \times$$
$$\left[\frac{\omega - \omega_{ng}}{(\omega - \omega_{ng})^2 + \Gamma_{ng}^2} \right] A_{ng}(Q_1, Q_2) \rho_{ng}^0.$$

(10.119)

We shall assume that light radiation can initiate transitions of charge carriers in a nanoparticle from the state $|n\rangle$ in quasicontinuous exciton states with an energy band $\hbar\Delta\omega_1$, which lie in the forbidden band below the bottom of the conduction band, and into quasicontinuous quantum-size states with an energy band of $\hbar\Delta\omega_2$ lying above the bottom of the conduction band. This allows us to move from discrete summation over states in (10.119) to integration, which gives the following expression for the refractive index of a nanoparticle [10]:

$$n(\omega, I) = n_0 + \frac{\hbar}{2} \sum_n A_{ng}(Q_1, Q_2) \rho_{ng}^0 \times$$
$$\left\{ g_1 \ln \frac{\left[\omega - (\omega_n - \Delta\omega_1) \right]^2 + \Gamma_n^2 (1 + I / I_s)}{(\omega - \omega_n)^2 + \Gamma_n^2 (1 + I / I_s)} + \right.$$

(10.120)

$$\left. g_2 \ln \frac{\left[\omega - \omega_n \right]^2 + \Gamma_n^2 (1 + I / I_s)}{(\omega - (\omega_n + \Delta\omega_2))^2 + \Gamma_n^2 (1 + I / I_s)} \right\},$$

where g_1 and g_2 are the reciprocal of the density of states for the selected transitions; $\Gamma_{ng} = \Gamma_{n}$.

As follows from (10.120), for nanoparticles in a field of light radiation, for low intensities $I < I_s$ the refractive index depends on the intensity and this dependence vanishes when $I \gg I_s$. The dependence of the refractive index on the radiation frequency is also observed and at $\omega \gg \omega_n$ also disappears. The peculiarities of such behaviour of nanoparticles are confirmed by the dependence shown in Fig. 30 for the refractive index and the absorption coefficient obtained for nanoparticles of Al_2O_3 and SiO_2 in nonpolar matrices. In Fig. 30 for comparison, the calculated dependencies shown with the dotted line are made using (10.120).

The low-threshold nonlinearity observed in liquid-phase nanocomposites can be used to control radiation in photonic devices.

One possible approach to the creation of such devices may be the process of collinear interaction of light rays with different wavelengths in a nanocomposite (the diagram is shown in the inset in Fig. 33), as capable of providing the greatest interaction length. This method of control allows you to create photonic devices in which light controls light. Figure 33 shows the experimental dependences of the intensity of the near-axis part of the laser beam with a wavelength of $\lambda = 532$ nm measured at the exit from the cuvette with a liquid-phase nanocomposite from its intensity at the input. The parameter in this series of dependences is the input power of a collinear beam

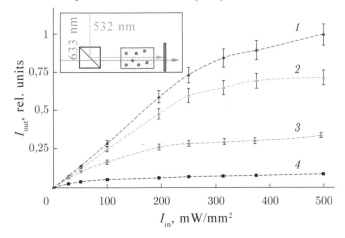

Fig. 33. Normalized experimental dependences of the intensity of the near-axis part of a laser beam with a wavelength $\lambda = 532$ nm, measured at the exit from a cuvette with a liquid-phase nanocomposite, on its input intensity, obtained for different radiation powers of a laser beam propagating with a wavelength collinear with it, $\lambda = 633$ nm: *1* – 0 mW; *2* – 500 mW; *3* – 1000 mW; *4* – 1500 mW. The inset is a diagram of the interaction of laser beams [37].

interacting with it with a wavelength of λ = 633 nm. As a liquid phase nanocomposite, a 0.3% solution of Al_2O_3 nanoparticles in immersion oil was used.

Since the wavelengths of the radiation are different, the interaction of the rays is not interference, but has a quantum nature. The mechanism of the process of interaction of rays can be explained as follows. In the collinear propagation of light rays with different wavelengths, the nonlinear change in the effective refractive index of the nanocomposite is determined by the intensity of the radiation with wavelengths λ = 633 nm and λ = 532 nm. So, for an input beam with λ = 532 nm, controlled by a light beam with λ = 633 nm, the magnitude of the change in the refractive index Δn due to a decrease in the population of the subzone of allowed levels caused by the absorption of green radiation with λ = 532 nm, becomes negative, which leads to a decrease in the intensity in the central part of the transmitted green beam due to its defocusing. And the higher the intensity of the control beam with λ = 633 nm, the lower is the intensity of the input radiation at which this process begins to be observed (Fig. 33).

References

1. Scholes G.D., Rumbler G.,Nature materials.2006. V. 5. P. 689–699.
2. Shik A.Ya., et al., Physics of low-dimensional systems, ed. A.Ya Shik. Saint Petersburg, Nauka, 2001.
3. Borisenko S.I., Physics of Semiconductor Nanostructures (textbook). Publishing house of Tomsk Polytechnic University, 2010.
4. Pedersen K. Quantum size effects in nanostructures. Aalborg University. 2006.
5. Seysyan R.P., Okno v mikromir, 2001. V. 2. P. 2–13.
6. Gurtov V.A., Osaulenko R.N., Solid State Physics for Engineers (textbook), ed. L.A. Aleshin. Tekhnosfera, 2007.
7. Gross E., Usp. Fiz. Nauk. 1962. V. 76. P. 433–447.
8. Hanemann T., Szabu D.V. Materials. 2010. V. 3. P. 3468–3517.
9. Suzdalev I.P., Suzdalev P.I., Nanoclusters and nanoclusters systems. Organization, interaction, properties. 2001. V. 70, No. 3. P. 203–236.
10. Kulchin Yu.N., et al., Moscow, Fizmatlit, 2011.
11. Landau, L.D., Lifshitz, E.M. Quantum mechanics. V. 3. Moscow, Fizmatlit, 2004.
12. Dzuba V.P., et al., Optics. 2014. No. 3. P. 22–37.
13. Pokutniy S.I., Fiz. Tekh. Poluprovod. 2006. V. 40, No. 3. P. 223–228.
14. Dzyuba V.P., et al., Advanced Material Research. 2013. V. 677. P. 42–48.
15. Dzyuba V.P.,et al., Journal of Nanoparticle Research. 2012. V. 14. P. 1208–1214.
16. Dzyuba V.P., et al., Fiz. Tekh. Poluprovod. 2014. P. 56, No.. 2. P. 355–361.
17. Kulchin Yu.N., Shcherbakov A.V., Kvant. Elektronika. 2008. V. 38, No. 2. P. 158–163.
18. Kupchak I.M.,, et al., Semiconductor Physics, Quantum Electronics & Optoelec-

tronics. 2006. V. 9, No. 1. P. 1–8.

19. Pokutniy S.I.,et al., Nanosistemi, nanomateriali, nanotekhnologii. V. 7, No. 4. P. 941–949.

20. Pokutnyi S.I., Phys. Express. 2011. V. 1, No. 3. P. 158–168.

21. Pokutnyi S.I., Tech. Phys. Lett. 2013. V. 39, No. 3. P. 233–235.

22. Dvurechensky A.V., Yakimov A.I., Fiz. Tekh. Poluprovod. 2001. Vol. 35, no. 9. P. 1143–1153.

23. Kulchin Yu.N., et al. Pis'ma Zh. Tekh. Fiz. 2010. V. 36, No. 21. C. 1–9.

24. Pokutnyi S.I., in: Adv. in Semiconductor Research. Physics of Nanosystems and Technological applications, Ed. D.P. Adorn and S.I. Pokutnyi. NOVA, New York, 2014.

25. Ledentsov N.N., et al. Fiz. Tekh. Poluprovod. 1998. Vol. 32, No. 4. P. 385–410.

26. Dzyuba V.P., et al. Fiz. Tekh. Poluprovod. 2011. V. 45, No. 3. P. 306–311.

27. Dzyuba V.P., et al., Journal of Nanophotonics. 2011. V. 5. P. 053528.

28. Milichko V.A., et al. Kvant. elektronika. 2013. V. 43, No. 6. P. 567–573.

29. Kulchin Yu.N., et al., Advanced Material Research, 2013. V. 677. P. 36–41.

30. Milichko V.A., et al., Applied Physics A: Materials Science & Processing. 2013. V. 11. P. 319–322.

31. Sheik-Bahae M., et al., IEEE J. Quantum. Electron. 1990. V. 26. P. 760–769.

32. Zeng H., Duan G., et al., Adv. Funct. Mater. 2010. V. 20. P. 561–572.

33. Kaftelen H., et al., Phys. Rev. B. 2012. V. 86. P. 014113.

34. Usui H., et al., J. Phys. Chem. B. 2005. V. 109. P. 120–124.

35. Akhmanov S.A., Nikitin S.Yu., Physical optics. Moscow, Nauka, 2004..

36. Shen I.R., Principles of nonlinear optics. Moscow, Nauka, 1989.

37. Kulchin Yu.N., et al., Pis'ma Zh. Tekh. Fiz. 2009. V. 35, No. 1. P. 1–7.

Index

T - #0098 - 111024 - C478 - 234/156/22 - PB - 9780367571603 - Gloss Lamination